The Soil-Plant System
IN RELATION TO INORGANIC NUTRITION

Other Related Books

A Manual on Conservation of Soil and Water	*USDA*
Agricultural Drainage Engineering : Field and Laboratory Manual	*Parmar, H.V.*
Biological Management of Soil Ecosystems for Sustainable Agriculture	*FAO*
Challenges and Strategies of Dryland Agriculture	*Rao, S.C.*
Chemistry of the Soil, 2nd Ed.	*Bear, F.E.*
Crop Production in Waterlogged Saline Soils	*Gupta, S.K.*
Diagnosis and Improvement of Saline and Alkali Soils	*Richods, L.A.*
Fundamentals of Soil Science	*Rathinasamy, A.*
Organic Agriculture in the Tropics and Subtropics	*Kopke, U.*
Research Methods in Plant Sciences: Allelopathy Vol.1 (Soil Analysis)	*Narwal, S.S.*
Reuse of Municipal Sewage and Sludge in Agriculture	*Ramulu, U.S.*
Salinity Tolerance in Plants: Methods, Mechanisms and Management	*Garg, B.K.*
Salt Affected Soils : Reclamation and Management	*Gupta, S.K.*
Soil and Plant Analysis	*Piper, C.S.*
Soil and Plant Analysis Laboratory Manual, 2nd Ed.	*Ryan, J.*
Soil and Water Conservation in Semi-Arid Areas	*FAO*
Soil Chemical Analysis	*Jackson, M.L.*
Soil Dynamics in Tillage and Traction	*Gill, W.R.*
Soil Salinity Assessment	*FAO*
Soil Survey Laboratory Methods Manual	*USDA*
Soil Survey Manual New Revised Edition	*USDA*
Standard Methods for Analysis of Soil Plant and Water	*Gupta, I.C.*
The Use of Saline Waters for Crop Production	*FAO*
Use of Saline Water in Agriculture	*Gupta, I.C.*

The Soil-Plant System
IN RELATION TO INORGANIC NUTRITION

MAURICE FRIED

Joint FAO/IAEA Division of
Atomic Energy in Food and Agriculture
International Atomic Energy Agency
Vienna, Austria

HANS BROESHART

Section of Agriculture
International Atomic Energy Agency Laboraroty
Vienna, Austria

Prepared under the direction of the American Institute of Biological Sciences for the Division of Technical Information, United States Atomic Energy Commission

United Book Prints

1st Reprint in India, 2015

Published by:

United Book Prints
4806/24, Ansari Road,
Daryaganj,
New Delhi - 110 002
Tel.: 011-41511055

Print: 2015

ISBN: 978-93-83692-12-5

Printed in India

FOREWORD

This monograph is one in a series developed through the cooperative efforts of the American Institute of Biological Sciences and the U. S. Atomic Energy Commission's Division of Technical Information. The goal in this undertaking has been to direct attention to biologists' increasing utilization of radiation and radioisotopes. Their importance as tools for studying living systems cannot be overestimated. Indeed, their applications by biologists has an added significance, representing as it does the new, closer association between the physical and biological sciences.

The association places stringent demands on both disciplines: Each must seek to understand the methods, systems, and philosophies of the other science if radiation biology is to fulfill its promise of great contributions to our knowledge of both the normal and the abnormal organism. Hopefully, the information contained in each publication will guide students and scientists to areas where further research is indicated.

The American Institute of Biological Sciences is most pleased to have had a part in developing this Monograph Series.

JOHN R. OLIVE
Executive Director
American Institute of Biological Sciences

PREFACE

This work is intended as a monograph on the soil–plant system in relation to the inorganic nutrition of plants. It can by no means be considered a textbook, recording in detail all points that affect this nutrition, e.g., very little coverage is given to such subjects as soil aeration, hydroponics, organic soils, etc., nor are all the nutrient elements discussed in detail in relation to each subject matter covered. It is rather an attempt to view the dynamics of the process of ion uptake in relation to those physical and chemical processes that must be considered both in understanding any observation made on the soil–plant system and in predicting the results of any stress placed on the system; such stress might, for example, be the removal of nutrients by the plant or the addition of nutrients by fertilization.

The book is divided into two parts: the first includes primarily, fundamental aspects of ion movement from the soil into and through the soil solution, then into the plant root, and finally into the shoot; the second includes the more practical aspects of the supply of nutrients to plants grown in the soil–plant system and how it can best be supplemented. The emphasis throughout is on an understanding of both problems and the considerations necessary to solve them rather than on the solution of any individual problem. References to the scientific literature are extensive. The authors have drawn heavily from their own research and experience, and, in addition, have attempted to present all other pertinent ideas together with their documentation. Nevertheless, this is not presented as a literature review or recording of facts, but rather as an over-all view as seen by the authors; finite conclusions are generally drawn.

Although the publication was partially supported by funds from the U. S. Atomic Energy Commission, no limitations were placed on the authors as to emphasis or subject matter coverage relating to isotope or radiation applications. Nevertheless, it should be pointed out that much of the understanding and principles developed both on the more fundamental and the applied aspects have involved isotope applications. This is particularly evident in the subject matter coverage of solid-phase–soil-solution relationships, movement of ions to the roots, into the roots (active or passive), and translocation to the shoot, the mobility of nutrients, laboratory, greenhouse, and field evaluation of soil nutrient supply and when, where, and what kind of fertilizer to apply.

The authors are greatly indebted to Professor H. Laudelout for his assistance in the preparation of the sections dealing with ion exchange and diffusion in soil systems. Thanks are also due to Professor G. H. Bolt for his help in the preparation of the section on cation exchange in soils. We are grateful to Professor R. A. Olson, Professor A. C. Caldwell, and Dr. R. E. Shapiro for their critical examination of the text and the many valuable suggestions made by them. Finally, we thank our wives for several years of patience; without their cooperation this manuscript could never have been completed.

October, 1967

MAURICE FRIED
HANS BROESHART

CONTENTS

PART I

Introduction

The soil–plant system together with energy from the sun is the primary source of food for land organisms and of food and fiber for man. The inorganic nutrition of plants grown in the soil–plant system is one of the main areas of effort where man has succeeded in appreciably affecting the quantity and quality of his food and fiber. Fertilizer consciousness by farmers and an enormous fertilizer industry are a result of the realization that man can exercise appreciable control over the amount and kind of agricultural produce yielded per unit area of land surface. In the United States alone the consumption of fertilizer nutrients in 1963 was 9,532,065 tons and the consumption is still increasing as shown in Table I.1.

TABLE I.1
TOTAL FERTILIZER NUTRIENT CONSUMPTION IN THE UNITED STATES FROM JUNE 1960 TO JUNE 1963[a]

Year	N	P_2O_5	K_2O	Total
1960–1961	3,030,788	2,645,085	2,168,533	7,844,406
1961–1962	3,369,980	2,807,039	2,270,537	8,447,556
1962–1963	3,903,629	3,092,070	2,536,366	9,532,065

[a] Data given in short tons. From Scholl *et al.* (1964).

In other areas of the world outside of Western Europe and the Far East very little fertilizer is used, although the trend is also upward (Table I.2). Actually, the same lack of utilization occurs in the countries in the Far East outside of Japan and Taiwan.

This lack of utilization in most of the land areas of the world is not because fertilizer is not required, but primarily is a result of high cost and other economic factors, including transportation facilities and the difficulty of bringing the appropriate information to the attention of the farmers. In spite of the general over-all need for fertilizer, there are local areas where too much of a particular fertilizer nutrient is being utilized, resulting in economic loss or even reduced yields and quality of the product.

Only with knowledge and understanding of the soil-plant system can man hope to exercise the kind of control that will lead to the most efficient method of high production under the many varied environmental conditions of soil and climate. The present volume is, therefore, divided into two distinct parts. The first (Part II) deals with soil–plant relationships in their more basic aspects, including the information and considerations necessary to understand the system. The second (Part III) deals with the more practical aspects of supplying nutrients to plants, including the research information and considerations necessary for final evaluation of the system.

TABLE I.2
WORLD FERTILIZER CONSUMPTION BY REGIONS[a]

Region	1945/1946	1949/1950	1954/1955	1959/1960
Europe (excluding USSR)	3372	6388	9191	23,800
USSR	680	1025	1718	2340
North America	2672	4024	5887	7130
Latin America	154	225	481	840
Near East	48	133	206	250
Far East	86	1014	1748	2530
Africa	115	181	307	400
Oceania	373	507	722	840

[a] Data given in thousand metric tons of plant nutrients. From FAO (1962).

In both parts the emphasis of necessity is on those nutrient elements for which most of the research information exists. This means that nitrogen, phosphorus, and potassium may receive the most attention. Nevertheless, the principles indicated and research information needed apply to all nutrient elements. One nutrient element is no more essential than another; it is only the likelihood of deficiency in the practical situation that may be different.

Since the emphasis is primarily on the principles involved, the soil is generally treated as a relatively homogeneous system. The soil is not generally homogeneous, however, particularly in its gross aspects, and any practical applications of the principles involved must, of necessity, take account of the heterogeneity *in situ*.

PART II

Soil-Plant Relationships in Ion Uptake

From the standpoint of practical agriculture, the inorganic nutrition of plants essentially concerns the uptake of nutrient ions from the soil into the plant system. In Chapters 1–4, the two components of this system, the soil and the plant, and their interrelationships are examined.

CHAPTER 1

Dynamics of the Soil-Plant System

As long as the plant is alive and its roots are in the soil, the soil-plant system with regard to inorganic nutrition is an open system in which nutrient ions are continuously removed from the system at one end (the solid phase) and accumulated at the other end (the plant), as shown in Eq. (1.1) where M is a nutrient ion.

$$\text{M(solid)} \rightleftharpoons \text{M(solution)} \rightleftharpoons \text{M(plant root)} \rightleftharpoons \text{M(plant top)} \quad (1.1)$$

The movement of ions from soil to plant top has been written as an equation not only to emphasize that all the processes taking place are physicochemical processes but also to suggest that knowledge of the concentrations of the reactants and the rate constants of each of the processes gives a rather complete description of the over-all process with the concomitant potential of process regulation.

Energy must be supplied to the system for the ion accumulation process to continue. In higher plants this energy comes from the absorption of quanta of light from the sun. To fully understand the process of inorganic nutrition in the soil–plant system, one must also understand this energy process (photosynthesis). For living higher plants, this energy process is quite capable of changing the concentration of reactants and the magnitude of the rate constants (as might other factors in the environment), yet the physicochemical description of the over-all process will not normally change. No attempt is made to describe this over-all energy process which itself is the subject of many books and review articles (Bassham, 1957; Hill, 1958; Rabinowitch, 1945 and 1956; Terrien, 1957; Bladergroen, 1960; McElroy, 1961).

In an open system, such as that described in Eq. (1.1), the amount of ion accumulated in the plant top in a given (short) unit of time will equal the net loss of ion from the soil. When all the intermediate reactions are occurring at the same rate, none of the reactants are changing in concentration and $dMi/dt = 0$. The system is then considered to be in a steady state condition.

Ion uptake is a continuous dynamic process and all the reactions suggested and implied in Eq. (1.1) are going on all the time. The solution of the kinetic equation involving even two consecutive reactions is extremely complex. Reactions involving more than two steps lead to even greater difficulties. Practical solutions normally necessitate certain approximations (Dixon and Webb, 1958; Laidler, 1958). The steady state approximation is one of the most useful. The ion uptake process may be treated as a steady state system which reacts to changes in concentration of one of the reactants or of one of the rate constants by the necessary adjustments in all other reactions to result in a new steady state level. A few characteristics of a steady state system, with particular reference to determining the rate of the over-all reaction by the steady state approximation, are listed below.

(1) In a system at steady state all reactions are going on at the same rate, i.e., $dMi/dt = 0$.

(2) If the reverse step of a process in a steady state system is very rapid compared with the rate at which this intermediate undergoes reaction to the final product, then the preceding reactions are essentially at equilibrium.

(3) The steady state approximation includes a concept of the rate-limiting step, i.e., in a sequence of consecutive reactions the slowest reaction (or reactions) determines the rate of all the reactions.

(4) The steady state approximation is particularly good when the intermediate components are very reactive and, therefore, present at very small concentrations.

(5) In a chain of reactions in a steady state system no reaction after the first irreversible reaction plays any part in determining the over-all rate.

(6) A zero-order reaction in a chain of reactions determines the over-all rate.

(7) If in a chain of reactions an irreversible reaction occurs before a zero-order reaction, a steady state is impossible.

The individual reactions making up this continuum of the ultimate accumulation of an ion initially present in the solid phase can be studied individually, but their significance to the accumulation process cannot be evaluated without consideration of the rest of the reactions in the continuum. It is in this context that the approximation of the rate-limiting step is particularly useful.

Chapters 2 and 3 of Part II deal with the solid and liquid phase of the soil system, including their interrelationships, and with the plant system, including the plant root and translocation of nutrients. Chapter 4 of this more basic part will discuss the over-all process in relation to the accumulated information on the individual reactions.

CHAPTER 2

Characterization of the Soil System

The soil system consists of three phases: solid, liquid, and gaseous. The gaseous phase is important to the metabolism of the plant and interacts with the liquid phase. From the standpoint of nutrient supply to plants in the soil–plant system, however, only the solid and liquid phases will be considered. A discussion of the effect of soil air on plant nutrition is given by Currie (1962), Wiegand and Lemon (1958), Lemon and Wiegand (1962), Russell (1952), and Williamson (1964).

1. The Solid Phase

The solid phase is the reservoir of most plant nutrients and also contains the active surface which determines the concentration of ions in the soil solution. In Eq. (1.1) it has been referred to as M(solid), but M(solid) itself is a complex, since the soil consists of many distinct chemical species containing in their lattices or on their surfaces the same nutrient element.

The abundance in the lithosphere of elements of agricultural significance is given in Table 2.1. These totals represent the maximum reservoir from which nutrients can be drawn other than the contribution of the atmosphere and hydrosphere to the total H, O, N, and C and, to a lesser extent, S. The elements occur in distinct chemical forms and may be divided conveniently into primary minerals, secondary minerals, the uncombined oxides and salts, and organic matter. The oxides may be of primary or secondary origin, while the carbonates are always of secondary origin. The primary minerals (i.e., those minerals present in the original magma, the chemical composition of which remains unchanged) are, in general, confined to the coarser fractions of the soil that are greater than 2 μ. The secondary minerals (i.e., those minerals with a new chemical structure due to the action of the external environment on the primary minerals) are usually confined to the finer fractions that are smaller than 2 μ. There is, of course, some overlapping in sizes and the uncombined oxides and carbonates exist in both fractions, both as precipitates, particularly in the larger fractions, and as coatings on both primary and secondary minerals. The organic matter which may exist in all states of decomposition, in all size distributions, and both

TABLE 2.1
ABUNDANCE IN THE LITHOSPHERE OF SOME OF THE ELEMENTS OF POSSIBLE AGRICULTURAL SIGNIFICANCE[a]

Atomic number	Element	Abundance (ppm)	Atomic number	Element	Abundance (ppm)
1	H	—[b]	23	V	150
5	B	10	25	Mn	1000
6	C	320	26	Fe	50,000
7	N	—[b]	27	Co	40
8	O	466,000	29	Cu	70
11	Na	28,300	30	Zn	80
12	Mg	20,900	33	As	5
13	Al	81,300	34	Se	0.09
14	Si	277,200	35	Br	2.5
15	P	1200	37	Rb	280
16	S	520	38	Sn	150
17	Cl	480	42	Mo	2.3
19	K	25,900	53	I	0.3
20	Ca	36,300	55	Cs	3.2
21	Sc	5	56	Ba	430

[a] From Goldschmidt (1954).
[b] Figure not given.

as entities and as coatings or even in chemical combinations with the mineral phase imposes further complexity upon the system.

Table 2.2 summarizes the elemental composition of mineral soils from the two climatic zones, tropical and temperate, as indicated from analysis in the literature. Soils containing as much as 50% free $CaCO_3$ may exist in dry areas. The chemical analysis of the mineral phase with

TABLE 2.2
ELEMENTAL COMPOSITION OF MINERAL SOILS FROM THE TROPICAL AND TEMPERATE ZONES

Element expressed as oxide	Climatic zone: Tropical (%)	Climatic zone: Temperate (%)	Element expressed as oxide	Climatic zone: Tropical (%)	Climatic zone: Temperate (%)
SiO_2	3–30	60–95	CaO	0.05–0.5	0.3–2
Al_2O_3	10–40	2–20	MgO	0.1–3.0	0.05–1
Fe_2O_3	10–70	0.5–10	K_2O	0.01–1.0	0.1–4
MnO	0.1–1.5	0.005–0.5	Na_2O	0.01–0.5	0.1–2
TiO_2	0.5–15	0.3–2	P_2O_5	0.01–1.5	0.03–0.3

the $CaCO_3$ removed falls in the temperate range, as designated in Table 2.2.

1.1. Primary Minerals

A qualitative description of the primary mineral abundance by species is given in Table 2.3 and the chemical composition of representative primary minerals is given in Table 2.4.

Clearly, the nutrient elements Ca, Mg, K, Fe, and, to a lesser extent,

TABLE 2.3
Primary Mineral Composition of Igneous Rocks[a]

Mineral	Chemical composition	Content (%)
Feldspar	Aluminosilicates of K, Na, Ca, Ba	60
Hornblende and augite	Aluminosilicates of Ca and Fe with other cations such as Na, Ca, Ti	20
Quartz	SiO_2	10
Mica	Aluminosilicates of K often combined with Fe and Mg	3
Hematite and magnetite	Fe_2O_3 and Fe—Fe_2O_4	3
Apatite	$Ca_5(PO_4)_3X$ (X = anion, e.g., F, Cl)	1

[a] From Vilenskii (1958).

TABLE 2.4
Approximate Chemical Composition (%) of Principal Primary Minerals in Soils[a]

Mineral	SiO_4	Al_2O_3	Fe_2O_3	FeO	TiO_2	CaO	MgO	K_2O	Na_2O	P_2O_5
Quartz	100	—	—	—	—	—	—	—	—	—
Orthoclase	62–66	18–20	—	—	—	0–3	—	9–15	9–4	—
Albite	61–70	19–26	—	—	—	0–9	—	0–4	6–11	—
Anorthite	40–45	28–37	—	—	—	10–20	—	0–2	0–5	—
Muscovite	44–46	34–37	0–2	0–4	—	—	0–3	8–11	0–2	—
Biotite	33–36	13–30	3–17	5–17	—	0–2	2–20	6–9	—	—
Hornblende	38–58	0–19	0–6	0–22	—	0–15	2–26	0–2	1–3	—
Augite	45–55	3–10	0–6	1–14	—	16–26	6–20	—	—	—
Olivine	35–43	—	0–3	5–34	—	—	27–51	—	—	—
Epidote	35–40	15–35	0–30	—	—	20–25	—	—	—	—
Apatite	—	—	—	—	—	54–55	—	—	—	40–42
Magnetite	—	—	69	31	—	—	—	—	—	—
Tourmaline	35–40	30–37	0–10	0–10	—	0–6	0–12	—	0–6	
Rutile	—	—	—	—	100	—	—	—		-
Ilmenite	—	—	0–10	50	—	—	—	—		

[a] From Bear (1953).

P are abundant in primary minerals. Notably lacking in abundance are the major plant nutrients N and S, and the minor nutrients, Mn, Cu, Zn, Mo, B, and Cl. Table 2.1 indicates that these latter elements, except for possibly Mn and Cl are lacking in abundance in the lithosphere as a whole.

1.2. Uncombined Oxides and Salts

The oxides, carbonates, and sulfates may contain nutrient ions such as Ca, Mg, S, and Fe. However, the carbonates when present are probably more important as pH buffers than as nutrients, and the Fe oxides when hydrates are more important as cementing agents between particles and as surface coatings, particularly on secondary minerals, than as suppliers of Fe as a nutrient.

The oxides of Fe and Al exist in many soils in both crystalline and amorphous forms. Crystallization of hydroxides is usually a slow process which leads to the formation of minerals such as goethite, hematite, gibbsite, and boehmite.

Oxides of Fe and presumably of Al also exist as coatings on the clay particles. Much more is known of the Fe oxides owing to the interest of the soil scientist in the nature of laterites (Swarajosunghkam *et al.*, 1962). The detailed work of Fripiat and Gastuche (1952) indicates that crystalline Fe oxides do not form in the presence of kaolinite unless more than enough Fe is present to "saturate" the kaolinite. This amounts to over 20% Fe_2O_3. Thus, the surface of kaolinite is coated to a greater or lesser extent with Fe, and the kaolinite in the so-called ground-water laterites apparently has its surface almost, if not completely, saturated. The amounts of these oxides are indicated by the data for Fe_2O_3 and Al_2O_3 in Table 2.5.

TABLE 2.5
Approximate Chemical Composition (%) of Principal Secondary Minerals in Soils[a]

Mineral	SiO_2	Al_2O_3	Fe_2O_3	TiO_2	CaO	MgO	K_2O	Na_2O
Kaolinite	45–48	38–40	—	—	—	—	—	—
Montmorillonite	42–55	0–28	0–30	0–0.5	0–3	0–25	0–0.5	0.3
Illite	50–56	18–31	2–5	0–0.8	0–2	1–4	4–7	0–1
Chlorite	31–33	18–20	—	—	—	35–38	—	—
Limonite	—	—	75–90	—	—	—	—	—
Diaspore	—	85	—	—	—	—	—	—
Gibbsite	—	65	—	—	—	—	—	—

[a] From Bear (1953).

Much less is known about the conditions that lead to the formation of Al oxides in the soil. The pH normally encountered in soil is too low to enable their crystallization. On the other hand, the presence of kaolinite, organic matter, and heavy leaching may facilitate crystallization. According to Herbillon and Gastuche (1961, 1962) and Gastuche *et al.* (1964), the formation of crystalline Al hydrates will be stimulated under well-drained conditions and low-salt content of the percolating water, further by a high temperature and conditions at which silica is mobile.

Little is known about the nature of the amorphous inorganic material in the soil, owing in part to the inadequate techniques available for analysis. [For a review see Mitchell *et al.* (1964).]

Aluminum and iron oxides and hydroxides will, depending on external pH and salt concentration of the ambient soil solution, dissociate H^+ or OH^- ions and can therefore adsorb cations and anions at negative and positive charged spots.

The oxide coatings, particularly Fe and Al, thus provide a reactive surface capable of retaining certain anions, chief of which among the plant nutrients is phosphate. It is also becoming apparent that many of the exchange properties of soil are due to contributions from oxide coatings (Coleman and Thomas, 1964; Schwertman and Jackson, 1963).

1.3. Organic Matter

The soil organic matter is derived almost wholly from plants, therefore, it not only initially contains essentially all of the mineral nutrients, but contains them in a somewhat similar ratio to their relative occurrence in growing plants. Marked changes take place, however, when residues are added to the soil. Most of the inorganic nutrient elements are leached from the plant residues except for some of the N, S, and P which are present in coordinated compounds. Most of the remainder of these elements are released relatively rapidly as decomposition proceeds. The C:N ratio generally narrows to approximately 10:1, and the C:P and C:S, to approximately 100:1.

Since the organic matter content of the surface soil is usually between 1 and 10% in mineral soils, estimates of total N, S, and P in the organic matter are possible. The N content of the organic matter can be approximated at 5%. Thus, for each 1% organic matter in the soil, an acre surface of approximately 6 inches (2,000,000 lb) contains approximately 0.05% N and 0.005% P and S. This amounts to 1000 lb of N and 100 lb of P and S.

Those soils that are high in organic matter are generally high because the decomposition rate is low (except for newly cleared virgin land)

and considering that the classic 100-bushel corn crop, as an example of crop needs, removes about 100 lb of N and 15 lb each of both P and S, the organic matter could not long be the major source of these elements to plants in a cropping system. Yet, for other than leguminous crops, the organic matter is the major native N source for a growing crop in any one crop season. The N contribution from other sources such as rainfall and nonsymbiotic N-fixing organisms is relatively small. A large proportion of the total P supply usually comes from the mineral fraction. The S supply in humid regions comes partially from the organic matter, often with major additions from the rainfall in the vicinity of industrial areas. Most soils in drier regions contain gypsiferous zones in the profile which are a primary S source which supplements the organic matter contribution.

The organic matter also provides a reactive surface which both adsorbs cations in exchangeable positions formed by COOH and OH groups and also may complex such ions as Fe, Mn, and even Ca and Mg. This adsorption and complexing of nutrients can be appreciable; indeed, in very sandy soils organic matter may represent both the immediate reserves of many plant nutrients in addition to nitrogen and the factor determining the concentration of many of the nutrient ions in the soil solution. The cation exchange capacity of the humic acid approximates 250 to 400 meq/100 g, which is threefold that of the montmorillonite-type secondary clays and 30- to 100-fold that of the kaolinite type.

1.4. Secondary Minerals

The principal secondary minerals present in the soil are shown in Table 2.5, together with their over-all composition.

The chemical composition of the primary minerals is important in terms of the original source of nutrient ions. The chemical composition of the secondary minerals is relatively unimportant in this respect. The secondary minerals are primarily responsible for many of the physicochemical properties of soils that affect plant nutrition. The dominant reactive clay minerals, including the kaolinites, montmorillonites, and illites, derive their reactivity not only from their fineness of subdivision and broken exposed crystal edges but also from isomorphous substitution in the lattice, resulting in a net negative charge of the clay particle. It is this net negative charge and exposed crystal surface that result in the ionic adsorption of cations, including the nutrient cations.

The common reactive clay mineral types found in soils are kaolin and halloysite, two-layer nonexpanding types; illite, a three-layer essentially nonexpanding type; and montmorillonite, a three-layer expanding type. The two-layer types of clay minerals consist of units of one layer

FIG. 2.1. View of kaolinite structure normal to the *a* axis. From Wear *et al.* (1948).

of silica tetrahedrons and one layer of alumina octahedrons. The three-layer types consist of units containing two layers of silica tetrahedrons bound together by an octahedral layer in which the dominant cation may also be Al (e.g., muscovite, montmorillonite) or in which the dominant cation may be divalent, such as Mg^{2+} or Fe^{2+} (e.g., saponite, hectorite). Figures 2.1–2.3 show an *a* axis view of the normal structure, while Figures 2.4–2.6 show a more schematic *b* and *c* axis view.

FIG. 2.2. View of one-half the unit cell of the muscovite structure normal to the *a* axis. From Wear *et al.* (1948).

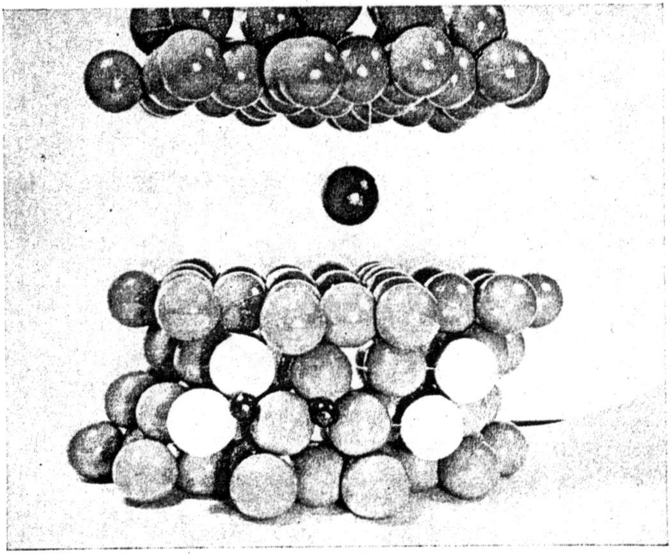

FIG. 2.3. Expanding lattice of montmorillonite structure. From Wear *et al.* (1948).

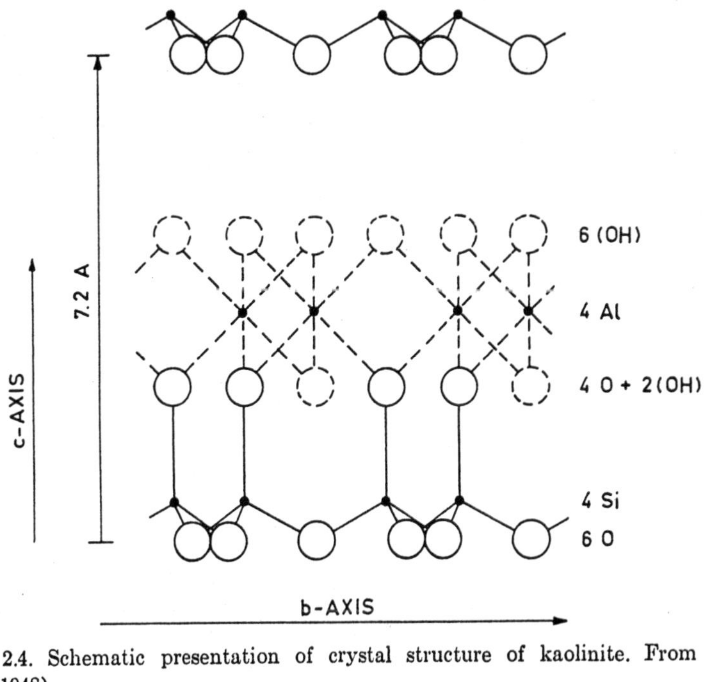

FIG. 2.4. Schematic presentation of crystal structure of kaolinite. From Wear *et al.* (1948).

The structures presented are all balanced structures and, as such, they have no intrinsic base exchange properties. Within the lattice, however, there are always substitutions in the three-layer minerals, such as Al and possibly even P, for Si in the tetrahedral layers and lower valence ions such as Mn, Fe, Li, Ni, Zn, etc., for Al in the octahedral layer. These substitutions give rise to exchange properties i.e., a net charge on the lattice resulting in an ability to adsorb ions. Lattice substitutions are presumably not common in the two-layer type minerals such as kaolinite, and most of the exchange properties of these minerals are

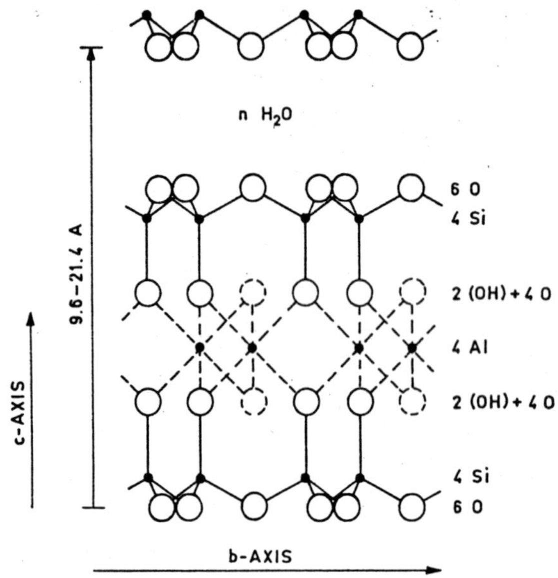

FIG. 2.5. Schematic presentation of crystal structure of montmorillonite. From Wear *et al.* (1948).

supposedly the result of the unbalanced structure at broken edges. However, the strong acid nature of freshly prepared H kaolinite (Chao and Harward, 1962) suggests some lattice substitution. Of course, broken edges also occur in the three-layer minerals, but they generally represent a relatively minor contribution to the total charge, as these minerals normally occur in soils.

In many of the three-layer minerals water can enter between the unit layers, giving these minerals an expanded lattice structure, e.g., montmorillonite may expand along the *c* axis from 9.6 to 21.4 A (Fig. 2.5) when Ca-saturated and much further in the presence of Na. It is in this interlayer that the exchangeable cations tend to congregate.

In the presence of K and ammonium ions, the lattice tends to collapse along the *c* axis, trapping these cations in a similar structure to that of the three-layer nonexpanding types such as muscovite (Fig. 2.6). In contrast, the two-layer minerals are nonexpanding types and, except when present in the structure, do not hold interlayer water or interlayer cations.

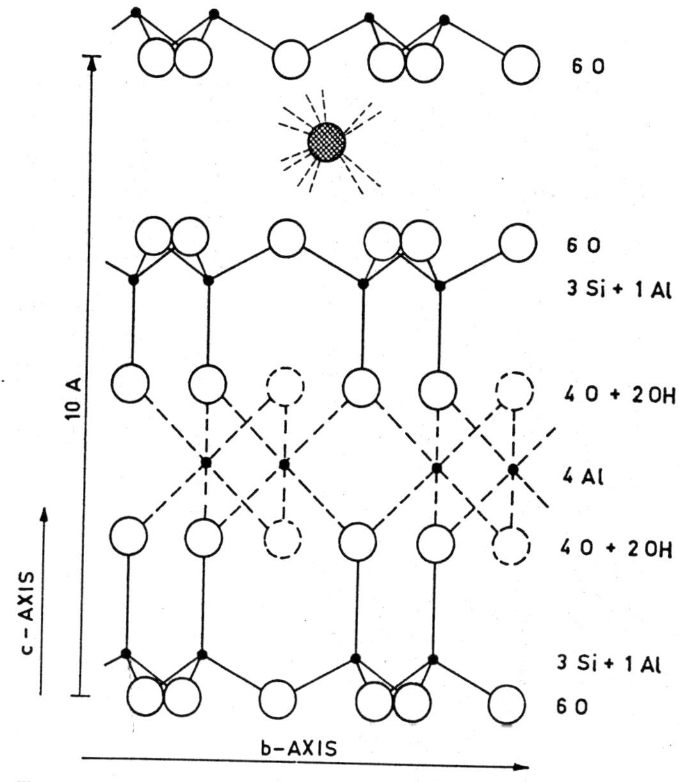

FIG. 2.6. Schematic presentation of crystal structure of muscovite.

The different clay minerals not only have differences in plate thickness but also marked differences in specific surface. Thus, the two-layer minerals may have a specific surface up to 50 m^2/g of clay, whereas the three-layer expanding type may go as high as 800 m^2/g and seldom go below 300 m^2/g, depending upon the relative proportion of expanding and nonexpanding types. The specific surface of the other three-layer minerals is between these extremes. The charge density also varies among the clay minerals. It is independent of pH for montmorillonite and illite and is approximately 3.5×10^{-7} meq/cm^2 for illite and 1×10^{-7} for montmorillonite (Lagerwerff *et al.*, 1959). The charge density of the two-layer minerals is pH dependent and is generally about 2×10^{-7} meq/cm^2. Thus, the surface charge density of clay minerals is about the magnitude of one elementary charge per square millimicron and does not fluctuate within wide limits. This charge density corresponds to an average distance between charged sites of about 10 A units, which is very nearly the average separation of ions in a normal solution of a 1:1 electrolyte (Robinson and Stokes, 1959). A normal solution is several orders of magnitude more concentrated than the soil solution, and this enormous discrepancy between the average separation of ions on the surface of soil colloids and in the soil solution may have a bearing on ion uptake by plants.

Thus, the capacity to hold exchangeable cations by the different clay minerals varies considerably, as shown in Table 2.6.

TABLE 2.6
CATION EXCHANGE CAPACITY OF CLAY MINERALS[a]

Mineral	Exchange capacity
Kaolinite	3–15
Halloysite	5–10
Montmorillonite	80–150
Illite	10–40
Vermiculite	100–150
Chlorite	10–40
Sepiolite-attapulgite-palygorskite	20–36

[a] Data given in milliequivalents per 100 g. From R. E. Grim, *Clay Mineralogy*, McGraw-Hill, New York, 1953. With permission.

The concentration of M(solid) reflects this difference in capacity to adsorb exchangeable cations. Those soils in which two-layer clay minerals predominate (e.g., lateritic type soils) typically contain relatively small amounts of exchangeable cations and have relatively small capacity to hold them. Those in which three-layer clay minerals predominate (e.g., chernozem soils) usually contain large amounts of exchangeable cations and have a relatively large capacity to hold them (chernozem soils also have a relatively large exchange capacity owing to the presence of organic exchange positions).

Broadbent (1955) has determined the relative contributions of organic material and the mineral fraction to the exchange capacity of five New York soils as shown in Table 2.7. Since the analytical method for determining the relative contribution involved destruction of organic matter with H_2O_2 to find the mineral contribution and destruction of the mineral component by HCl-HF treatment to find the organic matter contribution (both of which are subject to some error), the total exchange capacity is also listed.

A similar type of work was done in the USSR. It has been compiled and reported upon by Kononova (1961) as shown in Table 2.8.

TABLE 2.7
CATION EXCHANGE CAPACITY OF ORGANIC AND CLAY FRACTIONS COMPARED WITH VALUES FOR WHOLE SOILS[a]

Soil type	Organic matter content	Exchange capacity		Sum of clay plus organic matter	Amount due to organic matter (%)	Actual exchange capacity
		Clay	Organic matter			
Dunkirk	9.75	7.4	18.8	26.2	72	28.5
Honeoye A	6.57	3.2	13.3	16.5	81	19.5
Honeoye B	5.66	10.9	14.9	25.8	58	23.4
Ontario	7.34	2.7	13.7	16.4	84	20.1
Yates	5.70	3.4	9.4	12.8	73	14.8

[a] Data given in milliequivalents per 100 g of soil.

The results indicate that one can generalize that about 50% of the exchange capacity of humid region surface soils is due to organic matter. Considering this very high percentage contribution, relatively little has been done either to determine the nature of this exchange or the relationship between ions held by the organic material and those in the soil solution.

Broadbent and Bradford (1952) did attempt to determine the nature of the functional groups. They concluded that up to a maximum of 46% of the exchange capacity could be accounted for as imide N and phenolic and enolic hydroxyl, of which a maximum of 25% and more likely in a range below 10% was due to imide N. The other 36% was due to hydroxyl. Carboxyl presumably accounted for the remaining 54% and may even account for more. In a later study on a muck soil, Lewis and Broadbent (1961) concluded that much of the exchange capacity was due to phenolic and carboxyl groups. Fortunately, the synthetic resins have similar functional exchange groups. At present, the soil scientist can only infer from the resin literature many of the characteristics of the soil exchangeable cation reactions due to these groups.

TABLE 2.8

RATIO BETWEEN THE EXCHANGE CAPACITY AND HUMUS CONTENT IN SOILS OF THE USSR

			Exchange capacity (meq./100 g soil)		Total exchange capacity (%)	
Soil	Horizon	Humus determined by Tyurin's method	Mineral part[a]	Organic and mineral part	Mineral part	Organic part
Strongly podzolic	A_p[b]	2.82	7.22	10.08	71	29
	A_2	0.66	5.24	6.16	90	10
Moderately podzolic	A_p	3.24	7.07	12.56	56	44
	A_1	2.74	6.74	12.02	56	44
Slightly podzolic	A_p	5.29	14.65	26.54	55	45
	A_1	5.34	15.07	23.09	65	35
Slightly podzolic	A_p	6.13	13.69	35.29	39	61
dark-gray	A_3	4.97	15.46	34.09	45	55
Degraded chernozem	A_p	8.02	20.16	48.35	42	58
	A_1	5.50	23.24	47.88	48	52
Rich chernozem	A_p	11.10	23.82	62.64	38	62
	A_1	11.30	23.53	63.98	37	63
Ordinary chernozem	A_p	7.90	25.68	56.86	45	55
	A_1	5.96	25.00	52.91	47	53
Southern chernozem	A_p	4.95	24.66	42.50	58	42
Chestnut soil	A	2.07	8.60	16.68	65	35
Serozem	A	1.93	6.89	13.15	52	48

[a] The determination of the exchange capacity of the mineral fraction of the soil was carried out after the oxidation of organic matter with H_2O_2.

[b] Ploughed layer, A.

2. Liquid Phase

The liquid phase of the soil is, in general, not the reservoir of nutrient ions, except for those ions such as chloride and, to some extent, sulfate, that are not adsorbed by the solid phase or incorporated in the organic matter. Similarly in low organic matter soils it is not the reservoir for those ions, such as nitrate and sulfate, that are generally incorporated in the organic matter but, under usual conditions, are not associated with the mineral phase. In Eq. (1.1), the concentration of the ion in the soil solution was referred to as M (solution) and represents the intensity factor which, in the ion uptake system as represented in Eq. (1.1), drives the reaction toward ion accumulation by the plant.

If the plant takes up its ion from this solution, at any instant of time, the solution concentration of the ion at the root surface is the instantaneous supply. This concentration may change with time in a different way for each nutrient and each environmental situation. A continuous description of this solution concentration all during the growth of the plant would be the near-perfect description of soil nutrient supply over one growing season. The early recognition of the possible importance of the concentration of the nutrient ion in the soil solution to plant assimilation and growth has led to many attempts to determine its concentration.

In spite of the limitations concerning the accuracy of the measurement of the concentration of the ions in the soil solution (see Chapter 6, Section 1.1.1), a compilation of the values obtained is useful in reflecting the order of magnitude of the concentration of the various constituents.

TABLE 2.9
SOIL SOLUTION COMPOSITION[a]

Element	Range of all soils	Acid soil	Calcareous soil
Ca	0.5–38	3.4	14
Mg	0.7–100	1.9	7
K	0.2–10	0.7	1
Na	0.4–150	1.0	29
N	0.16–55	12.1	13
P	0.001–1	0.007	0.03
S	0.1–150	0.5	24
Cl	0.2–230	1.1	20

[a] From Burd and Martin (1923); Burgess (1922); Hoagland, Martin, and Stewart (1920); Morgan (1916); Pierre (1931); Burd (1948); Burd and Martin (1924); Eaton (1960); Pierre and Parker (1927); Vlamis (1953); and Reitemeier (1946). Data given in moles/liter $\times 10^3$.

Table 2.9 is a summary of the reported composition of the soil solution as compiled by Fried and Shapiro (1961). A glance at the table indicates that all of the major nutrient ions except P are normally present at concentrations approximating 10^{-3} to 10^{-4} *M*. Phosphorus is usually lowest in concentration, commonly about 10^{-5} to 10^{-6} *M*. The N figures are undoubtedly too high as most samplings of the soil solution are made on stored samples in the laboratory. Laboratory handling often results in a burst of production of soluble N. Often as much N is produced in the laboratory in only 2 to 3 weeks as might be produced in the field in months (Fitts *et al.*, 1955). The higher values for cations and often for S and chloride are usually associated with higher pH soils.

3. Solid Phase–Soil Solution Relationships

3.1. General Considerations

Solid phase-soil solution relationships involve not only considerations of the nature of the binding of nutrients to the solid phase, but also the quantities and concentrations associated with the two phases and the rate at which this reversible binding occurs.

3.1.1. *Ion Binding by the Solid Phase*

The interactions between the solid phase and the soil solution comprise several forms of binding. Part of these are electrostatic and are influenced primarily by the valence of the ion and the charge density of the solid phase. Others are influenced more by other forces (e.g., polarizability of the ion) which are specific for the ion involved. Although the specific forces are seldom absent, in the case of cations the interaction is still mainly governed by charge effects. For anions the interaction is somewhat different and chemical binding is sometimes dominant.

a. *Charge Effects.* The charge in clay minerals of the three-layer type is due primarily to isomorphous substitution. This results in a constant negative charge density on the surface of the crystal that is not affected by the composition of the outside soil solution.

In addition to isomorphous substitution, ionized SiOH and AlOH groups of two-layer clay minerals and on crystal edges of the three-layer clay minerals and FeOH or AlOH groups from coatings and gels may adsorb both cations and anions, depending on the composition and, in particular, the pH of the soil solution.

The SiOH groups at crystal edges will, under normal soil conditions have a negative charge because the pK value of the $SiO^- \rightleftharpoons SiOH$ reaction is approximately 9–10. Positively charged $SiO\!\begin{smallmatrix} \diagup H^+ \\ \diagdown H \end{smallmatrix}$ groups may occur only below a pH of about 3.5 (Bolt, 1957). For AlOH on the mineral edges, the pK value is difficult to estimate because the Al atoms are usually bound to silica atoms via an oxygen bridge, thus changing the pK value from that found in systems of AlOH in solution. For $AlO^- \leftrightarrows AlOH$ the pK of the dissociation is probably about 8–9 and for $AlOH \leftrightarrows AlOH_2^+$, about 4–5. The pK of Al and Fe hydroxide in the soil is probably near 5–6, which means that at a low pH, positive sites are present which may adsorb anions. The charge on the organic matter responsible for adsorbing ions is also primarily negative and a function of pH.

The adsorption of cations by dissociated groups may give rise to the formation of a diffuse double layer (cf. below) depending on the distances between the charges. It is doubtful, however, that the adsorption of anions to soil constituents has the nature of a double layer. Positive charges are not likely to be distributed homogeneously along the surface, and specific binding of anions in contrast to electrostatic attraction is probably much more important in normal soils (de Haan, 1965).

As a consequence of the negative charge, the composition of the soil solution in close proximity to the clay plates differs from the composition of the soil solution outside the sphere of influence of the negative electric field. The system consisting of a net negative charge on the surface and a layer of cations in close proximity to the surface of the clay is called an electric double layer. The simplest form of such an electric double layer is the charge distribution present in a system consisting of two parallel charged plates at close distance (i.e., so-called Helmholtz double layer).

However, this is a gross oversimplification with respect to the situation present in moist soils, since the layer of adsorbed cations is often diffuse, owing to their kinetic energy which causes them to diffuse into the ambient solution against the attraction forces of the clay plates. Inasfar as this distribution of the cations in a diffuse, double layer is dominated by electrostatic forces, it may be described by a simplified form of Boltzman's law*:

$$C = C_0 e^{-ze\psi/kT}$$

where C is the concentration of a cation at a point where the electric potential has the value ψ; C_0, the concentration of that cation in the equilibrium solution outside the sphere of influence of the solid phase (ambient solution); z, the valency of the ion; e, the elementary charge; ψ, the electric potential at C; k, the Boltzman constant; and T, absolute temperature.

The extent of the double layer in a given soil is a function of the valency of the cation and the total salt concentration in the ambient solution. Monovalent cations will be bound less than divalent cations and will thus swarm out further away from the clay plates than divalent cations. The lower the salt concentration of the ambient solution, the further away the double layer will extend.

On the other hand, close to the negatively charged surfaces of the

* In certain special cases (e.g., Na-montmorillonite) the above equation in combination with Poisson's equation of the electric field renders a fairly exact calculation of the ionic distribution of the double layer possible. In many other cases a substantial, if not dominant, influence of nonelectrostatic attraction forces prevents the simple calculation of this distribution.

clay, negative ions will be expelled and the anion concentration will be lower in the vicinity of the clay plates than in the ambient solution outside the electric field of the double layer (anion repulsion or negative adsorption).

In Figure 2.7 a schematic diagram is given of the distribution of cations and anions in diffuse electric double layers as affected by salt concentration and valency of the cations.

The difference in composition between double layers and the equilibrium soil solution has important implications for the supply of nutrients

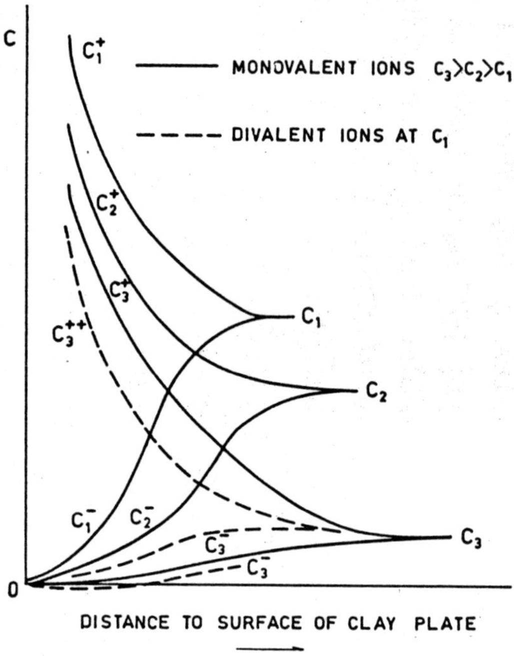

FIG. 2.7. The effect of ion charge and concentration in the ambient solution on the concentration of ions at various distances from a negatively charged clay plate.

to plants. The thickness of the double layers may vary from 200 A for a Na clay in a 10^{-3} N NaCl solution to 20 A in a 10^{-1} N NaCl solution. For divalent cations, the values are approximately 100 and 10 A, respectively. This means that in a heavy clay soil with a specific surface of 100 m^2/g and a double layer thickness of 50 A, 0.5 cm^3/g of the soil solution has a composition that differs from the equilibrium soil solution which is outside the sphere of influence of the negative charge. This may mean that at low-moisture contents almost all the ions in the soil solution are within the sphere of influence of the charge on the clay plates. As the plant roots are taking their nutrient ions from the same solutions, it is difficult to estimate intensity factors for such conditions.

b. *Specific Effects.* The strong and sometimes essentially irreversible binding between ions and the solid phase may be due to a high binding energy of the bond or to mechanical factors which prevent the movement of ions once they are bound.

Potassium and ammonium "fixation" by illitic clay minerals is probably the result of both electrostatic binding and mechanical hindrance. The K or NH_4 which is preferentially adsorbed as compared to Na is normally readily exchangeable. However, K and NH_4 may move between the layers of the mineral. Under certain conditions, particularly upon drying of the soil, the plates "condense," and because the lattice has collapsed and K and NH_4 ions have the right dimensions to remain "fixed" between the plates, they are very energetically bound and do not exchange with cations from the outside solution.

An interaction of a specific nature is that of H ion and clays. Unfortunately, although it was demonstrated in 1934 (Paver and Marshall, 1934) that H clays are unstable and spontaneously evolve toward Al saturated clays, this was largely ignored until relatively recently (Coleman and Craig, 1961; Davis, 1961; Harward and Coleman, 1964; Davis *et al.*, 1962; Chao and Harward, 1962; Frink and Peech, 1963; Chernov and Belyaeva, 1956, 1956a; Eeckman and Laudelout, 1958; Laudelout and Eeckman, 1961). Spontaneous transformation of H clay into Al clay apparently follows a parabolic rate law which may be interpreted as a surface interdiffusion of Al ions from the edges of the clay platelets against a countercurrent of H ions from the faces toward the edges, where they immediately react with the lattice, setting Al ions free as exchangeable cations (Laudelout and Eeckman, 1961). This complicates the study of H clays not only because they are unstable, but also because homoionic H clays cannot be obtained. A fraction of the total cation exchange capacity will always be occupied by Al ions, no matter how carefully or how rapidly the H clay is prepared.

Since the limitations on the stability and homogeneity of the hydrogen clays were not generally recognized, most if not all, of the older work is only of historical interest. The evolution of soil scientists' ideas about the nature of soil acidity is very clearly stated in a most interesting review article by H. Jenny, "Reflections on the Soil Acidity Merry-Go-Round" (1961).

3.1.2. *Intensity vs. Capacity*

In Eq. (1.1), M(soil) is the amount of a given ion associated with a definite amount of the solid phase which under conditions of no n' transfer of free energy is in equilibrium with M(solution), a defir quantity of the same ion species in a specific volume of soil solution. When a living plant root is imposed on this system, a reaction occurs between the ion in solution and the plant root, resulting in metabolic uptake of the ion by the root. Insofar as this reaction occurs the rate of the forward reaction representing ion uptake by the living root is a function of M(solution), the concentration of the ion in solution.

The concentration of a given ion in the soil solution represented by M(solution) is an *intensity factor* analogous to H ion concentration (pH) in the solution of a weak acid.

When the stress of a living plant root is placed upon the soil system, resulting in the removal of some of the ions from the soil solution, the relationship between M(soil) and M(solution) results in a shifting of ions from M(soil) to M(solution). This shifting of ions from solid phase to solution is a continuous process, continuing as long as ions are removed from solution by the plant roots.

The amount of a given ion in the solid phase that is associated with the same ion in the liquid phase is a *capacity factor* analogous to the molarity of an acid. The capacity factor is thus simply the size of the reservoir or pool from which the ion from the solid phase is withdrawn. The capacity factor must be multiplied by a rate constant to obtain a measure of the ability of the solid phase to continue to supply a given nutrient ion to the soil solution rapidly enough for plant use when the soil solution is placed under the stress of removal of the ion.

Relating these intensity and capacity factors to rates at which substrate is transformed is in the province of the thermodynamics of irreversible processes (Prigogine, 1955; de Groot, 1952; Denbigh, 1951). Although there may be no fundamental theoretical difficulty in applying these concepts to a highly heterogeneous system such as the soil and its solution, the experimental difficulties in determining the many parameters that occur in the equations are so great that this approach is of more theoretical interest than of practical significance.

A glance at Table 2.9 indicates that both the intensity and capacity factors are important for many ions that exist in the soil solution in such small quantities that the soil solution must be renewed frequently in order to supply the needs of the plant. This is particularly true of P, where renewal must apparently occur many times per day from the solid phase M(soil) in order to give the total uptake of P needed by plants (Fried and Shapiro, 1956; Parker, 1927).

3.1.3. *Rate Processes*

The extent to which the rate of transfer of a given ion between the solid phase and the soil solution may vary is extremely large. Ionic exchange reactions between the interface liquid–solid and the solution are practically instantaneous; when slower rates are observed, this may well be due to diffusion control of the exchange reaction. Borland and Reitemeier (1950) found the exchange of radiocalcium with the Ca adsorbed on clays was in equilibrium in half an hour, the shortest time period utilized, and Barbier *et al.* (1954) concluded that for Ca, and probably all exchangeable cations, the ions on the solid phase exchange with those in solution several times within the course of minutes.

An exception to this generalization occurs where geometrical considerations become important. Transfer between phases requires exchange with another ion. If the other ion present cannot exchange owing to restrictions other than charge, then exchange will not occur. The most common example, the K reaction with clays, is not by any means the only example. The surface ions of almost any crystal will exchange with the identical ion in solution and, indeed, the surface area of a crystal can be determined in this way (Paneth, 1922; Olsen, 1952). However, other ions of the same charge will not necessarily be exchanged. If the amount of exchangeable Ca is determined in a soil containing $CaCO_3$ and this is done with labeled Ca, the amount obtained does not necessarily reflect the amount of solid phase Ca that will go into solution in the presence of other cations. The Ca in the surface of the $CaCO_3$ exchanges with Ca in solution but not with other cations. It, therefore, does not represent plant-available Ca. Fortunately, the amount of this Ca even in soils high in free $CaCO_3$ is generally low relative to the exchangeable Ca. Similarly, the fact that phosphate on the surface of the soil exchanges with P^{32} in solution does not necessarily mean that this P will come into the soil solution when P is removed from the solution. In the case of surface Fe or Al phosphates in acid to neutral soils, this replacement by hydroxide is to be expected. In soils in which the phosphate system is primarily associated with Ca and a high pH predominates, the extent of this replacement by anions such as hydroxide, carbonate, or bicarbonate may be more limited.

Nutrients held on the surface of the soil particles (solid phase) by some form of covalent bonding do not necessarily transfer between phases at a rapid enough rate for them to represent a source of nutrient during the growing season. This could apply to such ions as Ca and Mg and probably applies to such ions as Zn and Fe held within the organic fraction.

The rate of phosphate transfer has been experimentally investigated. The results (Fried and Shapiro, 1961; Fried, 1960; Fried *et al.* 1957) suggested that at least for a fraction of the P, the rate of transfer between phases was at least 250 times as fast as the rate of absorption of P by plant roots. Other fractions of P and, for that matter, K and many of the minor elements will transfer too slowly to represent an appreciable supply of nutrient while the plant is growing, but over the years may make available to the plant an appreciable amount of total nutrient.

Transfers between the solid phase and the solution are not limited to the processes described above which involve only the interface liquid–solid or at least those parts of the solid phase which are fairly readily accessible to transfer processes to and from the solution.

Another class of transfer processes, namely, the movement in and out of the solid phase, is a process that is characterized by slow or very slow rates and may have, nevertheless, an important implication in the long-term supply of nutrients to the plant. One of the interesting contributions in that field is the study of Lahav and Bolt (1963) of the rate of self-diffusion of Ca in $CaCO_3$. Values as low as 10^{-19} cm^2 sec^{-1} were found for the self-diffusion coefficient. This shows that processes of this type will lead to the transfer of negligible amounts of materials, unless the specific surface of the solid-phase particles into which diffusion takes place is very large.

Similar observations were made for a transfer in the reverse direction by Nagelschmitt (1944) at Rothamsted. He studied the mineralogical composition of lime fragments isolated from a soil to which phosphate had been applied a century before. Some of the phosphate ions were found to have diffused into the lime particles and crystallized as apatite, which is known to supply a negligible amount of plant nutrient to the soil system.

3.2. Cation Adsorption

3.2.1. *Theory*

A first approach to a description of the relationship M(soil) ⇆ M(solution) may be based on equilibrium considerations. From these considerations certain "exchange" equations have been derived in the literature which will be mentioned below. Prior to this, however, some general remarks will be made about the thermodynamics of the system considered.

Treating cation exchange as a chemical reaction of the form:

$$\text{A soil} + B^+ \leftrightarrows \text{B soil} + A^+ \tag{2.1}$$

it is found that at equilibrium the (Gibbs) free energy of the system is at a minimum. Expressing this quantity in terms of the thermodyn[illegible] potentials (i.e., partial molar free energy) of the reactants, t[illegible] ts for the above homovalent reaction in

$$\mu(A^+) + \mu(\text{B soil}) - \mu(\text{A soil}) - \mu(B^+) = 0$$

The μ values can be written in the form:

$$\mu = \mu_0 + RT \ln a \tag{2.3}$$

in which μ_0 is the value of the thermodynamic potential of a reactant in a chosen standard state, and a is the activity of the reactant in the system. Introduction of Eq. (2.3) into Eq. (2.2) then yields, *at equilibrium:*

$$RT \ln \frac{(A^+)(B \text{ soil})}{(A \text{ soil})(B^+)} = -\ \{\mu_0(A^+) - \mu_0(B^+) + \mu_0(B \text{ soil}) - \mu_0(A \text{ soil})\} \tag{2.4}$$

in which the quantities in parentheses of the left-hand side indicate activities. The right-hand side of Eq. (2.4) is called the "standard free energy of the reaction," i.e., the change in free energy accompanying the reaction of 1 g mole of reactants, when all reactants are in their respective standard states. This quantity is a constant (per definition) and thus *at equilibrium the* "activity product" of the reactants, (A^+)-$(B \text{ soil})/(A \text{ soil})(B^+)$, is a constant, which is then called the thermodynamic equilibrium constant K.

Obviously the numerical value of K depends on the chosen standard states. Usually (although not necessarily) the standard state for the adsorbed ion species is taken as the homoionic form of the adsorber (i.e., the adsorbed ion as a mole fraction or equivalent fraction equal to unity).

Although completely unambiguous, Eq. (2.4), which is usually written as:

$$(A^+)(B \text{ soil})/(A \text{ soil})(B^+) = K\, e^{-\Delta\mu_0/RT} \tag{2.5}$$

cannot be applied in practice for predictive purposes unless the activity of the reactants is expressed in terms of experimentally accessible quantities. For the ion species in solution this is no specific problem. For the adsorbed species it is convenient to express its (*a priori* unknown) activity as the product of the (also unknown) activity coefficient f and the measurable equivalent fraction on the adsorber, N. This results in:

$$(A^+)N_{B^+}/N_{A^+}(B^+) = f_{A^+}/f_{B^+} \cdot K = K_c \tag{2.6}$$

Applying the above reasoning to the monodivalent exchange one finds:

$$(A^{++})N_{B^+}{}^2/N_{A^{++}}(B^+)^2 = f_{A^{++}}/(f_{B^+})^2 \cdot K = K_c \tag{2.7}$$

In the above equations the quantity K_c is neither known *a priori* nor constant. It constitutes, in effect, a formal notation, expressing that the experimentally measurable, complex ratio of solution activities and equivalent fractions adsorbed equals the product of the thermodynamic reaction constant with the ratio of the activity coefficients pertaining to the adsorbed phase (Gaines and Thomas, 1953). To go from here (i.e., toward working out an equation which lends itself to predictive use), one must either determine the said activity coefficients from experimental data or construct a theory which enables one to calculate these on the basis of certain assumptions specifying the precise nature of the bonding mechanism between ions and adsorber. Such a theory is therefore based on the adoption of a certain "model."

a. *The Experimental Approach.* Equilibrium data may be (and have been) collected on many soil systems, involving different types of clay minerals, with or without the presence of organic constituents. As has been shown by Thomas *et al.* (1954, 1954, 1955, 1956), Amphlett and MacDonald (1956, 1958), Martin and Laudelout (1963), Gilbert and Laudelout (1965), such data, if properly covering the entire range of the exchange equilibrium, may then be summarized by means of a plot of ln f_A and ln f_B as a function of N, in addition to the numerical value of ln K. Research workers have often limited themselves to giving the value of K_c as a function of N over the range of practical interest, unfortunately often using different units for the fraction adsorbed, thus making direct comparison impossible.

b. *The Model Theories.* Perhaps the only detailed model theory applied to cation exchange in soil is the application of the diffuse double-layer model in the simplified version of the Gouy Chapman theory (Erikssen, 1952; Bolt and Peech, 1953; Bolt, 1955; Bolt and Warkentin, 1958; Bolt, 1955; Bolt and Warkentin, 1956; Bolt, 1960; de Haan and Bolt, 1963; Bolt and Miller, 1955). Because of its detailed nature, this model theory renders possible the calculation of the absolute value of K_c for the mono-divalent exchange. At the same time, so many assumptions are involved in its usage that its "exact" value is rather limited. In fact, it could perhaps be considered as a standard(ized) model which makes possible an estimate of the magnitude of the valence effect in the case of mono-divalent systems, thus enabling one to distinguish numerically between these valence effects and the mentioned specific effects. As such it serves a useful purpose in connection with the experimental approach. The calculated value of K_c is indeed fairly close to the experimental value obtained for Na-Ca clays over certain ranges, indicating that for these ions the specific effects are perhaps of relatively small magnitude.

Other model theories are much less specific than the double-layer model and accordingly only predict trends in the variation of K_c as a function of N. The oldest one in this respect is Vanselow's theory, which assumes—as based on solid-solution theory—that the activity of the adsorbed state equals the mole fraction of the adsorbed species (Vanselow, 1932). This implies that K_c', i.e., the value of K_c when using mole fractions rather than equivalent fractions in Eq. (2.6), would be a true constant. The variation of K_c (based on equivalent fraction) is then easily calculated. Comparison with experimental data indicates that this constancy of K_c' is often roughly true.

A more detailed model is the one introduced by Davis, Krishnamoorthy, and Overstreet, which is based on certain statistical considerations as to the composition of two ions of different valence for surface positions (Jenny, 1936; Krishnamoorthy *et al.*, 1948; Barrer and Falconer, 1956). Their assumptions led to the prediction that a parameter K_c'', i.e., the value of K_c obtained when using "corrected" mole fractions, is a constant. On the basis of the data presented the above authors show that in certain systems K_c'' varies less with composition than does K_c' introduced by Vanselow.

The treatment presented by Gapon (1933) is of practical interest for the mono-divalent exchange (Bolt, 1955). As the underlying theoretical considerations appear to be rather weak, it will suffice to mention that according to this theory the parameter K_G defined as

$$K_G = N^+/N^{++} \cdot C_0^+/\sqrt{C_0^{++}} \tag{2.8}$$

should be constant. In this equation C refers to the solution concentrations and, in contrast to the previous K_c values, the equivalent fractions of both mono- and divalent ions appear as a first power. As experiments show K_G is usually fairly constant over a limited range (more specifically between 0 and 40% monovalent ion adsorbed) (Lagerwerff and Bolt, 1955). Especially since the U.S. Salinity Laboratory (1954) has used the Gapon equation as a means of expressing the results of extensive studies on cation exchange in saline and alkaline soils, the Gapon "constant" has been accepted rather widely for the prediction of the relation between adsorbed ions and solution composition in these soils.

The Donnan equation has also been used as a starting point for the description of ion exchange. This approach results in an equation identical with Eq. (2.6) except that K_c is now interpreted to be the ratio of the activity coefficients in the adsorbed state, K becoming unity because of the arbitrarily chosen common standard states for the ions in solution and in the adsorbed state. Since the theory offers no means to give any pre-estimate of the values of the activity coefficients nor of their constancy, its usage in practice does not differ from the experimental approach discussed before. A thorough and very objective treatment of the so-called Donnan potential has been given by Overbeek (1956).

In summarizing the situation, it may be stated that thermodynamic reasoning indicates how experimentally determined exchange data could be represented in an unambiguous manner. Model considerations show that in certain cases the experimentally accessible quantity K_c and its variants based on mole fraction or corrected mole fraction and the quantity K_G—which is related to K_c in an algebraically known manner—should perhaps be fairly constant over limited ranges. Depending on the required accuracy (when using exchange equations for predictive purposes) one may choose to use that model which gives a satisfactorily constant value for one of the exchange parameters discussed or take its variation with composition into account.

3.2.2. *Discussion of Experimental Data*

Ion exchange constants are summarized in Tables 2.10–2.12, together with the formulations that have been examined previously. Some of

TABLE 2.10
SUMMARY OF SOME ION EXCHANGE "CONSTANTS" USED IN SOIL SCIENCE LABORATORIES

Common name	Equation	Reference
Donnan	$\frac{A^+ \text{ Soil}}{\sqrt{C^{++} \text{ Soil}}} \cdot \frac{\sqrt{C^{++}}}{A^+} = 1$	Vageler and Woltersdorf (1930)
Mass action	$\frac{A^+ \text{ Soil}}{\sqrt{C^{++} \text{ Soil}}} \cdot \frac{\sqrt{C^{++}}}{A^+} = K$	Kerr (1928)
Gapon-Schofield-Eriksson-Bolt	$\frac{A^+ \text{ Soil}}{C^{++} \text{ Soil}} \cdot \frac{\sqrt{C^{++}}}{A^+} = K_1$	Gapon (1933); Eriksson (1952); Schofield (1947); Bolt and Peech (1953)
Langmuir	$\frac{\sqrt{A^+ \text{ Soil}}}{\sqrt{2C^{++} \text{ Soil}}} \cdot \frac{\sqrt{C^{++}}}{A^+} = b_1/b_2 = K_4$	Boyd *et al.* (1947)
Vanselow	$\frac{A^+ \text{ Soil}}{\sqrt{C^{++} \text{ Soil } (C^{++} \text{ soil} + A^+ \text{ soil})}} \cdot \frac{\sqrt{C^{++}}}{A^+} = K_2$	Vanselow (1932) · Krishnamoorthy and Overstreet (1950)
Statistical	$\frac{A^+ \text{ Soil}}{\sqrt{C^{++} \text{ Soil } (1\frac{1}{2}\, C^{++} \text{ soil} + A^+ \text{ Soil})}} \cdot \frac{\sqrt{C^{++}}}{A^+} = K_3$	Krishnamoorthy *et al.* (1948); Krishnamoorthy and Overstreet (1950)

TABLE 2.11
APPARENT EQUILIBRIUM CONSTANTS (VARIOUS INVESTIGATORS) FOR PAIRS OF IONS OF THE SAME VALENCE

Material	Initial ion	Complementary ion	Apparent constant (K)	Ref.
Utah bentonite	NH_4	Li	0.042	*a*
Utah bentonite	NH_4	Na	0.21	*a*
Yolo clay (montmorillonite)	NH_4	Na	0.30	*a*
Hanford clay (hydrous mica)	NH_4	Na	0.17	*a*
Aiken clay (kaolonite)	NH_4	Na	0.20	*a*
Utah bentonite	NH_4	K	1.29	*a*
Yolo clay	NH_4	K	1.01	*a*
Hanford clay	NH_4	K	1.15	*a*
Aiken clay	NH_4	K	1.85	*a*
Utah bentonite	NH_4	Rb	3.19	*a*
Yolo clay	NH_4	Rb	2.34	*a*
Hanford clay	NH_4	Rb	3.30	*a*
Aiken clay	NH_4	Rb	3.60	*a*
Utah bentonite	NH_4	Cs	14.60	*a*
Yolo clay	NH_4	Cs	8.40	*a*
Black earth	K	NH_4	0.733	*b*
Montmorillonite	K	Cs	13.1	*c*
Wyoming montmorillonite	K	Na	0.567	*d*
Utah montmorillonite	K	Na	0.306	*d*
Belle F. montmorillonite	K	Na	0.440	*d*
Wyoming montmorillonite	Li	Na	0.982	*d*
Wyoming montmorillonite	Na	K	1.755	*d*
Utah montmorillonite	Na	K	3.090	*d*
Belle F. montmorillonite	Na	K	2.231	*d*
Wyoming montmorillonite	Na	Li	1.109	*d*
Utah bentonite	Ca	Mg	0.92	*a*
Wyoming bentonite	Ca	Mg	0.606	*d*
Yolo clay	Ca	Mg	0.70	*a*
Utah bentonite	Ca	Sr	1.10	*a*
Utah bentonite	Ca	Sr	1.29	*e*
Vermiculite	Ca	Sr	1.37	*e*
Illite	Ca	Sr	1.04	*e*
Kaolinite	Ca	Sr	0.99	*e*
Yolo clay	Ca	Sr	1.35	*a*
Yolo clay	Ca	Ba	1.24	*a*
Utah bentonite	Sr	Ca	0.77	*e*
Vermiculite	Sr	Ca	0.76	*e*
Illite	Sr	Ca	0.93	*e*
Kaolinite	Sr	Ca	0.95	*e*
Wyoming bentonite	Mg	Ca	1.65	*d*

[a] Krishnamoorthy and Overstreet (1950).
[b] Antipov-Karatayev (1956).
[c] Faucher and Thomas (1954).
[d] Tabikh *et al.* (1960).
[e] Heald (1960).

the formulations mentioned in Table 2.10 for the sake of completeness have not been discussed previously, such as Kerr's expression or Langmuir's equation. Their interest is minor or historical only. It may be pointed out that most of the formulations presented are formally identical when they refer to homovalent exchange, and this greatly simplifies

TABLE 2.12
APPARENT EQUILIBRIUM CONSTANT OBTAINED BY VARIOUS INVESTIGATORS FOR PAIRS OF IONS OF DIFFERENT VALENCE

Material	Initial ion	Displacing ion	Exchange equation	Ref.
Utah bentonite	Ca	NH_4	Statistical	*a*
Black earth	Ca	NH_4	Mass action	*b*
Yolo clay	Ca	NH_4	Statistical	*a*
Hanford clay	Ca	NH_4	Statistical	*a*
Aiken clay	Ca	NH_4	Statistical	*a*
Black earth	Ca	K	Mass action	*b*
Montmorillonite	Ca	K	Gapon	*c*
Illite	Ca	K	Gapon	*c*
Utah bentonite	Ca	Cs	Statistical	*a*
Utah bentonite	Ca	Cs	Vanselow	*a*
Utah bentonite	Ca	Cs	Gapon	*a*
Utah bentonite	Mg	Rb	Statistical	*d*
Utah bentonite	Rb	Mg	Statistical	*d*
Utah bentonite	La	Cs	Statistical	*a*
Utah bentonite	La	Cs	Vanselow	*a*
Utah bentonite	La	Cs	Gapon	*a*
10 Soils	Mg	Li	Gapon	*e*
10 Soils	Ca	Li	Gapon	*e*
10 Soils	Ca	Na	Gapon	*e*
Illite	Ca	Na	Gouy	*f*
59 Soils	(Ca + Mg)/2	Na	Gapon	*g*
59 Soils	(Ca + Mg)/2	K	Gapon	*g*

[a] Krishnamoorthy and Overstreet (1950).
[b] Antipov-Karatayev (1956).
[c] Lagerwerff and Bolt (1959).
[d] Tabikh *et al.* (1960).
[e] Pratt and Grover (1964).
[f] Bolt (1959).
[g] Bower (1959).

the comparison of the results which have been tabulated in Table 2.11. This is unfortunately not the case when heterovalent exchange is considered. For this reason the numerical values found for heterovalent exchange by different investigators have not been tabulated in Table 2.12.

Although it is not the purpose here to develop ion exchange theory in its entirety, an explanation of these observations should help in utilizing this type of information for the prediction of the concentration of the nutrient cations in the soil solution.

The reason the apparent equilibrium constants for ions of equal valence are usually within an order of magnitude of 1 is a reflection of the electrostatic nature of the reaction. Apparently covalent bonding or complex formation is essentially absent. However, neither Al^{3+} nor H^+ comparisons are made in Table 2.12. Although many comparisons with H^+ available in the early liturature indicate weak acid formation, it was only recently realized that H-saturated clays very rapidly convert

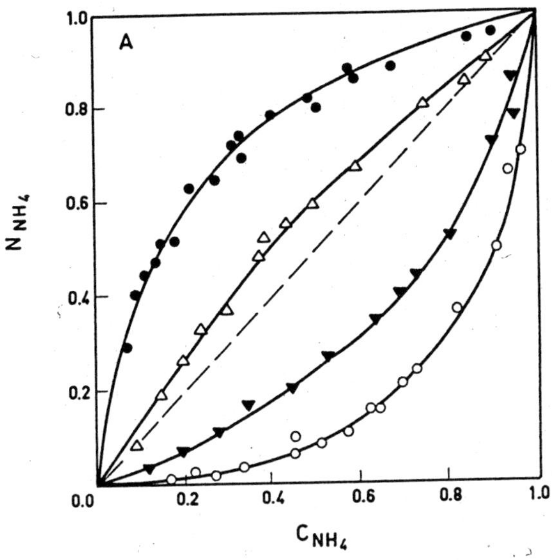

FIG. 2.8. Exchange isotherm of ammonium against (from top to bottom) sodium, potassium, rubidium, and cesium. From Martin and Laudelout (1963).

to H-Al clays on standing as indicated in Fig. 2.8. Thus, the older literature either dealing with H clays or involving the prior preparation of H clays is not subject to adequate interpretation.

Although the apparent equilibrium constant is in the order of magnitude of 1, it is seldom 1, indicating clearly that other factors in addition to charge affect the results. These are the polarizability of the ions, the size of the ions, the adsorption of solute, the swelling that occurs in some systems, the possible effect of the nature of an ion already adsorbed on the preference for a particular ion species, and the specific adsorption of K and NH_4 by micaceous clays. Martin and Laudelout (1963) show ion exchange isotherms that are typically obtained for two different ion species. They are reproduced in Fig. 2.8.

The apparent equilibrium constants for ions of equal valency are dimensionless. For ions of mixed valency, however, the apparent constant usually has dimensions and its actual numerical value is determined by the units used and the exchange equation assumed. Thus, if millimoles are used in contrast to moles, the constants will differ by a factor of 1000 in the Gapon formulation. The actual magnitudes, therefore, do not have the same meaning as they do with pairs of ions of the same valency. In comparing the results of different investigators using the same formulation, the values must be converted to the same unit basis. If different exchange equations are used, the same experimental data will clearly give different numerical (apparent) constants.

The reason for the preference of one ion for another is probably primarily due to the fact that the strength of electrostatic attraction depends chiefly on the ionic charge and the distance of closest approach. The higher valence ions are generally preferred over the lower valence ions. This preference increases with the dilution of the system and is strongest the higher the charge density of the clay. The strength of the binding between electrostatic charges varies inversely with the square of the distance between the charges. The distance of closest approach of cations to clays has been found to depend on the solvated radius. Thus, the distance of closest approach of solvated Cs, K, Rb, NH_4, Na, and Li to a cation exchange material as calculated by Boyd *et al.* (1947) was 2.6, 3.2, 4.1, 4.5, 5.2, and 6.9 A, respectively. This preference for ions with smaller solvated radius is interrelated with the swelling pressure which develops with swelling clays and which can be taken into account in quantitative treatments.

In a soil system where both many cations and many types of exchangers are always present, it is of primary importance that the constants be so closely related to mechanism that they are internally consistent, e.g., the constant experimentally determined for one pair of ions should be predictable from constants determined separately for each of these ions with a third ion in common, i.e.,

$$K_{\mathrm{B}}^{\mathrm{A}} = K_{\mathrm{C}}^{\mathrm{A}} \cdot K_{\mathrm{B}}^{\mathrm{C}} \tag{2.9}$$

Furthermore, these constants should vary as little as possible with the type of exchange material. This may sound like a vain hope, but it is certain that with full knowledge and understanding of the events the predictability of events can be accomplished.

A big step forward in this direction was made by Eriksson (1952) when he suggested that the charge concentration be taken into account in exchange equations. Although he had not measured the actual charge densities by assuming them proportional to exchange capacities (not necessarily valid), he was able by recalculation of data of Krishnamoorthy and Overstreet (1950) to equate a Utah bentonite and a Yolo clay (which is primarily montmorillonite) in a reaction involving a monovalent ion NH_4^+ and a divalent ion Ca^{2+}. Bolt *et al.* (1943), Bolt (1955), and Eriksson (1952) have estimated these charge densities as have Bower (1959) and Pratt and Grover (1964), but considering the nature of this breakthrough, relatively little work of this kind is in progress, including work on multiple ion systems such as occur in soils. It is evident that much information can be transferred from knowledge of pure resin systems, but the soil systems are unique, particularly in the nature of the charge, the nature of the exchange material, the reaction of the exchange material with water, and the complete mixture of exchange materials and ions that always occur in soil systems.

Although the complexity of the system from a physicochemical standpoint somewhat discourages the hope of practical utilization of basic approaches, this need not be the case. Exchange equations have been successfully used since the 1940's to delimit soil salinity problems and are now one of the main bases for recommendations with regard to these problems. Work with many arid regional soils has shown that there is an excellent correlation between the ratio of exchangeable Na to exchangeable Ca and Mg and the ratio in the saturation extract of the molarity of Na divided by the square root of the sum of the molarities of Ca and Mg divided by 2. This relation is equivalent to the Gapon–Eriksson–Bolt equation.

The potential use of ion exchange equations to predict the concentration of nutrient ions in the soil solution has not yet been fully realized. Yet, if one takes into account that the cation exchange material of most soils is at least 60% saturated with Ca or Ca plus Mg and that the predictions need not be based on equations that are applicable over the whole range of possible mixtures of cations on the soil cation exchanger, there appears to be enough information already present in the literature to make suitable predictions of the concentrations and ratio of cations that would be found in the soil solution. In fact, such predictions are now being successfully made (Bower *et al.*, 1957; Bower and Goertzen, 1958; Rible and Davis, 1955; Bower, 1959) even if not equally applied.

3.3. Anion Adsorption

The concentration of anions in the soil solution is affected by the nature of the solid phase. The over-all negative charge of the adsorption complex causes anions to be repulsed or "negatively adsorbed" in close proximity to the charged sites (Scofield, 1947). For systems with a constant charge density this negative adsorption can be described in parameters of the outside solution and the charge density of the adsorbing surface. Anion repulsion, however, is of little importance either in relation to capacity or intensity factors.

Anion binding is the most important phenomena which determines either M(soil), M(solution), or both. Positively charged sites in the soil adsorb anions. This binding is generally more difficult to describe than ion repulsion because of the existence of different binding mechanisms that are affected by pH and salt concentration of the soil solution. Anions can be classified into four categories based on the binding phenomena involved: (1) anion repulsion, (2) slight binding, (3) moderate binding, and (4) strong binding.

3.3.1. *Anion Repulsion (Negative Adsorption)*

The equivalent distance at which anions are excluded from a negatively charged clay surface with constant density of charge has been described by Bolt and Warkentin (1956, 1958) for systems containing single salts. de Haan (1965) derived expressions for anion exclusions of general applicability to most mixed systems containing mono- and divalent cations and mono-, di-, and trivalent anions. The equivalent distance of exclusion is expressed as a function of three variables: (1) a factor which is a function only of the equivalent fractions of the different ions in the equilibrium solution; (2) the total electrolyte level of the equilibrium solution; and (3) a factor which is related to the surface charge density of the colloid.

3.3.2. *Slight Binding*

Since the nitrate ion does not normally react with the solid phase, its concentration in solution at any instant is simply determined by the total amount of nitrate present divided by the total amount of water present, except at low moisture contents where negative adsorption may become important. A good yielding annual crop utilizes from 100 to 150 lb of N per acre per season. Therefore, a concentration in the soil solution of N of approximately 20 meg/liter would be adequate for the entire growth of the crop, assuming 20% soil moisture and no leaching losses. The soil solution composition is seldom this high (Table 2.9 and accompanying discussion) and replacement must take place either by fertilization or biological breakdown of the organic N. Replacement of N into the soil solution from the inorganic fraction of the soil is practically nil in most soils. Small contributions are made by rainfall additions (Black, 1957).

Since NO_3 is not adsorbed by the soil to any extent, it is a very mobile nutrient and the efficiency of utilization of the N in the soil solution is very high (Legg and Allison, 1959; Jansson, 1963; Walker *et al.*, 1956; Stewart *et al.*, 1963). Lack of utilization is normally due primarily to losses owing to leaching, denitrification, or microbiological competition and not to immobility of the nutrient in relation to plant roots or limited rate of uptake by the plant. Losses of N due to leaching while the crop is actively growing are highly unlikely, except under excessive irrigation (see Chapter 2, Section 4), while losses of N due to denitrification may be severe (MacVicar *et al.*, 1951; IAEA, 1964; Walker *et al.*, 1956; Fried *et al.*, 1958). Losses due to microbiological competition may also be temporarily severe when large additions of

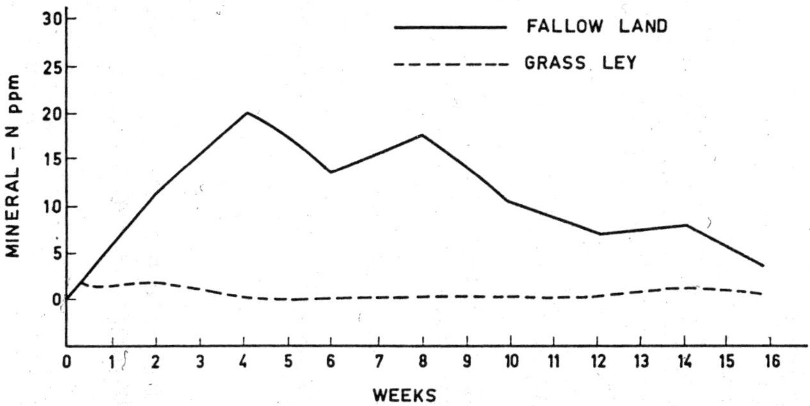

FIG. 2.9. Mineral nitrogen content in a clay soil on fallow and cropped parts of the land from May to August inclusive. From Harmsen and van Schreven (1955).

fresh organic matter with a wide C:N ratio are added to the soil system, but this N is eventually released upon death and decomposition of the microorganisms.

The native concentration of NO_3—N in the soil solution is, therefore, controlled almost entirely by the amount of N present in the organic matter and the rate at which this is turning over.

The graph in Fig. 2.9 indicates that this concentration in the field is often transient and is usually low in cropped soils. The low figures that occur periodically under fallow conditions could be related to heavy leaching by rainfall.

Nitrogen exists in the soil solution both as ammonium and nitrate ions. Since the presence of nitrate in solution is not a function of the physical chemistry of the system but primarily of the biology of the system, this constitutes a whole subject matter area not within the scope of this monograph. For further information see Alexander (1961), Waksman (1952), Harmsen and van Schreven (1955), Black (1957), and Fried *et al.,* (1958). However, one comment may be in order. Considering the major importance of N as a plant nutrient and the newer techniques available including N^{15}-labeling, chromatography, electrophoresis, etc., relatively little effort is being devoted to this subject matter area.

3.3.3. *Moderate Binding*

Sulfate retention by surface soils and other soils above pH 6.0–6.5 is generally either slight or nonexistent. Yet it may be pronounced in the case of acid soils high in Fe and Al oxides (Williams and Steinbergs, 1962; Chao *et al.*, 1963; Freney *et al.*, 1962).

In these soils, as with nitrate, the concentration in solution at any instant is simply determined by the total amount of sulfate present divided by the total amount of water, except at low-moisture contents where negative adsorption again begins to become important. However, sulfate retention can generally be demonstrated even on less retentive soils if suitable techniques are chosen (Berg and Thomas, 1959; Fang *et al.*, 1962; de Haan and Bolt, 1963).

In soils in which sulfate adsorption does not occur, sulfate is seldom supplied by the inorganic solid phase to the soil solution in any appreciable quantities except where solid-phase $CaSO_4$ may be present naturally, or by addition as an amendment or where sulfate may be an occlusion in the $CaCO_3$ (Williams and Steinbergs, 1962). The concentration of sulfate in the soil solution in these soils is determined almost entirely by the amount of S in the rainwater. This varies from area to area as a function of the atmospheric content which, in turn, is a function of the combustion of fossil fuels. This is normally a reflection of the presence of heavy industry, although fossil fuels for heating and as fuel in the railroad industry can make a substantial contribution.

Sulfur deficiency occurs where the amount of S in rainwater is very low, amounting to less than 10 lb per acre per year, such as in parts of Australia, New Zealand, Southeastern, Northwestern, and North Central (Minnesota) United States, Northern Nigeria, Ghana, East Africa, USSR, India, Brazil, Canada, and Lebanon (Williams and Steinbergs, 1962; Whitehead, 1964).

The N:S ratio in soil organic matter approximates 10:1, and the relative plant requirement also approximates 10:1. Thus, only when the organic matter is turning over fast enough to supply all the N needs of the plant will it also supply a high proportion of the S needs of the plant.

A good yielding annual crop from 1 acre of land contains from 10 to 80 lb of S. One acre plow-depth at 20% soil moisture contains approximately 400,000 lb of water. To have all the needs of the plant in the soil solution at one time would require a concentration in solution of approximately 90 ppm or 1×10^{-3} *M* sulfate concentration.

An analysis of the soil solution for sulfate in other than alkaline soils is often below the ability of present techniques to detect (Whitehead, 1964). Water soluble S in low S soils will be undetectable in many soils even where 5 ppm is easily detectable (Freney, 1958; Kilmer and Nearpass, 1960). Other extractants have been suggested in order to extract enough S to enable ready detectability (Kilmer and Nearpass, 1960; Chesnin and Yien, 1950; Bardsley and Kilmer 1963).

The two factors that primarily determine the concentration of S in the soil solution are the S in the rainfall and the reaction with the solid phase, if any, with the usual relatively minor additions from organic matter decomposition.

The concentration in the soil solution, M(solution), is a function of this amount as modified by the reaction with the solid phase. The nature of this reaction has been variously studied with only limited success. The suggested mechanisms for this adsorption have varied from adsorption or interaction with organic matter (Chao *et al.*, 1962), bridged to the solid phase through exchangeable Ca (Birch, 1951), adsorption on positive sites on the clay surface (de Haan and Bolt, 1963), adsorption by hydroxide replacement from surface sesquioxides (Jackson, 1963; Mattson, 1931), simultaneous Al hydrolysis cation replacement and sulfate adsorption (Chang and Thomas, 1963), positive charges developed at low pH levels on soils high in sesquioxides, (Chao *et al.*, 1963), substitution in the clay lattice (Schell and Jordan, 1959) to molecular adsorption (Ayres and Hagihara, 1953).

None of the mechanisms proposed satisfies all the observations. It was pointed out by Chang and Thomas (1963) that any proposed mechanism must take into account their observation that when a salt such as K_2SO_4 is added to a soil both cation and anion simultaneously disappear from solution. However, the adsorption of cation to planar surfaces and the adsorption of anions to positive charge groups in exchange for OH^- or other anions refer to entirely different mechanisms that are governed by the composition of the outside solution in a different qualitative and quantitative way. The stoichiometry as observed by Chang and Thomas is therefore likely to be highly fortuitous (de Haan, 1965).

A likely mechanism for this adsorption in mineral soils and subsoils is a replacement of sulfate of a hydroxyl associated with Al of Fe in a mechanism similar to phosphate adsorption followed by adsorption of the cation through a neutralization reaction. In this case, however, the affinity of the sulfate ion would be much less. If this were the mechanism, then the intensity of adsorption at a given pH should be predictable on the basis of an affinity constant or exchange coefficient (Fang, 1962).

3.3.4. *Strong Binding*

Phosphorus as orthophosphate reacts with practically all soils with an almost quantitative removal from solution. This is not surprising as soils characteristically have very reactive surfaces containing Fe, Al, Ca, etc., all of which form very insoluble phosphates. Thus, in a heterogeneous system such as soil containing many ions which will remove phosphate from solution, it would be surprising to find any appreciable quantity in solution.

The N:P ratio in soil organic matter is similar to that of S, i.e., approximately 10:1 and the relative plant requirement is a similar order of magnitude.

Theoretically, the P needs of the plant are always adequately supplied if the organic matter turns over fast enough to supply adequate N. Practically this is actually not valid. Two reasons for this lack of validity other than the much wider N:P ratio that is found in some soils are: (1) the P thus released reacts with the mineral fraction of the soil, thus decreasing the likelihood that this released P will be taken up by the plant and (2) P is a relatively immobile nutrient because of its low concentration in the soil solution, (see Chapter 2, Section 4). If the plant roots only derive their P from 1/10 or 1/25 of the soil volume, then only 1/10–1/25 of the released P will be in this feeding zone. Nitrogen released by the organic matter turnover neither reacts with the soil nor is immobilized. Thus, the efficiency of utilization of the released N and P during the growing season may differ by a factor of 25 to 100. In fact, in greenhouse tests the recovery of applied N as shown by data obtained with labeled N fertilizer was from 50 to 70% (Legg and Allison, 1959) in spite of the occurrence of NH_4^+ fixation. In contrast, the percentage utilization of soluble phosphate applied as P^{32}-labeled fertilizer is seldom above 10% and is often much lower (Dean *et al.*, 1948; Andersen *et al.*, 1961; McConaghy and Stewart, 1963). Of course, when the organic matter turnover is not sufficient to supply even the N needs of the plant, it would only represent a negligible supply of P during the growing season. Over a longer time period, however, organic matter may represent an appreciable source of P, providi- there is a net release of P (i.e., that microorganisms do n- -p the P in the organic matter at the same rate that it is bein- l). Since the C:P ratio in the soil tends to remain the sa- , this m there would then have to be an actual decrease in the organic matter content of the soil.

Considering all of the above factors, the organic matter is not normally a major source of P nutrition to the growing plant, although in local situations of very high organic matter soils or very sandy soils containing appreciable organic matter or other soils continuously decreasing in organic matter content, organic matter as a source of P may be temporarily important.

An examination of the soil solution data, as shown in Table 2.9, indicates a level of phosphate usually of the order of 10^{-5} to 10^{-6} *M*. A good yielding annual crop will contain from 25 to 50 lb of P per acre in the total plant material. Taking the figure of 20% moisture in the soil, it is readily seen that at any one time there is not enough P in the soil solution to supply the plant for more than a few hours. The soil solution usually must be renewed many times per day or even more than once per hour to supply the needs of the plant (Fried and Shapiro, 1956; Parker, 1927). The continuous renewal of the soil solution with phosphate and the level of phosphate maintained in solution characterizes the soil phosphate system. This renewal pattern varies markedly from soil to soil (Fried and Shapiro, 1956).

The attempt to describe the solid-phase phosphate supply in relation to solution characteristics has taken essentially three forms: (1) chemical potentials, (2) solubility products, and (3) surface adsorption equations. These descriptions are not intended to apply to the initial reactions that occur upon the addition of fertilizer salt to soil (Lehr *et al.*, 1959; Lindsay and Stephenson, 1959; Lindsay *et al.*, 1959).

a. *Chemical Potentials.* Schofield (1955) suggested that "quite general considerations point to the chemical potential of monocalcium phosphate as the function most likely to give a numerical index of the condition in the soil which mainly controls the availability of phosphate." This reasoning is based on the stated working hypothesis "that availability of soil phosphate is mainly determined by the appropriate chemical potential and by its rate of decrease with phosphate withdrawal." As performed by Aslyng (1954) and described by Schofield (1955), phosphate potential is defined as $\frac{1}{2}pCa + pH_2PO_4$ in a 0.01 *N* $CaCl_2$ solution. Interestingly enough, since the measurement is in standard 0.01 *N* $CaCl_2$ concentration, $\frac{1}{2}pCa$ is essentially constant and the only variable is the logarithm of the reciprocal of the molar $H_2PO_4^-$ ion in solution, i.e., $pH_2PO_4^-$. This is an intensity factor and is essentially the soil solution H_2PO_4 concentration when the soil contains 10^{-2} *M* Ca^{2+}. Since, as shown in Table 2.9, the soil solution concentration of Ca is more generally closer to 10^{-3} *M* but often goes as high as 10^{-2} *M*, the choice of 1[illegible]

M is within the range encountered in soils. Not only is the use of phosphate potentials to characterize plant nutrient supply attendant with difficulty, but even using phosphate potentials as a soil characterization may not be rewarding. As a means of characterizing the phosphate sys-

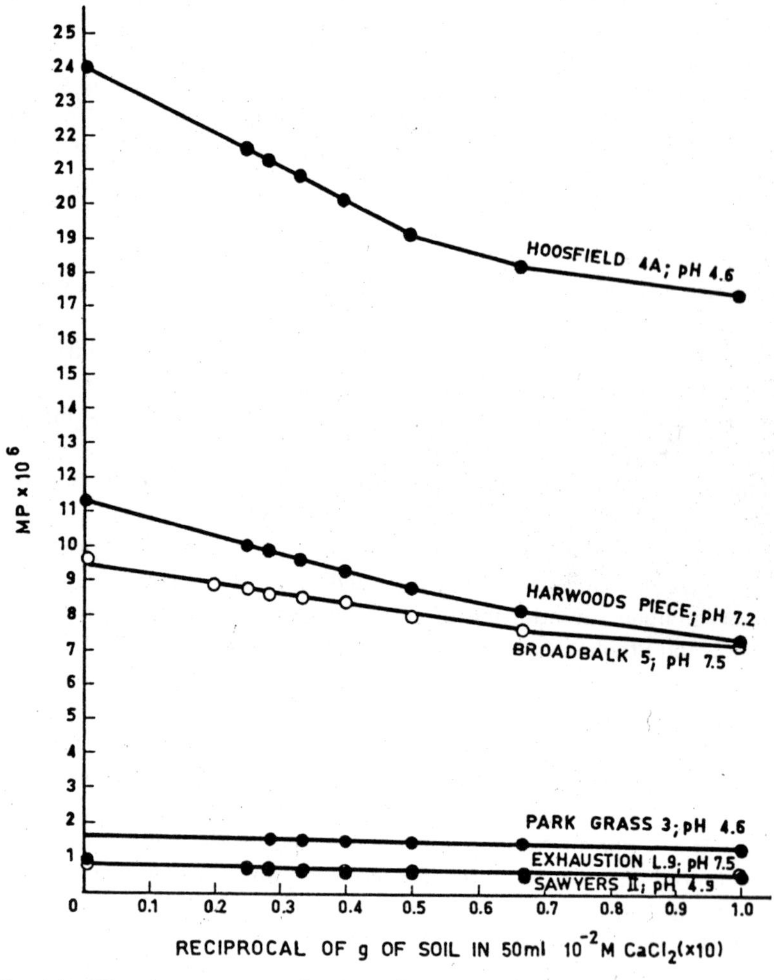

FIG. 2.10. Phosphate concentration in 10^{-2} M $CaCl_2$ for different liquid–soil ratios. From Aslyng (1954).

tem of the soil, one of the major difficulties is the effect of soil–water ratio on the monocalcium phosphate potentials (Aslyng, 1954; Larsen and Court, 1960; Wild, 1959; Fordham, 1962). Typical results are presented in Fig. 2.10. Since $\frac{1}{2}pCa^{2+}$ is constant at all soil-solution ratios (as well as the pH), the monocalcium phosphate potential varies directly with the concentration of phosphate. Aslyng (1954) suggested an extrapolation procedure to get around this difficulty, whereas Fordham (1962) suggested an interpolation procedure. Other environmental effects, such as the effect of other cations, anions, and organic matter, are known to affect the phosphate concentration in solution, but may or may not affect the monocalcium phosphate potential. There is no question but that an empirical method can be set up to measure the monocalcium phosphate potential of soils that will vary less than the phosphate ion concentration. Since the solid phosphate phase is definitely not monocalcium phosphate (the potential being a measure of the intensity factor only and does not take into account the capacity factor) and since there is no particular indication that phosphate is taken up by the plant as monocalcium phosphate, it is difficult to see how monocalcium phosphate potential as such is an improvement over the measurement of phosphate ion concentration in the soil solution. Thus, although phosphate potentials have been measured by many investigators (Henze, 1963; Ulrich, 1961; Mattingly *et al.,* 1963; Gough and Beaton, 1963; Scheffer and Ulrich, 1960; White and Beckett, 1964) and can possibly be standardized to measure a reproducible soil property which might be useful for categorizing soils in a chemical framework, it neither characterizes the soil in a plant nutrition framework nor helps in the understanding of the nature of the continuous soil supply of P to plants.

b. *Solubility Products.* Solubility products and even mineralogical identification have been used to characterize the soil phosphate system (Clark and Peech, 1955; Cole and Jackson, 1950; Cole and Jackson, 1951; Cole and Olsen, 1959; Olsen *et al.*, 1960; Haseman *et al.*, 1950; Kittrick and Jackson, 1954, 1955, 1956; Nagelschmidt and Nixon, 1944; Burd, 1948; Ghani and Islam, 1946; Swenson *et al.*, 1949; Jackson, 1963). The basic principle of this characterization is simply that if one could identify the solid phase and determine its solubility product, the concentration in solution could be predicted at all times from a knowledge of the concentration of the associated cation. Although the principle is simple, carrying it out is not so simple for two basic reasons. First, no one has yet been able to identify with certainty the existence of a phosphate salt as the dominant species influencing the soil solution concentration of phosphate and, second, instead of finding a constant concentration of phosphate in solution with dilution, as might be expected for the equilibrium solution of a well-defined phosphate salt, a marked change of phosphate concentration occurs with dilution for many soils, as shown in Figs. 2.10 and 2.11. Although it is evident that the soil solution concentration can change with dilution even in the presence of a specific compound of phosphate, the inability to use solubility criteria, the inability to identify the solid phase, and the inability to predict the soil solution concentration is a severe hindrance to the utility of solubility product criteria for describing the system. In addition, incongruent solubility processes are prominent in soils and adsorptive interactions modify the kind of substances and their solubility. Nevertheless, a combination of phosphate potential and solubility product criteria can be used to distinguish one soil phosphate system from another, as

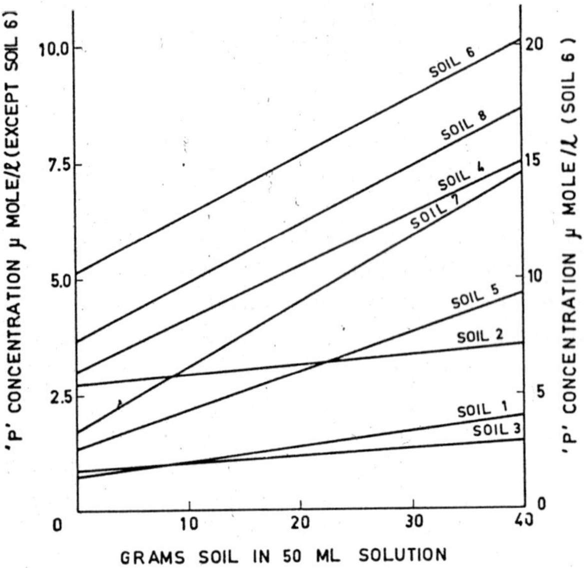

FIG. 2.11. Variation of "P" concentration with soil–solution ratio. From Larsen and Court (1960).

demonstrated by Aslyng (1954), Cole and Olsen (1959), Clark and Peech (1955), and others, such as Gough and Beaton (1963) and Olsen *et al.* (1960). This is shown in the solubility diagram in Fig. 2.12. Clark and Peech (1955) show such a diagram with pH $-$ $\frac{1}{2}$pCa as abscissa and $pH_2PO_4 + \frac{1}{2}pCa$ as ordinate. The lines represent the specific chemical compounds—monocalcium phosphate, dicalcium phosphate, hydroxylapatite, and $CaCO_3$ to various CO_2 pressures. It is only necessary to extract the soil solution and to determine Ca ion concentration, pH, and phosphate ion concentration to locate a soil on this diagram. It can then be said that a particular soil may be characterized by that point on the diagram. It is, however, quite another thing to be able to say that putting a stress on the system, such as adding or removing Ca or phosphate, will move the point along the line. This has yet to be demonstrated. An observation of interest in this connection is that of Cole and Olsen (1959) who have shown a constant dicalcium phosphate activity for three soils that were varied in their monocalcium phosphate and $Ca(OH)_2$ potentials. Although the dicalcium phosphate activity was constant under these conditions, it was not constant at

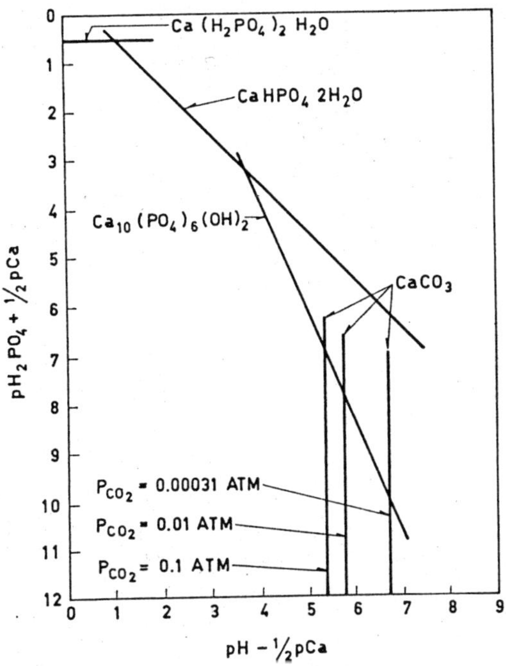

FIG. 2.12. Solubility diagram for the calcium phosphates. From Clark and Peech (1955).

different soil-solution ratios and not equal to the theoretical value. Nevertheless, calcareous soils often fall within the solubility limits of hydroxylapatite and dicalcium phosphate, although not always (see Aslyng, 1954); Gough and Beaton (1963) have even reported points above the dicalcium phosphate line for the two calcareous soils they investigated. Clark and Peech (1955) point out that for natural or acid soils the points invariably fall below the hydroxylapatite line, suggesting the presence of a less soluble phosphate compound. They have postulated iron and aluminum phosphate and present the solubility diagram for the aluminum phosphates which is reproduced here as Fig. 2.13. If the data were available, a similar diagram could be made for ferric phosphate.

The solubility diagrams are useful physicochemical descriptions of pure chemical phosphates such as might be necessary for process predictions in the fertilizer industry, but it is doubtful whether locating a soil on one of these diagrams bears any necessary direct relationship to plant nutrient supply. In distinguishing one soil from another as to

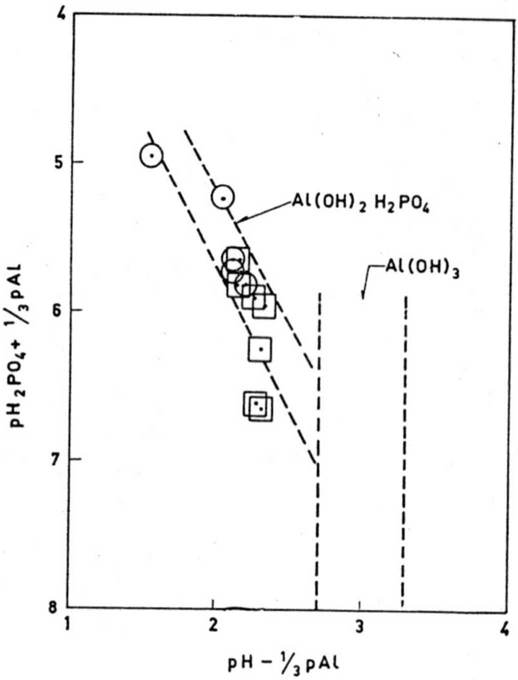

FIG. 2.13. Solubility diagram for the aluminum phosphates; ⊙, data from Cole and Jackson; ⊡, present experimental data. From Clark and Peech (1955).

its dominant phosphate chemistry, however, it may be useful to say this soil falls on the dicalcium phosphate solubility line as compared to another that may fall on a dihydroxyferric solubility line. To some extent, this is therefore an improvement over a chemical description of the soil in terms of monocalcium phosphate potentials.

c. *Surface Adsorption Equations.* The third means of characterizing the solid-phase supply in relation to solution characteristics is by the use of surface adsorption equations (Davis, 1935; Metzger, 1940; Scarseth, 1935; Fried and Dean, 1955; Kurtz *et al.*, 1946; Russell and Low, 1954; Fried and Shapiro, 1956; Shapiro and Fried, 1959; Schofield,

1947; Antipov-Karatayev, 1956; Cole *et al.*, 1953; Olsen and Watanabe, 1957; Dean, 1949; de Haan, 1965). While this characterization does not necessarily identify the solid phase, it does have the advantage of describing the system with a set of constants which, when known, can be used to predict the effect of placing a stress on the system.

Kurtz *et al.* (1946) were able to describe phosphate adsorption by the Freundlich equation which contains two empirical constants and can

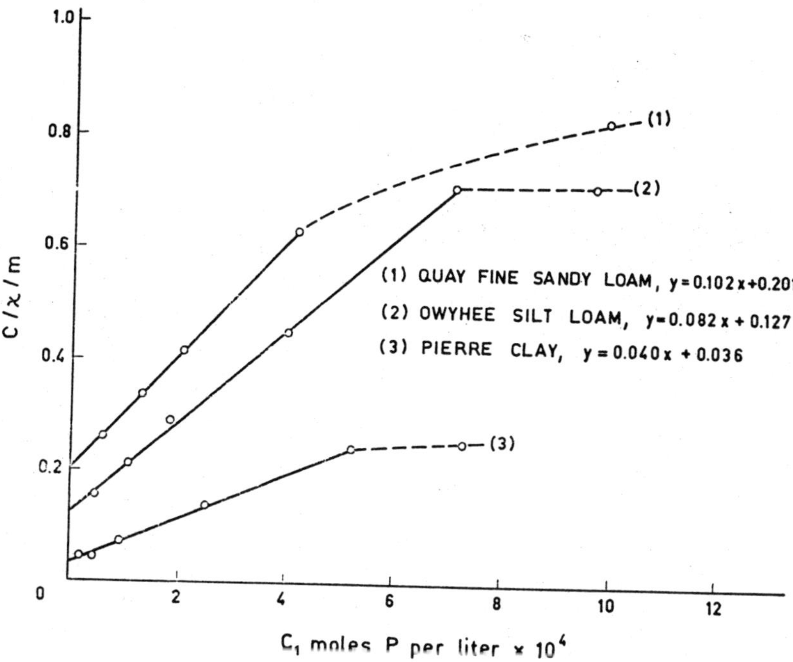

FIG. 2.14. Phosphorus data for three soils plotted according to the Langmuir isotherm. From Olsen and Watanabe (1957).

predict the amount adsorbed from the solution concentration. Actually, the results of these authors deviated from the Freundlich adsorption line when the concentration went below 1 ppm. Olsen and Watanabe (1957) and Fried and Shapiro (1956) were able to describe calcareous and acid soils, using the Langmuir equation which also contains two constants and can predict the amount adsorbed from the concentration in solution. However, these two Langmuir constants have additional significance: the one describes an adsorption maximum and the other is related to the energy of adsorption.

The results of Olsen and Watanabe (1957) are presented in Fig. 2.14

for three soils. They represent a plot of the data in a linear form:

$$c/(x/m) = 1/Kb + c/b \tag{2.10}$$

where c is the equilibrium P concentration; K, a constant related to the bonding energy of the adsorbent for the adsorbate; b, the adsorption maximum and x/m, amount of P adsorbed per unit weight of soil.

Thus, from the slope $1/b$ and the y intercept $1/Kb$ both b and K can be calculated. Interestingly enough, the adsorption maximum was found to vary with different soils in the range of 10 to 33 mg P per 100 g of soil or 200 to 660 lb per acre, and the P concentration in solution at that maximum was over 10 ppm. These are values seldom reached in soils, although they are possible in the neighborhood of the fertilizer band.

Using mass action considerations, an expression formally analogous to the Langmuir isotherm was derived by Shapiro and Fried (1959) on the assumption that phosphate adsorption was primarily an exchange with another ion such as hydroxyl.

The final equation is:

$$\text{P(solid)} = -K_p\text{P(solid)}/p + \text{P(solid)}_{\text{max}} \tag{2.11}$$

where P(solid) is the amount of P adsorbed by the soil; P, the concentration of P in solution; $\text{P(solid)}_{\text{max}}$, the amount of P adsorbed by the soil when all the adsorption site is saturated with P; and K_p, apparent dissociation constant.

In this form of the equation, a plot of P(solid) vs. P(solid)/P(solution) should give a straight line for a site with uniform points of adsorption, i.e., a constant energy of adsorption where the slope of the line is the apparent dissociation constant, K_p, and the y intercept is $\text{P(solid)}_{\text{max}}$, the maximum amount of P adsorbable by that site. Figure 2.15 gives an example of what would theoretically be obtained for a soil with a maximum adsorption by a particular site of 10×10^{-6} moles P per gram of soil, 200 lb of P per 2,000,000 lb of soil, and an apparent dissociation constant, K_p, of 5×10^{-6}. These two constants, $\text{P(solid)}_{\text{max}}$ and K_p, completely define the line. If a soil could be characterized by this line, then, knowing the two constants, the amount in solution and the amount on the surface can be calculated for any quantity of phosphate associated with this adsorption reaction without first knowing the distribution of the P between the solid and the solution. Of course, the amount on the surface could easily be calculated by knowing the amount in solution or vice versa. Since this is a polar coordinate-type plot, the constant concentration lines radiate from the origin and all points on the same line have the same concentration. Thus, at 50% saturation, 5×10^{-6} moles of P adsorbed per gram of soil, the equilibrium concentration in solution would be 5×10^{-6} *M* phosphorus. At 85% saturation it would be 3×10^{-5} *M* phosphorus (approximately 1 ppm) and at 40%, 3.3×10^{-6} *M* phosphorus (approximately 0.1 ppm).

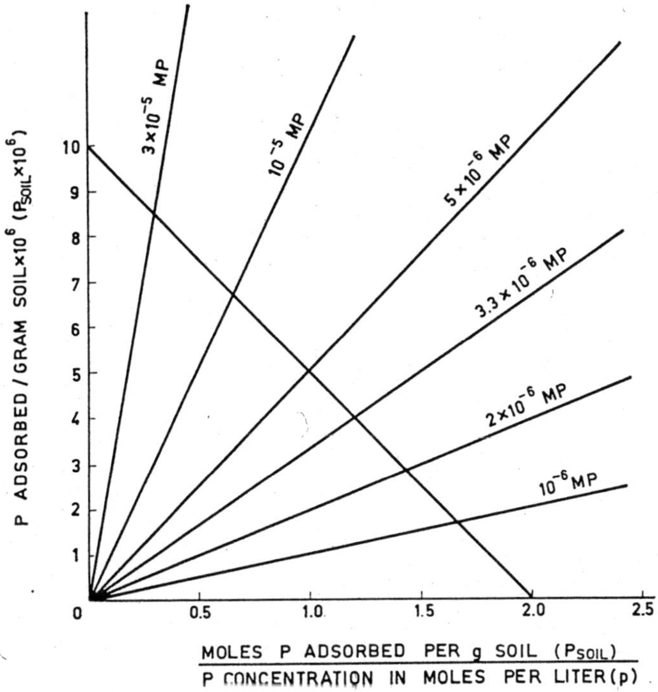

FIG. 2.15. Theoretical plot of phosphorus adsorbed vs. phosphorus adsorbed divided by the concentration for a soil with a maximum adsorption at one site of 10×10^{-6} moles per P per gram of soil (200 lb of P/2,000,000 lb) and an apparent dissociation constant, *Kp,* of 5×10^{-6}. From theoretical diagram of Shapiro and Fried (1959).

These are values obtained for a theoretically drawn curve. What is obtained for real soils under practical conditions? Only one publication in the literature has plotted phosphate data in this way (Shapiro and Fried, 1959).

Fried (1959) and Heald (1960) have plotted adsorbed Na and adsorbed Ca according to the same type of relationship. For Na and Ca the data obtained can be of the same form as shown in the theoretical single site picture given in Fig. 2.15. The one published figure for soil phosphorus, however, which is reproduced below (Fig. 2.16) shows a hyperbolic function that can be broken into the summation of two lines, as suggested and done by the authors. The same authors showed other justifications for assuming that two different reactions were controlling the soil phosphate system, one that dominated at relatively high concentrations of P in solution and the other that dominated at much lower P concentrations. Thus, to fully describe the soil system four constants would seem necessary, two maxima and two apparent dissociation constants. Actually, if the example given is typical, and the results given

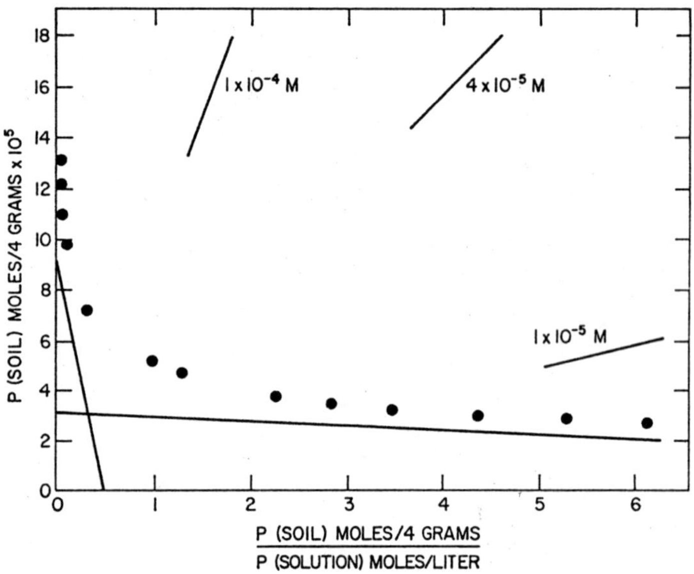

FIG. 2.16. The effect of P concentration on the uptake of P by 4 g of Caribou loam. The uptake from solutions of various P concentrations is plotted as a function of uptake divided by the equilibrium P concentration. From Shapiro and Fried (1959).

for three other soils suggest they are, the lower site usually dominates at a concentration less than 1 ppm P in solution, the concentration usually found in the soil solution. Of course, in the neighborhood of the fertilizer band, where higher phosphate concentrations predominate owing to movement of P away from the band, the upper site would be more significant. (The initial fertilizer–soil interaction must reasonably be described in other terms). That this can be of real utility and appears significant was clearly demonstrated by the same authors, but unfortunately only for one soil. They showed that varying the soil–solution ratio by a factor of 400 with the same total P present or varying the total P by a factor of 40 at the same soil–solution ratio only served to move the point up and down the same curve. In other words, everyone of the 28 points tested could have been calculated from just the knowledge of the constants.

More recently, the same approach was used by de Haan (1965) to determine the apparent dissociation constants and maxima for the adsorption of phosphate on clays. He also found that two different reactions were controlling the clay-phosphate system and his results are summarized in Table 2.13. Four constants again were adequate to describe these clay-phosphate systems.

TABLE 2.13
REACTION CONSTANTS AND MAXIMUM VALUES FOR THE ADSORPTION OF PHOSPHATE ON CLAYS[a]

	Reaction constants (μg P/ml)$^{-1}$		Adsorption (μg P/g clay)	
Clay	b_1	b_2	P_{m_1}	P_{m_2}
Montmorillonite (osage)	0.9	0.006	30	260
Illite (winsum)	2.7	0.045	20	460
Illite organic matter removed	2.7	0.042	140	800
Kaolinite (drybranch)	0.8	0.005	15	?

[a] From de Haan, Thesis (1965).

Actually the determination of anion adsorption should involve the correction for anion repulsion (negative adsorption) (de Haan and Bolt, 1963). This means that the net P adsorption found from the determination has to be corrected for negative adsorption to find the amount that is actually adsorbed. The data of de Haan quoted above were corrected by him before plotting. The extent of this correction may be fairly large as indicated in Fig. 2.17 (taken from de Haan, 1965), showing the relationship between phosphate adsorption and solution concentration as affected by the presence of organic matter. The values before and after correction are recorded. The data not only indicate the magnitude of this correction, but also that the presence of organic matter can appreciably reduce adsorption, presumably by partially saturating the adsorption sites.

Since this method of approach is only a mass action derivation that is analogous to a Langmuir isotherm and Hitchcock (1926) long ago showed the possible formal identity of the two, data in the literature that have been plotted as a Langmuir isotherm can be replotted this way. While there is essentially no mathematical difference, there is a difference for graphical purposes of the visual weighing of the points.

The data given by Cole and Olsen (1959) in their Table 2 for soil clays and Wyoming bentonite, are replotted in Fig. 2.18. Although the authors state adsorption maxima of 53, 59, and 81 mg P per 100 g for the various clays, all showed higher amounts associated with the solid phase at the highest P concentrations. Figure 2.18 shows clearly that at least two sites of adsorption were present and the adsorption

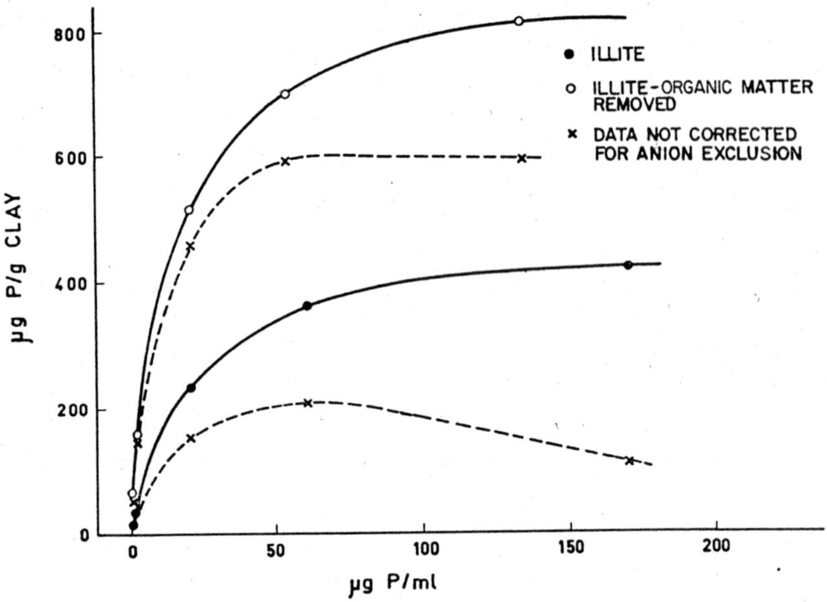

FIG. 2.17. Phosphate adsorption isotherm for illite, before and after removal of organic matter. From de Haan, Thesis, 1965.

maxima given by the authors was higher than the adsorption maxima for the site dominant at low concentrations that would be obtained if the curve were broken into two components. The slopes of all three curves were essentially the same, showing that the apparent dissociation constants were similar in these three clays. Interestingly enough, they are the same order of magnitude as those reported by Shapiro and Fried (1959) for three entirely different soils. Again, at concentrations of P usually present in the soil solution, i.e., 1 ppm, the lower site of adsoption appears to predominate. If a generalization can be made with regard to the apparent dissociation constants of soils, or even groups of soils that differ only in their maximum capacity for adsorption at a given site, then per cent saturation of the site would determine the concentration in solution. The use of per cent saturation as a useful criterion was demonstrated by Cole and Olsen (1959) and Olsen and Watanabe (1957), providing the maximum sorption capacity is determined by the proper adsorption equation and not by the so-called anion exchange capacity determination. Heald (1960) obtained the same adsorption

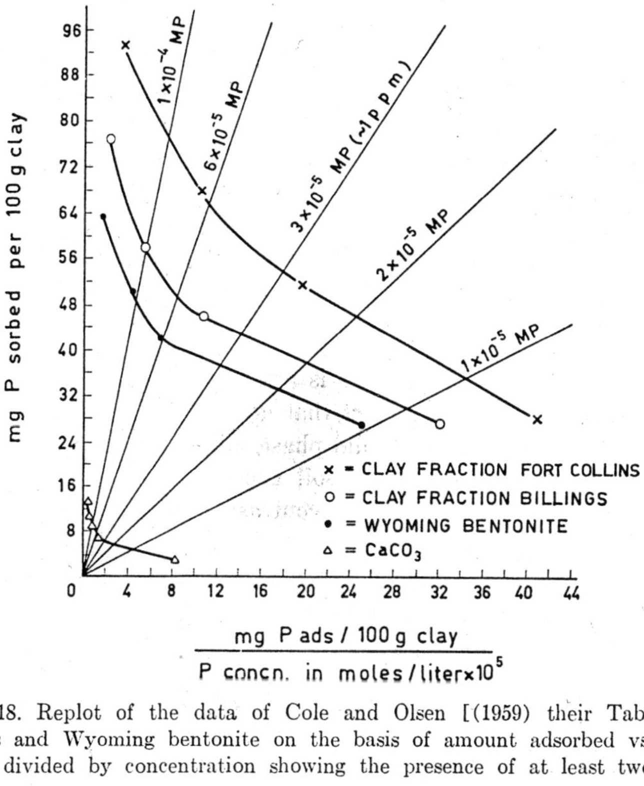

FIG. 2.18. Replot of the data of Cole and Olsen [(1959) their Table 2] for soil clays and Wyoming bentonite on the basis of amount adsorbed vs. amount adsorbed divided by concentration showing the presence of at least two sites of adsorption.

maximum for Ca using adsorption equations and ammonium acetate leaching. Perhaps the anion exchange capacity is more closely related to the $P(\text{solid})_{max}$ obtained from the sites that predominate at higher phosphate concentrations.

There is one difficulty attendant on this determination when dealing with real soil systems, and that is the difficulty of determining the amount already adsorbed on the surface of the soil. In most cases, this has been ignored and the amount removed from solution simply plotted as the amount adsorbed. This can be one of the reasons for the deviation (e.g., the results of Kurtz *et al.*, 1946) at lower P concentrations that was previously noted (p. 47). One means of overcoming this difficulty is by making a reasonable estimate of the amount of surface P present, using the principle of P^{32} exchange in calculating the E value (discussed in Chapter 6, Section 1.3.1). This correction is important, for it is just at these amounts of adsorbed P that the soil characteristics are most usefully described.

Of the three means of describing the soil phosphate system in relation to solution concentrations, (1) phosphate potential, (2) solubility diagrams, and (3) adsorption equations, the adsorption equation appears to be the most useful description for purposes of plant nutrition. The adsorption equation depends only upon the ion of interest and can predict the response of a stress upon the system caused by the addition or removal of the ion under consideration. They apply primarily to the soil system at equilibrium or pseudo-equilibrium and not to the initial reactions occurring right at the fertilizer band.

The quantitative description of the solid-phase soil–solution relationships of cations could be generalized, but that of anions require, in general, individual treatment. This is primarily because charge and size of cations are the primary factors that govern the relationship between one cation and another in this solid-phase solution system. Those anions, however, that are bound by the soil generally reflect both a sharing of electrons and covalent bonding in contrast to the electrovalent bonding of the cations. Thus, the association of the anions with the solid phase are infinitely more complex than cations and not suitable for general treatment. Nevertheless, the individual anions can be quantitatively treated and, indeed, must be so treated in order to predict M(solid) and M(solution) and the relationship between them for any given situation.

4. Movement of Ions to the Root

The movement of ions in soils under agricultural conditions occurs primarily through the water phase. In considering the movement of ions in soils, two general aspects are pertinent: movement that accompanies water flow in response to gravitational potential and movement caused by gradients of ions owing to the presence of plant roots. When water moves, ions move with it. Movement by diffusion also occurs, but only over much shorter distances.

4.1. Movement of Ions Not Adsorbed by the Soil

The movement of those ions not appreciably adsorbed by the soils is governed primarily by the movement of water. Of the nutrient ions this includes NO_3^-, Cl^-, and in most but not all surface soils SO_4^{2-}. The nutrients that are not adsorbed by soil tend to leach through the profile. The restraining forces, however, on the soil water are considerable, particularly as a consequence of both evaporation and transpiration which tend to keep nutrients in the root zone. In addition, during the growing season, any applied nutrient cannot move farther than the rainfall moves.

The movement of the rainfall into the profile in general is predictable, for water generally moves in a water front, displacing but not bypassing water already present. Therefore, when evaporation or evapotranspiration exceeds precipitation, there is no net movement downward. Since an average figure for evapotranspiration for a growing crop is about 0.2 inch per day, rainfall must exceed this approximately 6 inches per month to result in net downward movement. The distance moved will also depend upon the water-holding capacity of the soil, which varies from quite low in sandy soils to rather high values in heavy soils.

Under normal humid-region conditions, rainfall per month may approximate 3 to 5 inches and the soil on the average can hold one-third of its volume in the form of water. The movement per month in these soils from an initial starting point will be a maximum of 9 to 15 inches under these conditions, not counting evapotranspiration. In a month roots will normally grow that far, if not farther, and considering the upward movement due to evaporation and transpiration, including the utilization of water by the growing plant, the nutrient is not likely to be leached out of the zone of influence of the plant. For sandy soils with much poorer moisture-retaining properties, and for high rainfall incidence as may occur in the tropics, the leaching of nutrient during the growing season may occur to some degree. For heavier soils and drier conditions not only is leaching not a problem but salt accumulation may even become serious. Even under average soil and rainfall conditions, however, leaching can be serious, considering that the soil may not have continuously growing crop cover and considering that the annual rainfall under average humid-area conditions is about 25 to 50 inches.

4.2. Movement of Ions Adsorbed by the Soil and in the Soil Solution

The gross movement of those ions which can be adsorbed by soils is also governed to a large extent by water movement. The concentration of the ion in the soil water, however, and hence the amount of the ion that moves along with the water, is governed by the physicochemical relationship between the ion in solution and on the solid phase, and the presence of the other ions in the system. Of the nutrient ions this includ-

all the cationic forms, phosphates, borates, zincates, molybdates, etc., or essentially all nutrients except N in the form of nitrate, Cl in the form of chloride, and to some extent S in the form of sulfate. Thus, even if water movement is predictable, the movement of adsorbed ions may vary by several orders of magnitude depending upon the relationship between the ion in the adsorbed phase and that in solution. A quantitative prediction of this relationship is fundamental to prediction and control of the fate of nutrient and other ions in soil.

A typical humid-region soil may have 10 meq of adsorbed Ca, 1 meq of adsorbed Mg, 0.25 meq of adsorbed K, 0.1 meq of adsorbed Na, etc. per 100 g of soil. At the moisture equivalent this amount of soil may contain 25 ml of water or an order of magnitude of some 10–25 μeq of each of these cations in solution. Thus at the moisture equivalent, only $\frac{1}{10}$ to $\frac{1}{100}$ of the cation is in solution at any given time and only this amount can move with the water stream. At lower moisture contents the total amount, and thus the proportion in solution, is generally lower.

For phosphate the relationship is even more extreme. The amount of relatively rapid exchangeable P as determined by P^{32} exchange (see Chapter 6, Section 2.3) may be in ordinary soil, about 1.5 to 2.0×10^{-4} moles per 100 g with an order of magnitude of 0.1 μeq in the associated soil solution. Thus only $\frac{1}{1500}$–$\frac{1}{2000}$ of the phosphate is in solution at any given time, and only this can move with the water movement.

The movement of ions by diffusion, in contrast to movement with the water stream, only occurs over very short distances during a growing season and follows the diffusion laws.

Even when concentration gradients are high, as in the neighborhood of the fertilizer band, the movement of an ion, such as phosphate, is limited to a maximum of a few centimeters (Heslep and Black, 1954; Benko, 1962).

The diffusion coefficient of ions in pure water is very accurately known (Washburn, 1929). If the Einstein formula

$$\sqrt{\overline{\Delta x^2}} = \sqrt{2Dt}$$

is used to estimate the root mean square of the movement of an ion in soil or clay pastes, 1 day may be substituted for t in the formula and the values of the diffusion coefficients of Table 2.14 substituted for D. The order of magnitude of the diffusion distance in soils would then be less than 1 cm per day. At normal soil solution concentrations, the concentration gradients cannot be high and diffusional movement is even more limited. Although the movement of nutrient ions in soil

TABLE 2.14
SUMMARY OF REPORTED APPARENT DIFFUSION COEFFICIENTS IN SOILS AND ANALOGOUS SYSTEMS

Ion species	D (cm²/sec)	Medium	Other factors	Ref.
P	$<5 \times 10^{-9}$	Soil	Moisture 20%	*a*
P	$2 - 4 \times 10^{-11}$	Soil	Autoradiographs	*b,c*
P	$\sim 1.5 \times 10^{-14}$	Soil	Extracted with resin	*d*
P	$0.9 - 6.7 \times 10^{-7}$	Soil	Calculated, not determined, soils and moisture	*e*
Rb	$6.4 - 44 \times 10^{-8}$	Soil	Moisture	*f*
Rb	$1.2 \times 10^{-7} - 5.5 \times 10^{-12}$	Soil	Moisture and bulk density	*g*
Rb	1.4×10^{-9}	Soil	Autoradiographs	*b,c*
Mo	$0.5 - 8.4 \times 10^{-7}$ (Dp/b)	Soil	b in Dp/b is a correction factor for interaction with soil	*h*
Cl	$5 - 55 \times 10^{-7}$	Soil	Moisture, texture	*i*
Cl	$1.2 - 10^{-6}$	Clay paste	Salt concentration, $10^{-1} - 10^{-2}\,N$	*j*
Cl	1.2×10^{-6}	Clay paste	Salt concentration, $10^{-1} - 10^{-2}\,N$	*j*
NO_3	$0.3–1.3 \times 10^{-5}$	Silt	Moisture	*s*
NH_4	1.4×10^{-6}	Bentonite clay	No diffusion at 25% relative humidity	*k*
K	1.4×10^{-6}	Bentonite clay	No diffusion at 25% relative humidity	*k*
Li	$5–15 \times 10^{-6}$	Bentonite clay	—	*t*
Na	$4–10 \times 10^{-6}$	Bentonite clay	—	*t*
Na	3×10^{-6}	Bentonite clay	—	*u*
K	2×10^{-6}	Resin	—	*l*
Na	1.8×10^{-7}	Dowex 50	—	*m*
Na	$1 - 9.4 \times 10^{-7}$	Exchange resin	—	*n*
Ca	3×10^{-7}	Exchange resin	—	*o*
Sr	3×10^{-8}	Exchange resin, amberplex membrane	—	*p*
Ba	$1 - 4.5 \times 10^{-7}$	Vermiculite	—	*q*
Rb	$6 - 90 \times 10^{-7}$	Glass beads, moisture 75 μ and 200 μ	—	*r*

[a] Lewis and Quirk (1964).
[b] Vasey and Barber (1963).
[c] Barber *et al.* (1963).
[d] Vaidyanathan and Talibudeen (1962).
[e] Olsen *et al.* (1962).
[f] Patil *et al.* (1963).
[g] Graham-Bryce (1963).
[h] Lavy and Barber (1964).
[i] Porter *et al.* (1960).
[j] Dutt and Low (1962).
[k] Husted and Low (1954).
[l] Gregor *et al.* (1951).
[m] Merriam *et al.* (1952).
[n] Boyd and Soldano (1953).
[o] Spiegler and Coryell (1953).
[p] Lopez-Gonzales and Jenny (1958).
[q] Keay and Wild (1961).
[r] Klute and Letey (1958).
[s] Romkens and Bruce (1964).
[t] Mokady and Low (1966).
[u] Kemper and van Schalk (1966).

by diffusion has been the subject of many investigations, the relative importance of this process in soil–plant nutrition is yet to be determined.

Considering a cube of unit volume in a solution in which the solute is not at a constant concentration and neglecting possible activity effects, the flow of solute per unit time in a given direction through face AB of the cube, which is normal to the concentration gradient, is proportional to the concentration gradient as shown in Eq. (2.12)

$$\Phi = -D\,\partial c/\partial x \tag{2.12}$$

where Φ is the flux; c, the concentration; and x, the distance. The minus sign indicates that the flow occurs in the direction opposite to concentration gradient $\partial c/\partial x$. The proportionality constant between the concentration gradient and the flux of matter through a unit surface normal

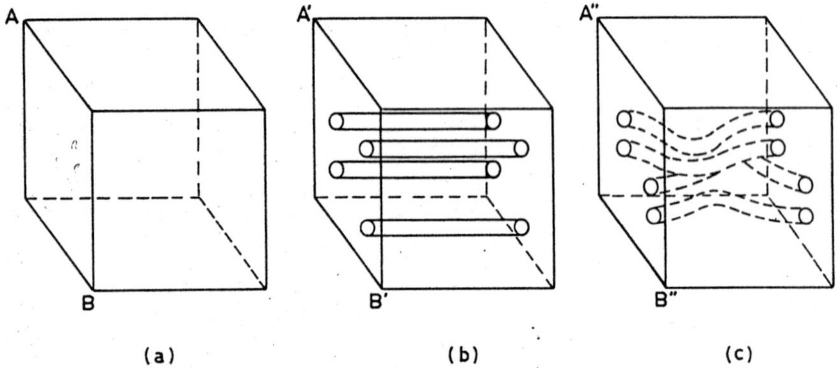

FIG. 2.19. The relation between porosity and tortuosity of a porous medium: (a) liquid (porosity = 1); (b) porous solid (porosity = q), no tortuosity; (c) porous solid at porosity q with tortuous and interconnecting channels.

to it, D, is the diffusion coefficient. Its dimensions in the cgs system are $cm^2\ sec^{-1}$. Equation (2.12) embodies Fick's first law of diffusion. If, instead of a homogeneous solution, the solid is porous and contains liquid in the pores, there would be no difficulty in applying Fick's first law if the pores were parallel to the concentration gradient. Defining the porosity as the ratio of pore space in a given volume of porous solid to that of the given volume, the reduction of the diffusion cross section may be set equal to q and Fick's first law would become

$$\Phi = -qD\,\partial c/\partial x$$

The situation is now as depicted in Fig. 2.19, illustration A″ re the diffusion process will take place along tortuous and intercon channels. Further, the porous solid, which is the soil, is made up of an assembly of solid particles, the geometry and the mode of packing of which will determine the tortuosity of the pores at a given porosity.

A further complexity of the problem is that the total pore space of the soil is not completely filled with the soil solution. Part of the porous volume must be filled with soil air to ensure the normal growth of plants. Since movement of cations is restricted to the solution, movement rates will be affected by the amount of water and the thickness of the film. The diffusion of soil gas through the soil is equally complicated by the presence of the liquid phase.

Finally, since the soil surface is electrically charged, its counter ions will be able to move along it when a concentration gradient is present, and this transport will be superimposed on the bulk diffusion process in the pores.

Even though the difficulties of a study of this process seem insuperable, more and more attention has been devoted to its study as evidenced by the results compiled in Table 2.14.

Soil scientists were not the first to study this problem. An identical situation occurs in the field of oil-well logging data evaluation. There an attempt is made to use the measured conductivity of rock formations in order to evaluate their porosity. This determines the use of the oil reservoir and of the oil saturation of this porosity. An abundant literature exists on the movement of cations in these systems. This precise problem faces the soil scientist if soil air is substituted for oil content.

Fick's first law can be used to evaluate the rate of change of concentration due to the diffusion process:

$$\partial c/\partial t = \partial/\partial x(D\, \partial c/\partial x) \tag{2.13}$$

or considering D as a constant

$$\partial c/\partial t = D\, \partial^2 c/\partial x^2 \tag{2.14}$$

Equation (2.14) is Fick's second law of diffusion, referring to an unidimensional process with a concentration independent diffusion coefficient.

In the case of the interdiffusion of cations, constancy of the electroneutrality requires that D be concentration dependent (Plesset *et al.*, 1958). Integration of the partial differential equation representing Fick's second law gives the amount of solute which was diffused from a source under the given initial and boundary conditions, provided D is known or may be calculated from experiments.

Two different approaches have been used to evaluate some of the factors discussed that influence the mobilities of cations in soils, namely, electrical conductivity measurements and the self-diffusion of cations in soil pastes and suspensions.

The influence of type of clay, nature and concentration of electrolyte in the solution, nature of the counterions, concentration of the clay, temperature on the electrical conductivity of clay suspensions and pastes are the subject of many investigations (Fricke, 1924; Burger, 1919; Van Olpen, 1957; Dakshinamurti, 1960; Street, 1956; 1960; 1963; Van Olphen and Walssman, 1958; Cremers and Laudelout, 1966; Low, 1958; Dutt and Low, 1962), whereas the self-diffusion of cations in clays has been studied by Lai and Nortland (1961, 1962).

The information may be summarized as follows:

When the porosity of the clay suspension diminishes from the value 1 (i.e., pure solution), the mobility decreases at certain salt concentrations and increases below a certain salt concentration level. Thus, at a certain electrolyte concentration, the clay suspension and the solution in which the clay is suspended have the same electrical conductivity. This has been termed the isoconductivity point by Dakshinamurti (1960) since it is independent of the porosity. The decrease of mobility is related to increased tortuosity and the restricted diffusion area when the clay concentration is increased, whereas the increase of mobility with increased clay concentration at low-salt content is due to a surface conductivity or diffusion effect which is favored by an increase in the number of contact points among the clay particles.

The problem of numerically evaluating the tortuosity effect is simplified if the so-called labyrinth or formation factor (F) is introduced. It is defined as the ratio of the diffusion coefficients or of the electrical conductivities of the soil as a whole to that of the soil solution. If Ks is the electrical conductivity of the soil or clay gel, the porus spaces of which are totally filled with an electrolyte solution, the conductivity of which is Ke, then

$$F = Ks/Ke \tag{2.15}$$

and similarly for the self-diffusion coefficients Ds and De

$$F = Ds/De \tag{2.16}$$

Actually F is an expression of the labyrinth factor only if the surface conductance or diffusion effect is eliminated. This is feasible and the value of F corrected for the surface mobilities effect can then be compared with the value that can be calculated from theoretical considerations as shown by Fricke (1924) and Burger (1919).

The agreement between the calculated value of F and those observed is generally very good, especially since the value of the simple parameter, which has to be adjusted for fitting the theoretical curve to the experimental data, may be checked by independent measurements on the influence of the clay particle axial ratios on the viscosity of their suspensions.

Once the factor F is known for a given soil paste, it can be used for calculating the surface diffusion effect. The latter can also be measured by working on salt-free soils or clay gels.

When the surface electrical mobility of a cation is known, its surface diffusion coefficient may be calculated by the Nernst-Einstein formula. The agreement between the calculated diffusion coefficients and those which have been measured is fairly good. By tracer techniques, Lai and Mortland (1962) found self-diffusion coefficients ranging from 3.6 to 6×10^{-6} cm^2 sec^{-1} for Na^+ in Na bentonite gels, with clay contents ranging from 8 to 16% by weight. The self-diffusion coefficient of Cs^+ in the same clay varied between 4 and 7×10^{-7} cm^2 sec^{-1}. Cremers and Laudelout (1966) calculated for these two ions from electrical conductivity measurements in montmorillonite gels 1.5 to 3.7×10^{-6} cm^2 sec^{-1} for Na^+ and 2.4 to 5.6×10^{-7} cm^2 sec^{-1} for Cs^+.

The last figures were quoted not only to show the agreement between two very different methods for estimating the surface mobilities of cations, but also to show that the sequence of surface mobilities for alkali metal cations is the reverse of what occurs in solution. The bulk mobilities increase from Li and Cs, i.e., from the largest solvated volume to the smallest. The surface mobilities decrease from Li to Cs, i.e., from the smallest crystallographic volume to the largest.

The diffusion coefficients of ions in the soil system are still a matter of conjecture. Table 2.14 lists some of the values gathered from the literature for the apparent diffusion coefficients in soils and artificial systems. In all cases the apparent diffusion coefficient of the ion is lower than that in pure water. The various reasons for this, among others, are the smaller cross-sectional diffusion area, since the particles physically take space, the tortuosity or increased path length, the effect of the interaction with a charged solid phase, the discontinuity of water films, and the changes in the properties of water near the particle surface. The values in Table 2.14 indicate how little information is available on the actual or apparent diffusion coefficients of ions in the soil system. Only four values are available for the diffusion coefficient of P and they differ, at the extreme, by a factor of 10^7. The highest value as given by Olsen *et al.* (1962) is actually a calculated value based on the assumption that the diffusion coefficient of P in soil is related to its diffusion coefficient in water as that of chloride in soil is related to its diffusion coefficient in pure water.

Actually, the values in Table 2.14 are not comparable. Olsen's estimate

(Olsen *et al.*, 1962) assumed diffusion took place under the driving force of the concentration gradient in solution [as did Shapiro *et al.* (1960) in their calculations]. They then treated separately the capacity of the soil to renew this concentration. Separately treating the capacity of the soil to renew the soil solution concentration more clearly separates one of the actual factors involved in determining the diffusion coefficient of ions in soils. It may involve the additional assumption, however, that the dissolution of ions off the soil surface, i.e., the rate at which M(soil) becomes M(solution), is faster than diffusion in the liquid phase. The results of Fried *et al.* (1957) and Shapiro and Fried (1959) indicate that this may be a valid assumption for P, but these rates, in general, have not been studied on soil systems. For ionically bonded nutrient elements the rate of dissolution might be expected to be the same order of magnitude as diffusion, but for covalently bonded nutrient elements or for so-called lattice-fixed elements this would not apply. For those nutrient ions held between the clay plates of an expanding-type lattice the measurement would have to be made.

Many other investigators working with soils have used the total surface plus solution ion present per unit volume of soil as the driving force in their calculations. This then attributes higher concentration gradients to the system in which the ions moved, resulting in smaller calculated values for the diffusion coefficient. Any comparison of apparent diffusion coefficients must take this into account.

Similarly, only three Rb diffusion coefficient values, one chloride, and one Mo are at present recorded for soils. The other values in Table 2.14 are for pure clay systems, resin systems, and even glass bead system representing a porous medium.

It is, of course, impossible to generalize from the limited results in Table 2.14, particularly since the diffusion coefficient falls rapidly with decrease in soil moisture and at very high moisture tensions no diffusion takes place (Husted and Low, 1954; Tshapek, 1963). Nevertheless, even at low moisture tensions the diffusion coefficient as measured by various techniques gives relatively low values compared to those in pure solution.

A glance at Table 2.14 suffices to indicate the differences of opinion still existing concerning the diffusion coefficient of ions in the soil system under normal growing conditions. Although most nutrient ions and water have a diffusion coefficient of about 10^{-5} cm^2 per second in pure water (Washburn, 1929), one of the values for P in Table 2.14 is approximately 1.5×10^{-14} cm^2 per second. This is an extreme value and the authors themselves point out that this determination using resin may be mechanistically rate limited by a different process (Vaidyanathan and Talibudeen, 1962). Nevertheless, the other values of P, from 4×10^{-11} to

6.7×10^{-7}, and of some cations in soils, from 10^{-9} to 10^{-7} cm^2 per second, are apparently reasonably determined values. Klute and Letey (1958) found that for glass beads at tensions as low as 30 to 100 cm of water (i.e., well above the moisture content at the moisture equivalent of 1000 cm of water) the apparent diffusion coefficient of Rb decreased approximately 25-fold, (i.e., almost two orders of magnitude). Soil compaction was shown by Graham-Bryce (1963) to decrease the diffusion coefficient as much as tenfold at equivalent moisture tensions. The moisture tension not only has an extreme effect, even resulting in a complete cessation Low, 1954), but also results in marked changes in diffusion coefficient of one or more orders of magnitude at different moisture tensions Graham-Bryce (1963); Patil *et al.*, 1963; Husted and Low, 1954; Low, 1962.

TABLE 2.15

NUMBER OF MOLES MOVED PER SECOND THROUGH A CROSS SECTIONAL AREA OF 40 CM^2 AS AFFECTED BY THE THICKNESS OF THE BOUNDARY AND THE CONCENTRATION GRADIENT[a]

Boundary thickness (cm)	Soil solution concentration (moles/liter)			
	1×10^{-6}	1×10^{-5}	1×10^{-4}	1×10^{-3}
1×10^{-5}	2×10^{-11}	2×10^{-10}	2×10^{-9}	2×10^{-8}
1×10^{-4}	2×10^{-12}	2×10^{-11}	2×10^{-10}	2×10^{-9}
1×10^{-3}	2×10^{-13}	2×10^{-12}	2×10^{-11}	2×10^{-10}
1×10^{-2}	2×10^{-14}	2×10^{-13}	2×10^{-12}	2×10^{-11}
1×10^{-1}	2×10^{-15}	2×10^{-14}	2×10^{-13}	2×10^{-12}

[a] If the diffusion coefficient is 1×10^{-8} cm^2/sec and the concentration gradient assumed to be one-half the soil solution concentration.

Even salt concentration has been shown to have a small but still measurable effect (Dutt and Low, 1962; Letey and Klute, 1960). Under normal soil conditions and soil-solution concentrations, where the concentration in the soil solution for cations and anions in highest concentration could vary from 10^{-3} to 10^{-6} moles/liter and using a value of D of approximately 10^{-8} cm^2/second, the amount of ions diffusing can be calculated. The results of such a calculation are presented in Table 2.15 for different thickness of boundary and for an area of 40 cm^2. This area was chosen because it represents the approximate equivalent planar surface are of 1 g of roots (fresh weight) if the diameter of the root is assumed to be 1 mm.

Actually if these values are applied to roots of 1 mm diameter, cylindrical diffusion considerations would raise the number of moles by a factor of 4.

Shapiro *et al.* (1960) approximated an uptake rate of P of 3×10^{-11} moles per second per gram fresh weight from their experimental data on corn grown in soil [a value of the same order of magnitude may be calculated from the data of Sayre (1948)]. The soil solution concentration was 3×10^{-6} moles per liter of solution. Olsen *et al.* (1963) approximated only about $\frac{1}{100}$ of this rate of uptake from their experimental data with three soils. These values can be associated with the same values in Table 2.15 to show what combination of boundary thickness and soil solution concentration are necessary to supply this rate of uptake.

At a soil solution concentration of 3×10^{-6} M, if diffusion had to supply ions at a rate of 3×10^{-11} moles per second, the ions could not diffuse from farther away than 3×10^{-5} cm. The volume of soil involved would be 120×10^{-5} ml and, taking into account the density of soil of approximately 1.6, this amounts to approximately 192×10^{-5} g of soil. One gram of soil with sufficient P may have some 50 ppm of P on its surface or 1.6×10^{-6} moles per gram of soil or approximately 300×10^{-11} moles in the 192×10^{-5} g of soil in the diffusion volume. This is the total supply and it would be exhausted in 100 sec. If the necessary rate of supply is only one-tenth as great, the diffusion distance and, therefore, volume would be tenfold as great, giving a total supply for 1000 seconds. If the necessary rate of supply is only one-tenth as great and the diffusion coefficient instead of 10^{-8} is tenfold greater (i.e., 10^{-7}), the total amount of the surface in the diffusion volume would be enough for 10,000 sec or almost 3 hr. Using cylindrical diffusion over a short distance might increase the value some fourfold. It is not reasonable, however, to consider exhausting all of the soil as this would bring the concentration in solution to too low a value to maintain the presupposed rate of uptake. This would tend to at least cancel this fourfold factor.

There are so many "guestimates" in these calculations that they are not intended to suggest that diffusion is not responsible for supplying a major portion of the P needs of the plant. They are intended to point out that the experimental data are sorely lacking to make proper calculations of the importance of diffusion, that diffusion can only involve very short distances when the soil solution concentration of an ion is very low, such as with P, and that the necessary diffusion component of the ion uptake continuum of plants grown in soil is not necessarily rate limiting to the over-all process.

When the concentration in the soil solution is much higher, such as 10^{-4} or 10^{-3}, as might occur with most of the major cations, then the diffusion volume, even if the ion has a diffusion coefficient of 10^{-7} or 10^{-8}, will be increased by a factor of 100 to 1000 and the diffusion process may be capable of supplying the requirements of the plant root. At these higher concentrations, however, direct movement with water moving under forces set up due to transpiration is appreciable, as shown by the following considerations.

An acre of corn growing in the field utilizes approximately 20 acre-inches of water (Nelson, 1956; Doss *et al.*, 1962) of which at least one-half or 10 acre-inches is lost by transpiration. Similar magnitudes of transpiration have been shown for other crops grown in the humid region, e.g., 10–16 acre-inches of transpiration per year with wheat at Rothamstead (Buck, 1961).

Since this water taken up by the plant will always reach the root surface containing the nutrient ions in the soil solution, there is a normal flow of ions to the root by this mechanism. A substantial crop, such as a 100-bushel corn crop, utilizes 25–50 lb of P, 100 to 200 lb of K, and intermediate amounts of Ca and Mg (Nelson, 1956). If the soil solution contains approximately 20 ppm of P (6×10^{-4} *M*), 100 ppm of K (2×10^{-3} *M*), 40 ppm of Ca (1×10^{-3} *M*) and 40 ppm Mg (1.7×10^{-3} *M*), then the 10 acre-inches of soil-solution water that will be transpired by the plant contains within it the total amount of these elements needed by the plant.

This description is somewhat oversimplified since ion uptake is continuous (i.e., day and night), whereas transpiration is discontinuous, and the ion uptake pattern as a function of age of plant is not necessarily the same as that of water consumption. Nevertheless, considering that roots are continuously moving into new areas and can remove appreciable amounts of nutrient from the diffusion volume, the above direct contribution of transpirational water is not unreasonable. Under these conditions of nutrient concentration, diffusion of the nutrient ion is essentially unimportant and movement with the bulk of the water is the primary factor bringing sufficient nutrient to the plant. This does not mean that the ions move into the plant with the water. The nutrient ions and water move into the plant independently, reflecting entirely different physicochemical processes; nevertheless, the water moving to the root carries ions in it. A portion of these ions is taken up by the plant by an independent mechanism. The remainder is left behind, at the root surface or on the soil particles in the vicinity of the root, where they represent a subsequent source of supply.

A glance at Table 2.9 indicates that Ca and Mg are usually at high enough concentration in the soil solution to have the needs of the plant brought to the surface of the root along with the transpirational water. Potassium often is at a high enough concentration, but P essentially never is. Under conditions of K deficiency, the amount of K in the water (i.e., the amount reaching the plant root and not the concentration) is too low. Vlamis (1953) has clearly shown that if the solution concentration of K can be maintained at the root surface even at very low levels (Williams, 1961) adequate uptake can be maintained.

Under these conditions, a diffusion volume of soil must be supplying the K and P to the plant. This diffusion volume might be expected to be larger for K than for P since the concentration is, normally, at least 100-fold greater and the concentration gradient should also be that much greater. In fact, Vasey and Barber (1963) and Barber *et al.* (1963) have clearly shown that this diffusion volume is greater for Rb and, by analogy, for K. The calculations made previously, however, have indicated that diffusion alone is probably too slow to supply the needs of the plant. Indeed, when one considers the marked effect of soil moisture level on the diffusion coefficient and the normal situation of low moisture in the close proximity to plant roots, diffusion alone can hardly supply all the needs of the plant.

A diffusion equation has been used to describe the movement of water at all moisture contents (Gardner and Mayhugh, 1958). Depending on the moisture content, the soil water diffusivity is 5×10^2 to 5×10^5 times greater than the apparent diffusion coefficient of P or K in soil. If the rate of water movement is sufficiently rapid, then the diffusion of ions under a concentration gradient is overwhelmed by the movement of the liquid phase. Additionally, the heterogeneous character of the velocity of water movement will cause so-called "hydrodynamic dispersion" (Day and Forsythe, 1957).

The diffusion volume must be continuously recharged with the nutrient ions. This recharge must come from mechanisms that involve water movement. Shapiro *et al.* (1960) showed that diffusion did not limit the P supply to corn and soybeans because of the offsetting effect of the soil-water transport factor on ion renewal at the plant root. These mechanisms of water movement include mass flow under forces due to transpiration and turbulence, evaporation and gravitation, and even temperature differentials. Much more water, containing ions, undoubtedly moves through the soils diffusion volume around the roots than the amount of water taken up by the plant and eventually transpired.

The relative importance of diffusion and water movement in supplying nutrient ions such as K and P cannot be categorically stated. Both can be considered rate limiting but not at the same time. The diffusion volumes from which the plant derives the nutrient is determined by diffusion characteristics. The recharge of this diffusion volume is determined by mass flow considerations. When the capacity factor, M(solid), is very high, the diffusion volume can be replenished without as much need for mass flow; or alternatively, where water movement is high, such as under irrigation in hot, dry areas, the capacity factor need not be as great for the so-called relatively immobile nutrients. In other words, an increasing water flow increases the effective volume from which a plant root feeds, while a higher diffusivity increases the diffusion volume from which the root may withdraw nutrient ions.

The movement of ions in soil takes place under the influence of both diffusion and mass flow. Under a given set of environmental conditions in each soil there is a diffusion volume, that is, a volume of ions moving under the influence of diffusion is delivering ions to the plant root surface at the same rate as the plant takes up the ions. This diffusion volume is determined by the diffusion coefficient of the ion in the soil system and the concentration gradient, the concentration gradient having as its maximum the concentration of the ion in the soil solution. The diffusion volume also must take into account the internal capacity of the soil. By internal capacity is meant the capacity within the diffusion volume to maintain the concentration gradient, e.g., the amount of exchangeable ion in the diffusion volume. This can be taken care of by using a modified diffusivity or preferably by applying a correction factor as suggested by Olsen *et al.* (1962, 1963). When the capacity factor is the same and the rate of uptake of the ions is the same, the diffusion volume is directly proportional to the diffusivity and the concentration gradient. This is valid for all ion species adsorbed or nonadsorbed. Similarly, the diffusion volume is inversely proportional to the requirement of the plant for nutrient, since if the nutrient requirement rate is low, more time is available for diffusion, and ions can thus diffuse for greater distances and still meet the needs of the plant.

The diffusion volume of 1 g fresh weight of roots can be calculated from these relationships as shown in Eq. (2.17).

$$\text{Diffusion volume} = \{[D/b\, \Delta C\, t/Q(Ar + Rr)]^2 - (Rr)^2\} \cdot 1/(Rr)^2 \qquad (2.17)$$

where D is the diffusivity (more properly diffusion coefficient); b, a correction factor due to the capacity of the system within the diffusion volume to renew the ions in the water phase; ΔC, the concentration gradient; Q, the amount of nutrient taken up in time t; t, time; Ar, the surface area of 1 g weight of root; and Rr, the radius of the root.

The diffusion volume is, therefore, the volume which if continuously supplied with nutrient ions will be capable of delivering these ions to the plant by diffusion. This diffusion volume can, of course, be supplied through the medium of mass flow of water. The water containing these ions need never reach the root surface. It needs only to deliver ions into the diffusion volume for those ions to be effective in the nutrition of the plant.

It is not quite valid to suggest, as has Barber (1962), that ions not associated with the soil particles move in the mass flow of water, whereas those associated with the particles do not. All ions move both by diffusion and mass flow. When water is not moving, then all ions are moving by diffusion only. When the diffusion volume does not contain enough ions for the needs of the plant, then the diffusion volume must be recharged by mass flow. The amount of ions brought into the diffusion volume by mass flow is a function of the concentration in the soil solution and not a function of whether the ion is adsorbed or not.

The outstanding contribution made by Barber and his associates (Barber *et al.*, 1963; Vasey and Barber, 1963; Barber, 1962; Lavy and Barber, 1964; Barber and Walker, 1961; Walker and Barber, 1961; Evans and Barber, 1964; Evans and Barber, 1964a; Place and Barber, 1964) to define, by radioautographs, the areas of depletion of nutrient ions by a living root have not defined by what mechanism this area was depleted. Since the diffusion volume is proportional to the concentration gradient, it would be much greater for K, N, or Ca than for P. If this diffusion volume does not contain enough nutrient ion to satisfy the needs of the plant, then ions must move into this diffusion volume by mass flow in order to obtain adequate plant growth. The radioautographs indicate the total area of depletion both by mass flow and diffusion.

For ions which reach the plant root in amounts too high to be utilized by the plant, the diffusion and mass flow processes must be effective in reverse, but the same principles would still apply.

CHAPTER 3

The Plant System

1. Passive Entry

1.1. Free Space

When the ion M reaches the plant root, three events have been suggested as possibly occurring. These are: (1) penetration of the root by passive movement into the free space (F.S.) of the root; (2) adsorption into external or internal root surfaces; and (3) active or metabolic accumulation.

Any conclusions with regard to the penetration of the ion into the root must be consistent with root structure. The following diagrams give an indication of the structures of a plant root that must be taken into account.

Figure 3.1 is the cross section of a primary root of about 0.6 mm

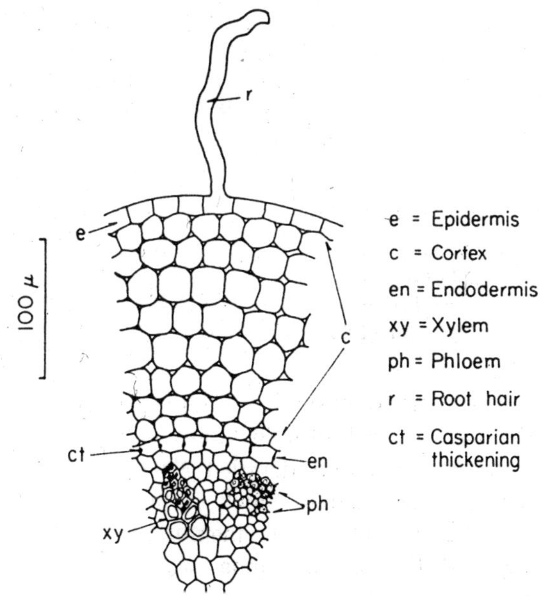

Fig. 3.1. Transverse section of a primary root. From Briggs *et al.* (1961).

in diameter, a common size for many plant roots. Intercellular spaces do exist between the cortical cells; however, they are small and do not continue into the endodermis. Furthermore, to reach the xylem or phloem, ions must traverse a distance of at least 0.2 mm if they move in a straight line but presumably much farther through a very tortuous path.

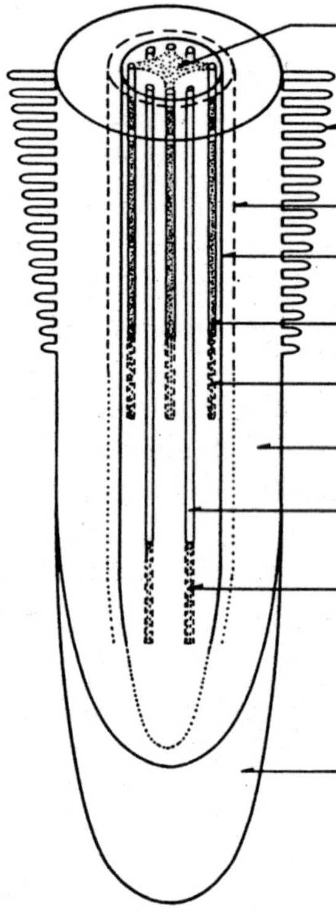

FIG. 3.2. Median longitudinal section through root apex (diagrammatic). RH, Root hair; End, endodermis; Per, pericycle; Pro X, protoxylem; I Pro X, immature protoxylem; Co, cortex; Pro Ph, protophloem; I Pro Ph, immature protophloem; RC, root cap. From Steward (1959).

Figure 3.2 is a diagrammatic longitudinal representation of a root showing the root cap and lack of phloem and xylem near the root tip.

Figure 3.3 shows the actual detailed structure of the cell and the relationship between the vacuole which accumulates ions and the cytoplasm, and also the tonoplast, the membrane separating the vacuole and cytoplasm, and the plasmalemma, the membrane separating the cytoplasm from the cell wall.

Although any conclusions with regard to penetration of the nutrient

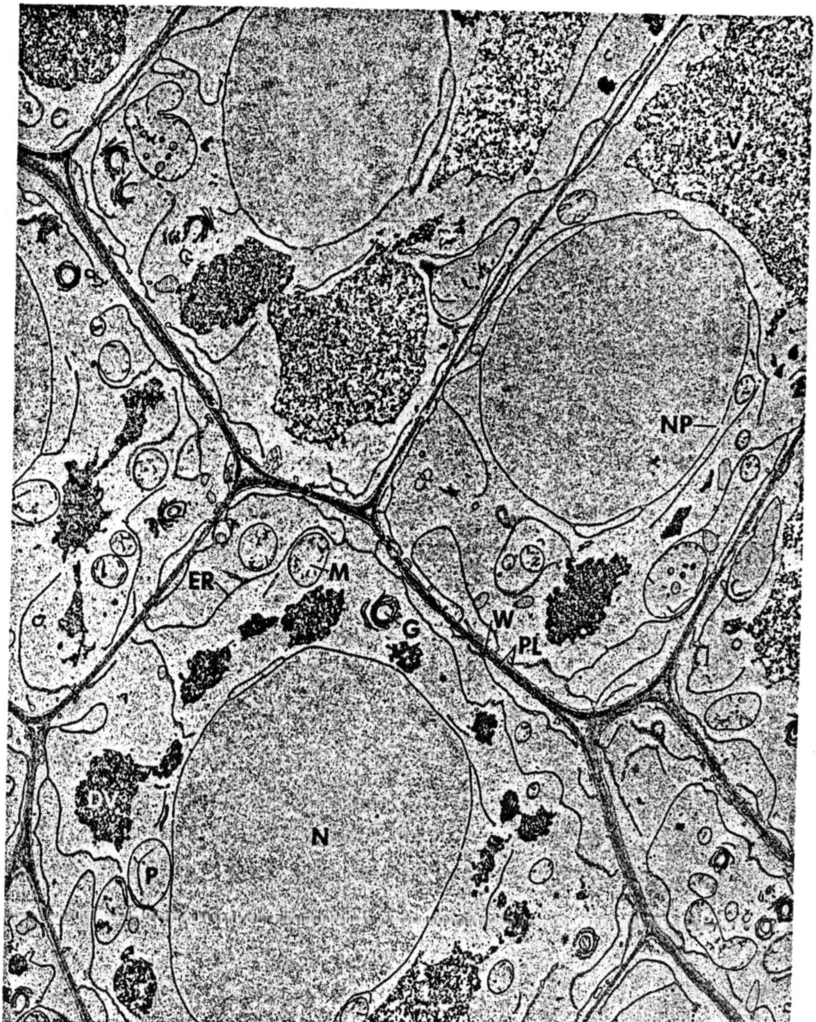

FIG. 3.3. Electron micrograph showing a cross section of a plasmolyzed root meristem (*Eichhornia crasipes*). DV, developing vacuole; ER, endoplasmic reticulum; G, golgi body; M, mitochondria, N, nucleus; NP, nuclear pore; P, plastid; PL, plasma membrane; V, vacuole; and W, wall. (Courtesy of Howard J. Arnott, University of Texas, Austin.)

ion must be consistent with root structure, direct observations of this penetration are limited by an inability to locate water-soluble ions at the cellular level at a given instant or time. Dyes have been used to simulate ion transport pathways, but their usefulness is questionable since they involve the assumption that the dye and the much smaller charged nutrient ions move in the same path and by the same mechanism. A few attempts have been made utilizing microautographic procedures, but the paucity of the literature is surprising since freeze-drying methods for preparation of animal tissue are routine and easily adaptable to this problem. Branton and Jacobson (1962) applied this technique quite successfully to the penetration of iron into pea roots, even showing that classic fixing methods by means of chemicals such as those employed by Possingham and Brown (1958) were not reliable. Lüttge and Weigl (1962), Weigl and Lüttge (1962), Ulrich *et al.* (1962), and Weigl and Ziegler (1962) also prepared sections using freeze-drying techniques. Hassbrock *et al.* (1962) indicated a carefully worked-out procedure for very thin sections that should be generally applicable for soluble ions. However, even the freeze-drying method has the disadvantage of ion movement during the drying process. When tissue is dried the ions must move. They cannot be left in the middle of a vacuum. With quick freezing and subsequent drying by sublimation this movement should be minimized even if not completely eradicated. The direct examination of the quick frozen tissue without drying is less subject to criticism, but is difficult to do experimentally. Fried and Franklin (1961) worked out just such a procedure, and suitable sections of barley roots were obtained. Unfortunately, suitable radioautographs of roots after ion uptake by nutrient ions using this procedure were never published. Nevertheless, some preliminary radioautographs were obtained by the same authors in cooperation with Noggle and Fleming of the same laboratory. A similar technique was used by Gahan and Rajan (1964) in studying longitudinal sections of *Vicia faba* after a 14-hour uptake period of $^{35}SO_4$.

Figure 3.4 is a radioautograph of an approximately 16 μ-thick section of a barley root that had received labeled sulfate ion for 10 minutes and shows a concentration of the ion both in the neighborhood of the epidermis and endodermis similar to that obtained by Weigl and Lüttge (1962) who used a fixing technique. This is not intended as any proof of the distribution of sulfate ions since the authors do not feel enough radioautographs were obtained for definite conclusions to be drawn. It is only shown to indicate that experimentally sections can be made and autographed of water-soluble ions by quick freezing and direct preparation of tissue slices.

Assuming that the only ion of interest is one that is finally actively metabolized by the root, the following points with regard to passive entry and adsorption become pertinent. (1) Does passive movement into the so-called free space occur in plant roots? (2) Does adsorption (3) If they occur, must either or both passive movement into and adsorption precede ion accumulation in the plant root?

or both passive movement into free space and adsorption do precede ion accumulation in the plant root, do these preliminary steps hinder, enhance, or have no effect on the rate of accumulation of a given ion species?

A consideration of these factors requires certain clearly stated definitions. As stated by Briggs *et al.* (1961), "The concept of free space was introduced to describe the phase in the cell or tissue into which solutes move relatively freely." They further state that "the free space . . . cannot be measured without being affected by the method

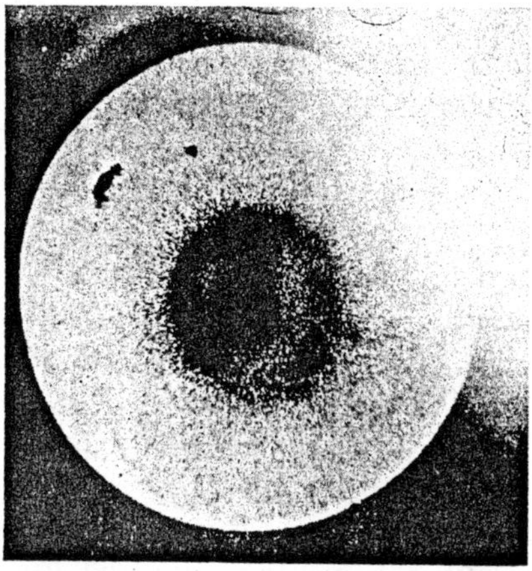

FIG. 3.4. Radioautograph of 16 μ section of a barley root that received labeled sulfate ion for 10 minutes. Section prepared by direct examination of quick frozen tissue. Outer dark area is in the region of the epidermis. Central dark area covers the central cylinder including the endodermis.

of measurement." Since at least one of these authors was initially responsible for the introduction of this concept, their definitions are most pertinent. They follow:

"Free space (F.S.) is defined as that part into and through which the solute and solvent from the external solution move readily; clearly there may be subdivisions of the free space based on different degrees of accessibility to solute."

"Since free space cannot be measured directly the apparent free space (A.F.S.) is estimated and is precisely defined (Briggs and Robertson, 1957) as follows: The apparent volume of the free space equals the amount of solute in the F.S. divided by the amount of solute per unit volume of external solution when free space and external solution are in equilibrium."

Unfortunately, none of these definitions are basic enough to be inde pendent of the method of measurement. Even free space is a relative term referring to accessibility. Once it is realized that these are empirical definitions that do not necessarily refer to an actual volume but rather to what might be termed an "equivalent" volume, much of the confusion can be eliminated.

The free space has been further broken down into water free space (W.F.S.) and Donnan free space (D.F.S.). The water free space is that portion of the free space in which the measurement of A.F.S. would give the same value for electrolytes and nonelectrolytes. The Donnan free space (D.F.S.) again according to Briggs *et al.* (1961) is "the portion of the F.S. where the concentration of an ion with a charge of one sign is smaller while that of the ion of opposite charge is greater than the external concentration." This definition, of course, applies whether the distribution of the ions is a so-called Donnan distribution or whether other distributions of the ions caused by the charge effects are more appropriate.

Keeping the definitions in mind and realizing the empirical nature of the determinations still does not eliminate the real possibility of making use of this concept. Nevertheless, one inconsistency in definitions is apparent. Free space is defined in terms of degree of accessibility, whereas apparent free space is defined in terms of equilibrium. This does not appear to be compatible.

Table 3.1 (occurrence of A.F.S.) lists some of the values obtained for A.F.S., together with the plant species and literature source.

In compiling Table 3.1, an attempt was made to omit values in which a large amount of adsorbed ion was clearly present, but there was no way to eliminate all values affected by adsorbed ions. In fact, one of the semantic difficulties is that W.F.S. and D.F.S. are defined in terms of a space, whereas A.F.S. is defined in terms of an operational procedure, the amount of solute in the free space divided by the concentration of solute in the external solution, (i.e., an apparent space). In other words, F.S. which cannot be determined has been divided into two compartments, D.F.S. and W.F.S., but A.F.S. which can be determined has not been so divided. The problem is clear when one tries to apply the working definition of A.F.S. to the D.F.S. (i.e., dividing the amount of solute in the D.F.S. by the concentration of solute in the external solution). The high concentration of ions at the surface owing to the charge of the root surfaces can give apparent volumes for D.F.S. which are greater than the volume of the tissue itself.

Many of the values in Table 3.1 are supposed to approach the value for W.F.S. if it is assumed that whenever D.F.S. is negligible, A.F.S. equals W.F.S. This distinction between W.F.S. and D.F.S. is very poorly delineated for there is a clear difference of opinion as to the assumptions

TABLE 3.1
SUMMARY OF REPORTED VALUES OF APPARENT FREE SPACE

Plant species	Whole plant or excised	Method	Equivalent volume of root (%)	Ref.
Barley	Excised	$H_2PO_4^-$, SO_4^{2-}, SeO_4^-, Ca^{2+}	22–25	*a*
Barley	Excised	HPO_4^{2-}	20	*b*
Barley	Whole	HPO_4^{2-}	31	*b*
Barley	Whole	Na^+, Rb^+	7.4–9.4	*c*
Barley	Whole	Mannitol Cl^-	23–33	*d*
Barley	Excised	Mannitol Cl^-	21–47	*d*
Wheat	Excised	Mannitol Cl^-	25–32	*d*
Wheat	—	NO_3^-	24	*e*
Wheat	—	SO_4^{2-} India ink	11–14	*f*
Wheat	—	Cl^-, $H_2PO_4^-$	24–33	*g*
Wheat	—	SO_4^{2-}, $H_2PO_4^-$	18–20	*h*
Wheat	—	Br^-	10–14	*i*
Wheat	Whole	SO_4^{2-}	27.5	*j*
Wheat	Excised	SO_4^{2-}	18	*j*
(*Agropyron elongatum* host) wheatgrass	Excised	Mannitol Cl^-	25.2	*d*
Maize	—	$H_2PO_4^-$, SO_4^{2-}	6	*k*
Maize	Whole	Mannitol Cl^-	12–26	*d*
Maize	Excised	Mannitol Cl^-	19–20	*d*
Pea	Whole	Mannitol Cl^-	4.5– 7.3	*d*
Pea	Excised	Mannitol Cl^-	7–15	*d*
Pea	—	Cl^-	7– 8	*l*
Pea	—	Rb^+	7–10	*m*
Bean	—	—	13	*n*
Bean	Whole	Mannitol Cl^-	5.6–10.6	*d*
Bean	Excised	Mannitol Cl^-	6–16	*d*
Sunflower	—	SO_4^{2-}	15	*o*
Cotton	Whole	Mannitol Cl^-	16–20	*d*
Milo	Whole	Mannitol Cl^-	10.6	*d*

[a] Epstein (1955).
[b] Jacobson *et al.* (1958).
[c] Waisel (1962).
[d] Bernstein and Nieman (1960).
[e] Kihlman-Falk (1961).
[f] Ingelsten and Hylmö (1961).
[g] Butler (1953).
[h] Butler (1959).
[i] Cseh and Böszörményi (1961).
[j] Kylin and Hylmö (1957).
[k] Weigl (1963).
[l] Hylmö (1953).
[m] Brouwer (1959).
[n] Hope and Stevens (1952).
[o] Pettersson (1961).

that can be made. A common assumption is that since the surfaces of roots are negatively charged, negative ions are not adsorbed to root surfaces but on the contrary are repelled. Thus, assuming negative adsorption is negligible, a measurement of A.F.S. with a negative ion should give a good approximation of what might be obtained by a nonelectrolyte. Thus, Khilman Falk (1961) obtained A.F.S values of about 24% using nitrate ion and 50–70% using K on the same tissue under the same conditions. According to the Briggs definition of A.F.S., this latter value is a perfectly valid determination of A.F.S. Yet even the assumption that anions are not adsorbed can be challenged by other data in the literature, e.g., Hagen *et al.* (1957) clearly show a phosphate intercept of 12×10^{-10} moles per gram barely roots for a time curve obtained from 2×10^{-6} *M* phosphate solution even after the roots were washed three times with distilled water, presumably washing away some if not most of the free space phosphate. If A.F.S. is calculated from this data, it amounts to 60%, clearly showing that this phosphate is indeed adsorbed and does not exist in so-called free space. In fact, 10^{-2} *M* NaN_3 reduced this intercept to one-third of its initial value, indicating that the phosphate was not only adsorbed, but was rather specifically related to some metabolic root function. The above authors believe this phosphate to be specifically adsorbed to carrier sites. In another figure they showed an intercept of 9×10^{-10} moles per gram of barley roots for a time curve obtained from 1×10^{-6} *M* phosphate (see Fig. 3.15). This, upon calculation, gives an A.F.S. of 90%. Again this intercept was reduced by 10^{-1} *M* β-hydroxybutyrate to approximately one-tenth of the original value, showing marked specificity of the adsorption as related to root function. Epstein (1955) also demonstrated a labile binding of sulfate by showing that measurements of A.F.S. with sulfate were concentration dependent if made at concentrations lower than 10^{-2} *M*. The specificity of this anion binding was evidenced by an inability to remove the effect with high concentrations of either KCl or KH_2PO_4. However, K_2SeO_4 at equivalent concentrations, successfully eliminated the effect even at lower concentrations consistent with the demonstrated SeO_4–SO_4 competition (Epstein, 1956).

Although the stated definitions of F.S., A.F.S., W.F.S., and D.F.S. do cause some semantic difficulties, there is no point in adding to the confusion by even further definitions. There are already other definitions to choose from, such as those of Jacobson *et al.* (1958), Conway and Downey (1960), Epstein (1955), Cowie *et al.* (1949). All definitions aside, there are clearly negatively charged surfaces in a root, and a space has been inferred from much experimental evidence. Admittedly, the size of this space is in doubt. It can conceptually be visualized, however, as one which the solute has access to without the application of metabolic energy. If the solute is a charged ion, its concentration will be changed in the vicinity of the negative charge, and this access volume can further be subdivided into that part affected by the charge and that part that is unaffected by the charge.

The weight of evidence suggests that a so-called F.S. exists in intact plant roots. In measuring its magnitude, however, it should be realized that plant roots are very delicate structures and are easily damaged. Such an action as blotting roots or other gentle handling for weighing or removing excess water invariably results in an outpouring of P and presumably other cell contents. It should also, therefore, result in permitting free penetration from the outside solution. In an experiment using methylene blue and intact plants, even though there was no physical handling of the roots, one of the 24 sets of plants allowed the dye to freely penetrate, whereas the other 23 sets of plants successfully kept the dye out (Fried, unpublished). Thus, despite all precautions, one set of roots was damaged, even though the roots were never physically handled directly (as the technique only involved handling whole plants and moving them from the reserve tank to the test solution). In addition, the experiments on A.F.S. always involve the assumption of somehow being able to separate the solution in the free space from the external solution by such techniques as blotting, centrifugation, etc. These are assumptions and not facts. Levitt (1957) raised another note of caution concerning the overestimation of the size of space involved and the suggested conclusions often made that part of this space was in the cytoplasm. Indeed, the estimates of Ingelsten and Hylmö (1961) listed in Table 3.1 for A.F.S. of wheat roots are downward revised estimates made from the same laboratory after Levitt's clear demonstration of necessary precautions. Another note to consider concerning the meaning of the apparent free space is that if the ions fully penetrated the first layer of cells and no further, volumes obtained would be similar to those noted in Table 3.1 when the measurement did not involve adsorbed ions. In fact, an explanation offered by Jackson *et al.* (1962) for their results comparing phosphate accumulation by intact roots with mitochondria from those roots and the relationship of accumulation to respiration was to suggest that only the first layer of cells (i.e., only a portion of the cells) was involved. If free space extended past this layer of cells, initial accumulation would also have to extend that far.

If it is assumed that a free space of some 10 to 20% exists, however, then the number of moles of the different ions in the free space can be calculated from the concentration of the ion usually found in the soil solution (Table 2.9). For example, at 10^{-3} M concentrations of Ca and Mg, 1 g of roots should contain in A.F.S. some $1–2 \times 10^{-7}$ moles of Ca or Mg, which is approximately the amount of Ca (or Sr) absorbed in 30 minutes from 10^{-3} M solutions by barley roots (Epstein and Leggett, 1954; Fried *et al.*, 1958). At $10^{-4} M$ K, 1 g of roots should contain in A.F.S. some $1–2 \times 10^{-8}$ moles of K, which is approximately the amount of K or Rb absorbed from 10^{-4} *M* solutions by barley roots in 5 to 25 seconds (Fried *et al.*, 1958; Epstein *et al.*, 1962). At 10^{-6} *M* P, 1 g of roots should contain in A.F.S. some $1\text{–}2 \times 10^{-10}$ moles of P, which is approximately the amount of P absorbed from 10^{-6} *M* solution by barley roots in 10 seconds, (Hagen *et al.*, 1957; Noggle and Fried, 1960) by alfalfa roots, in 20 seconds and by millet roots in 5 seconds (Noggle and Fried, 1960). These illustrations only serve to suggest that with some ions and at some concentrations prevalent in the soil solution, accumulation may be too rapid to allow for equilibration with A.F.S. where the ions must move appreciable distances, such as 0.2 mm, to reach the endodermis and even further to reach the xylem. These considerations are only suggestive, as calculations of this kind must be based on diffusivity coefficients that have been very little investigated. The value thus far obtained has varied from 9×10^{-7} for Rb entering cells of beet discs (Briggs *et al.*, 1961, p. 98), to "the order of $\frac{1}{100}$ of that which prevails in water for nonelectrolytes, and it is even less for some electrolytes, e.g., phosphates" (Steward, 1955).

Although Kramer (1957) pointed out that the presence of free space solved some conceptual problems of how ions are accumulated by each cell and are transported or translocated between cells, the importance of free space to ion uptake in respect to its effect on the rate of ion uptake is yet to be demonstrated. Two main effects are potentially possible. If negatively charged surfaces are present, there might be an appreciable change in the concentration of M(solution) in the vicinity of an active uptake process. A predictable change particularly in the relationship between monovalent and divalent ions might be expected with the accompanying change in the relative rate of uptake of these ions. These ionic charge effects should be similar to those discussed in Chapter 2, Section 3 involving the relationship of cations between solution and solid phase where the solid phase is negatively charged. The effects on the concentration of M(solution) in the apparent free space if present are not appreciable enough to affect the rate of ion uptake (Fried *et al.*, 1958; Kahn and Hanson, 1957; Lagerwerff and Peech, 1961; Shealtiel, 1958; Viets, 1944). The concentration of M(solution) in the vicinity of the root would also be different if diffusion into the root "free" space was rate limiting to the accumulation process. If diffusion were rate limiting, ion uptake should be directly proportional to concentration. Generally, it is not. The high temperature coefficients of ion accumulation for monovalent cations (Hoagland, 1944; Jacobson *et al.*, 1957), monovalent anions (Honert, 1936; Honert and Hooymans, 1955; Jacobson *et al.*, 1957), and phosphate (Honert, 1933; 1936) also indicate that diffusion is not rate limiting for these ions. The diffusivity of monovalent cations and anions is not very different from that of such divalent ions as Ca and sulfate (Washburn, ed., 1929). Since the turnover rates of Ca and sulfate are much slower than for monovalent cations and anions (Dijkshoorn, 1959; Fried *et al.*, 1958; Kahn and Hanson, 1957), diffusion is even less likely to be rate limiting for Ca and sulfate. Furthermore, diffusion effects are likely to be more pronounced, the lower the concentration of the ion M(solution) in the vicinity of the root. Of the major nutrient elements, P occurs in solution at the lowest concentration. Yet, Jackson and Hagen (1960) found that at $10^{-5}M$ phosphate, the rate of P^{32} appearance was constant in the five major products of phosphate uptake starting at 15 seconds and continuing for at least 40 minutes, indicating that new sites of incorporation did not become available as diffusion time increased. Thus, if "free" space does exist in living plant roots to any appreciable extent, it does not appear to affect the concentration of M(solution) in the vicinity of the sites of uptake.

1.2. Adsorption

Adsorption of ions on plant roots occurs whether the roots are dead or alive. However, most of the important questions about the nature and extent of this adsorption remain unanswered. The previous section has already mentioned the lack of knowledge concerning anion adsorption. Other questions such as amount, mechanism, and location of the adsorption phenomenon are largely unanswered. Much of the available information deals with dead roots. The relationship between dead and live roots is essentially unknown.

1.2.1. *Dead Roots*

The adsorption of ions by dead or damaged plant roots is called either cation-exchange capacity of roots (Drake *et al.*, 1951; Crooke, 1958; Wiersum and Bakema, 1959; Crooke *et al.*, 1960; Heintze, 1961; Keller and Deuel, 1957; Mouat and Walker, 1959; Crooke and Knight, 1962; Mitsui and Ueda, 1963; 1963a; Vose, 1963; Asher and Ozanne, 1961; Tarabrin, 1961; Arthur Wallace, 1963; Smith and Wallace, 1956; 1956; 1956a; McLean *et al.*, 1956) or anion-exchange capacity of roots (Peterburgskii and Nelubova, 1962, 1963).

The clear indication by Williams and Coleman (1950) of the occurrence of the phenomenon of cation exchange by plant roots led Drake *et al.* (1951) to suggest that the "differences in the ability of plants to take up cations from the soil are largely controlled by the cation-exchange capacity of the plant root and the valence of the cation." Over 30 species of plants were examined by these authors and later authors have added substantially to this list. Drake *et al.* further suggested that this explained the relative requirements of different species of plants for monovalent and divalent cations. This was strongly supported by other authors, including Smith and Wallace (1956, 1956a), Wallace (1963), Stahlberg (1955), Wiersum and Bakema (1959), Mouat and Walker (1959), and Asher and Ozanne (1961). Machado (1957) even uses it as the principal factor to explain the development of coffee trees and other associated plants. Others (e.g., Crooke, 1958; Keller and Deuel, 1957; Crooke *et al.*, 1960; Crooke and Knight, 1962) confined themselves to a study of the phenomenon and some of the correlations with observed plant properties without ascribing an active role to the cation-exchange capacity. Thus Crooke and Knight showed that "the cation exchange capacity of roots is positively correlated with the content in the tops of (a) the total cations, (b) the ash, (c) the excess base, and (d) the total trace elements." Cation-exchange capacity also showed a positive correlation with protein content and dry matter in the ether extract and negative correlation with crude fiber. It is a big step between showing a correlation and ascribing cause and effect. In fact, much evidence has accumulated indicating that factors other than cation-exchange capacity of roots actually control ion uptake by living plants and suggesting that the cation-exchange capacity of roots does not control the relative uptake of individual cations. Most of this evidence has been accumulated with living root systems and will, therefore, be discussed in the following section. But one might even here mention statements such as those by Vose (1963) who, after extensive work with dead roots involving determinations of cation-exchange capacity, suggested that some "regulator other than cation-exchange capacity and Donnan distribution laws must be postulated for the selective assimilation of monovalent and divalent cations by varieties of a species grouped within a relatively narrow range of cation-exchange capacity." Indeed the cation-exchange capacity of roots does not appear to be directly related to ion uptake.

1.2.2. *Living Roots*

Living roots adsorb cations. This has been clearly demonstrated by many investigators including Epstein and Leggett (1954), Fried *et al.* (1958, 1961), Leggett and Stolzy (1961), Waisel (1962), Jackson and Adams (1963), Marckwordt (1963), Barber and Russell (1961), and Scheuring and Overstreet (1961).

Figure 3.5 gives an illustration of this type of adsorption. It represents a 35 minute time curve of Na uptake by barley roots at 10^{-}, 10^{-4}, 10^{-3}, and 10^{-2} *M* concentrations of Na. Since uptake is a linear function of time, the time curve can be extrapolated to zero time, giving the amount adsorbed. Two other explanations for this intercept would be that (1) this is not adsorbed cation but water free space (W.F.S.), and (2) that there is a very rapid initial adsorption followed by a slower absorption. That the intercept does not represent W.F.S. is clear. To make the simple calculation, at 10^{-4} *N* Na, the intercept is 2×10^{-7} moles per gram fresh weight of roots. If this were W.F.S. it amounts

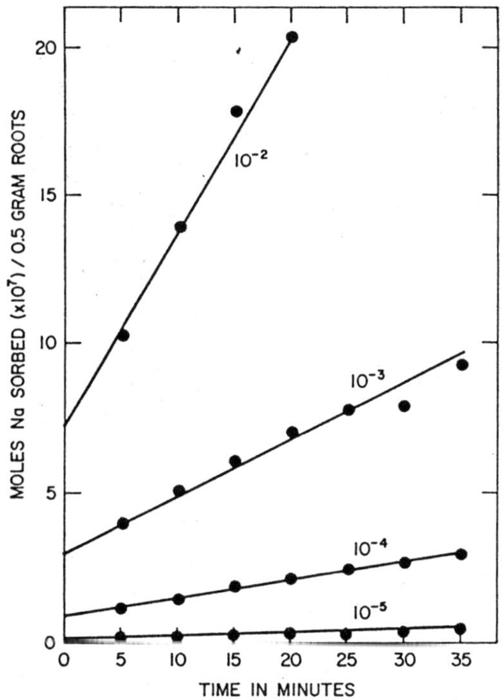

FIG. 3.5. Sodium uptake by excised barley roots as affected by the concentration of sodium. From Fried *et al.* (1958).

to 200% of the space occupied by 1 g of roots. The intercept presumably does not represent a rapid initial absorption since the ions responsible for this intercept will exchange with the identical ions in solution. Epstein (1964) clearly demonstrated this relationship by showing that the cation thus taken up could not be removed by water, but could be exchanged off by an isotope of the same ion. Epstein therefore rinsed his roots in the unlabeled isotope of the same element, Sr, in order to characterize the Sr uptake system of the plant.

Figure 3.6 shows the results obtained by Esptein, when after a 1-hour period of uptake of labeled Sr by barley roots, he compared washing the roots with water for 3 hours to washing the roots with unlabeled Sr. The amount removed by exchange was essentially equal to the intercept of the time curve. As shown in Fig. 3.7, Fried *et al.* (1958) demonstrated similarly the exchangeability of the Ca intercept by building a 2.5 hour time curve, points spaced 15 minutes apart, with and without washing with unlabeled Ca for 0.5 hour. Over 95% of the intercept

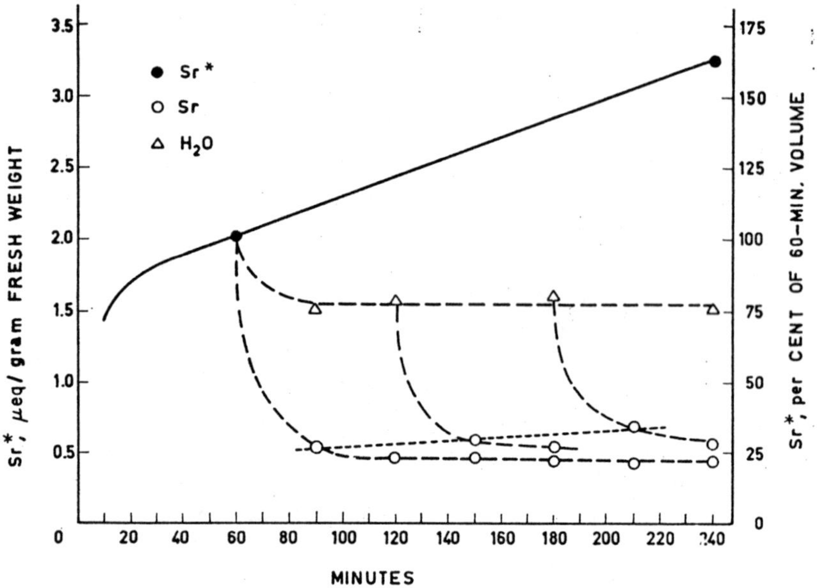

FIG. 3.6. Uptake of Sr* and loss of exchangeable Sr* to water and solutions of unlabeled Sr. Increase, with time, of nonexchangeable (absorbed) fraction of Sr* in roots held by water (dotted line). From Epstein and Leggett (1954).

was thus removed. Although this is not positive proof of a lack of absorption, since exchange could conceivably be metabolically mediated, the high isotopic dilution that ions such as Ca would presumably be subject to after absorption would mitigate against such a marked loss of the labeled Ca by metabolically mediated exchange.

It should be pointed out that this intercept is not always apparent, for the ability to experimentally demonstrate an intercept depends on the relative size of the intercept compared to the amount of ion taken up. For example, if the adsorption of Na in Fig. 3.5 remained unchanged but the rate of Na absorption was increased tenfold or more, then the intercept would represent less than 10% of the uptake in 5 minutes, which would not be detectable in such an experiment involving biological material. If the rate of adsorption of Na remained the same and the amount adsorbed decreased by one-tenth, the amount adsorbed would not be detectable. Even if the amount adsorbed and the rate remained

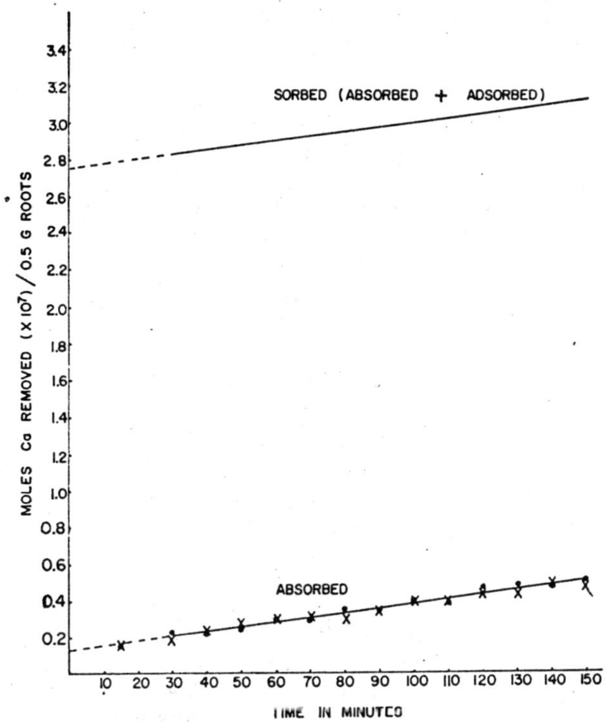

FIG. 3.7. The effect of time of immersion on the uptake of Ca by 0.5 g of excised barley roots. Upper line represents absorbed plus adsorbed Ca. Lower line represents only absorbed Ca. ● = Roots immersed in substrate for entire time period; × = roots immersed in substrate for 15 minutes. After rinsing, roots were then placed in a saturated atmosphere for the remainder of the time period. From Fried *et al.* (1958).

the same but the time points were at hourly intervals, the first being after 1 hour, the extrapolation would be too long and the amount adsorbed would be too low in proportion to the amount absorbed to be able to detect the adsorption experimentally. This, plus faulty techniques, undoubtedly explains the few observations on lack of adsorption of cations (Epstein *et al.*, 1962) and may even explain the inability to detect the adsorption of anions on living roots.

1.2.3. *General Discussion*

Cations are adsorbed on both living and dead roots, but the adsorption phenomenon is clearly different in living roots than in dead roots.

The so-called cation-exchange capacity of roots, i.e., adsorption capacity of dead or damaged roots, is some tenfold that of the amount of cation that living roots would be capable of adsorbing at high concentrations of cations as inferred from the intercepts of ion uptake—time curves (Fried *et al.*, 1958).

Furthermore, adsorption of cations on these dead or damaged roots is nonspecific, following in general terms the valence laws and lyotropic series just as with other cation-exchange materials such as clays and resins (Keller and Deuel, 1958); e.g., Keller and Deuel found the order of replacement $Li < Na \leqq K < H = Mg < Zn < Ba < Ca < Cu$. On the other hand, the adsorption of cations by living roots has been shown to be very specific. An example of this is shown in Figs. 3.8 and 3.9. Figure 3.8A clearly shows that the divalent cation Mg had no effect on the adsorption of Rb, as indicated by the lack of effect on the intercept, whereas the monovalent cation K had a marked effect on the intercept. Figure 3.8B shows that neither the divalent cation Ba nor the monovalent Na had any effect on adsorption, and Fig. 3.8C shows that although methylene blue has a marked effect on ion accumulation it had no effect on the amount of Rb adsorbed.

Figure 3.9 shows what a marked effect a neutral molecule like mannose or the presence of oxygen can have on the amount of K adsorbed. These are clearly effects that are characteristic of living tissue that do not occur in dead tissue.

Since the cation adsorption by dead roots is not apparent in living roots, the cation-exchange capacity of dead roots is a *result* of accumulated metabolic activity and *not a cause* of differentiation in the active transport mechanism. In fact, if any appreciable nonspecific adsorption of cations occurred, this might be expected to hinder the active transport processes by preventing the ion from reaching a site of active accumulation. Surely nonspecific cation adsorption is neither the first step in the ion uptake process nor the control mechanism for the ratio of ions absorbed by the plant. In fact, Broeshart (1962) demonstrated that uptake of cations was unaffected by the ratio of the cations adsorbed on the plant root, but was markedly affected by the ratio of these cations in the outside solution.

A possible meaning of the specific adsorption of ions that occurs with living roots is mentioned in Section 2.2 of this chapter.

The discussion of free space and adsorption phenomenon of living

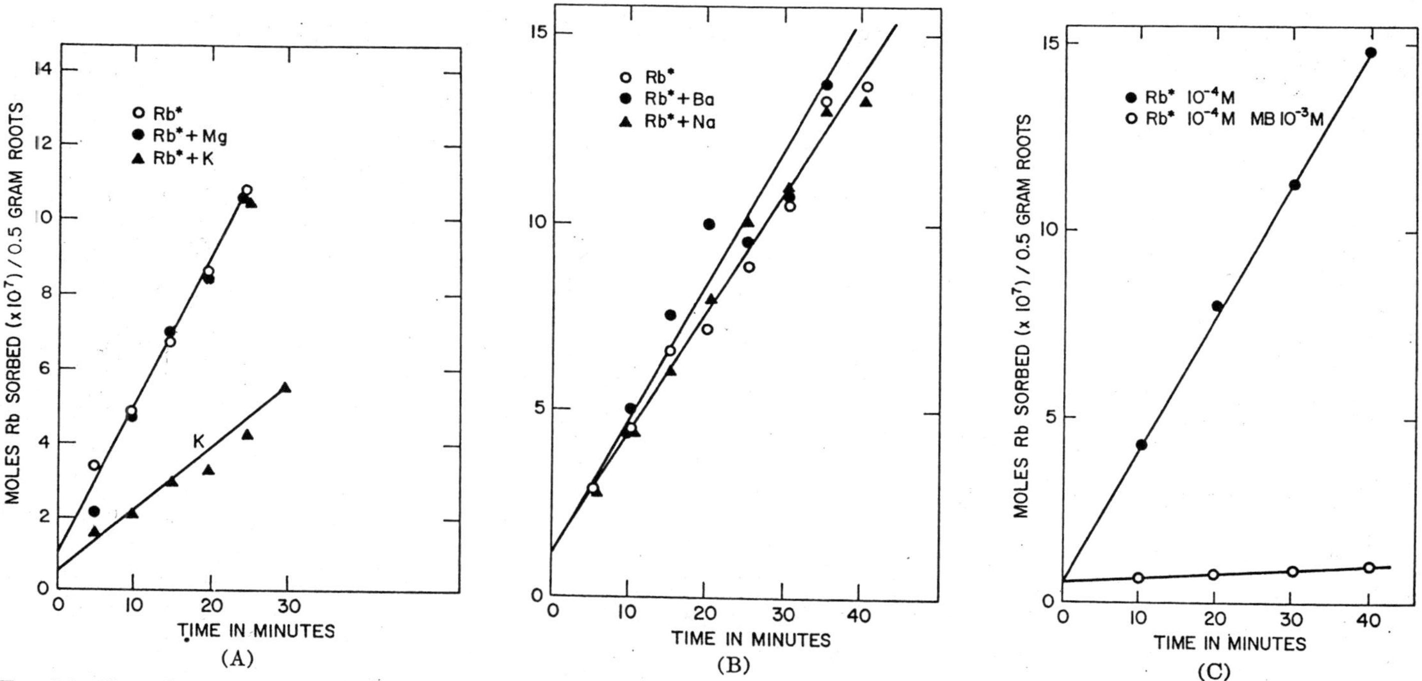

FIG. 3.8. The effect of added cations on the uptake of Rb by 0.5 g of excised barley roots. (A) The effect of 10^{-3} *M* Mg or 10^{-3} *M* K on Rb uptake from a 10^{-3} *M* RbCl solution; (B) the effect of 10^{-3} *M* Na or 10^{-3} *M* Ba on Rb uptake from a 10^{-3} *M* RbCl solution; (C) the effect of 10^{-3} *M* methylene blue on Rb uptake from a 10^{-4} *M* RbCl solution. From Fried *et al.* (1958).

and dead tissue has been confined to plant roots and has not referred to the much more extensive literature with storage tissue and other biological material such as chlorella, bacteria, yeast, leaves, etc. This is simply because free space is a physical phenomenon which may very

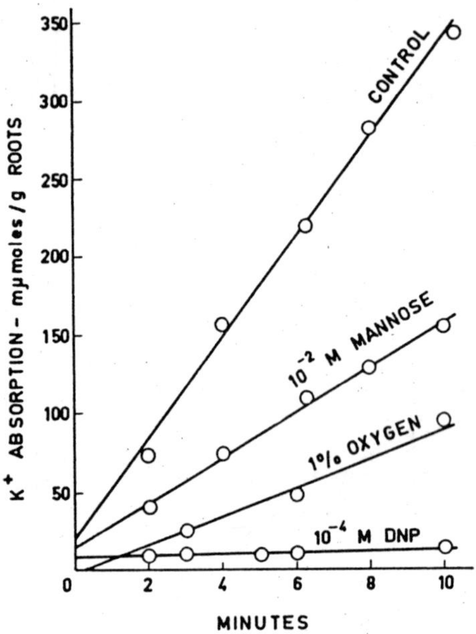

FIG. 3.9. K^+ absorption at pH 5 from 10^{-5} M KCl with 10^{-3} M NaCl. From Jackson and Adams (1963).

well be different for different tissues and adsorption as a passive process may not be consistent in the different tissues.

2. Active or Metabolic Entry into the Root

2.1. Carrier Hypothesis

Active accumulation of ions is most commonly described by the so-called "carrier" theory, i.e., accumulation results from the ion combining with a carrier, followed by the ion carrier combination going through a metabolic step requiring energy which results in the ion being deposited in or associated with the internal metabolic system of the plant. Various similar models consistent with this hypothesis may be visualized (van den Honert, 1937; Osterhout, 1936; Rosenberg, 1948; Jacobsen *et al.*

1950; Epstein and Hagen, 1952; Florell, 1957). A simple model by Fried and Shapiro (1961) following Epstein and Hagen (1932) is shown below

$$\mathrm{R} + \mathrm{M} \underset{k_{-1}}{\overset{k_1}{\rightleftharpoons}} \mathrm{MR} \tag{3.1}$$

$$\mathrm{MR} \underset{k_{-2}}{\overset{k_2}{\rightleftharpoons}} \mathrm{M(accumulated)} + \mathrm{R'} \tag{3.2}$$

where M is the ion, R the carrier containing a specific binding site, and MR an intermediate that goes through an active process resulting in the deposition of the ion across an energy barrier. Equation (3.1) describes the partial saturation of the binding site. Equation (3.2) describes the metabolic requiring step and, therefore, refers to the active accumulation process which is rate limiting. Since the back reaction for the accumulating ion is relatively slow, Eq. (3.2) under ordinary circumstances can be considered essentially irreversible insofar as active ion accumulation is concerned.

If Eqs. (3.1) and (3.2) or a variation of these equations describe ion accumulation, the mathematical expression of these equations must quantitatively describe the process. Furthermore, ion uptake would reflect the rate constants and the concentration of reactants including carrier concentration. Changes in ion uptake must, therefore, be due to changes either in the concentration of the reactants or changes in the rate constant. Assuming that k_{-2} is negligible or the experimental conditions are such that only accumulation of the ion M in the plant is being recorded, the following equation should hold:

$$dMa/dt = v = [\mathrm{MR}]k_2, \qquad (\mathrm{Ma} = [\mathrm{MR}]k_2t) \tag{3.3}$$

where v is the rate of accumulation of actively accumulated ions and Ma is the amount of ion accumulated in time t. If the concentration of M in solution is maintained essentially constant (this can be done by using a large enough volume of bathing solution and vigorous stirring) and the plant root maintains a constant carrier concentration, then a plot of Ma vs. t should give a straight line (steady state). If the concentration of M is increased, the maximum accumulation rate, $V_{\max}$, would occur when the concentration of MR was equal to the total concentration of carrier present, $[\Sigma\mathrm{R}]$, i.e., when $[\mathrm{MR}] = [\Sigma\mathrm{R}]$

$$V_{\max} = [\Sigma\mathrm{R}]k_2 \tag{3.4}$$

Furthermore, at steady state where the rate of accumulation of M, i.e., dMa/dt is a constant, indicating that the concentration of M is also constant, and the change in MR concentration is, therefore, zero, i.e.,

$$d[\mathrm{MR}]/dt = 0 = [\mathrm{R}][\mathrm{M}]k_1 - ([\mathrm{MR}]k_{-1} + [\mathrm{MR}]k_2) \tag{3.5}$$

from which it follows that

$$[\mathrm{R}][\mathrm{M}]/[\mathrm{MR}] = (k_{-1} + k_2)/k_1 = K_M \tag{3.6}$$

where K_M is a new constant, relating the various rate constants, and is analogous to the Michaelis–Menten constant (Michaelis and Menten, 1913) of an enzyme-substrate complex. Applying the conservation equation

$$[\Sigma\mathrm{R}] = [\mathrm{R}] + [\mathrm{MR}] \tag{3.7}$$

From Eq. (3.7)

$$[\mathrm{R}] = [\Sigma\mathrm{R}] - [\mathrm{MR}] \tag{3.8}$$

From Eq. (3.3)

$$[\mathrm{MR}] = v/k_2 \tag{3.9}$$

From Eq. (3.4)

$$[\Sigma\mathrm{R}] = V_{\mathrm{max}}/k_2 \tag{3.10}$$

Substituting in Eq. 3.8

$$[\mathrm{R}] = V_{\mathrm{max}}/k_2 - v/k_2 \tag{3.11}$$

and substitution of Eq. (3.9) and (3.11) in Eq. (3.6)

$$\frac{(V_{\mathrm{max}}/k_2 - v/k_2)[\mathrm{M}]}{v/k_2} = K_M \tag{3.12}$$

and

$$v = -K_M \cdot v/[\mathrm{M}] + V_{\mathrm{max}} \tag{3.13}$$

2.2. Biological Constants

This development or model has shown that the ion accumulation part of the ion uptake process can be characterized by essentially three constants, ΣR, k_2, and K_M, where ΣR is the total concentration of effective carrier, k_2 is the specific reaction rate constant of the rate-limiting metabolic step, and K_M equals $(k_1 + k_2)/k_1$, a constant equal to or a function of the dissociation constant of MR, the ion carrier combination, analogous to K_M, the Michaelis–Menten constant.

In this simple picture describing the kinetics of ion uptake, K_M, which is truly a constant, is shown as involving only two rate constants. The accumulation process may actually involve a sequence of more than two reactions. Although the kinetics will still describe the rate-limiting step, K_M may involve both other rate constants and the constant concentration of other reactants in the nonrate-limiting step (Jackson *et al.*, 1962). However, the surprising equvalence found by

Noggle and Fried (1960) for the K_M's of three different plant species (at both sites of phosphorylation) that differed markedly in their rates of phosphate uptake suggest that if other rate constants and reactants are involved in the K_M, they are not easily changed. All environmental effects on ion uptake can be described by their effect on the three biological constants, ΣR, k_2, and K_M, and the measurement of the effect of specific treatments on these constants can be utilized to aid in

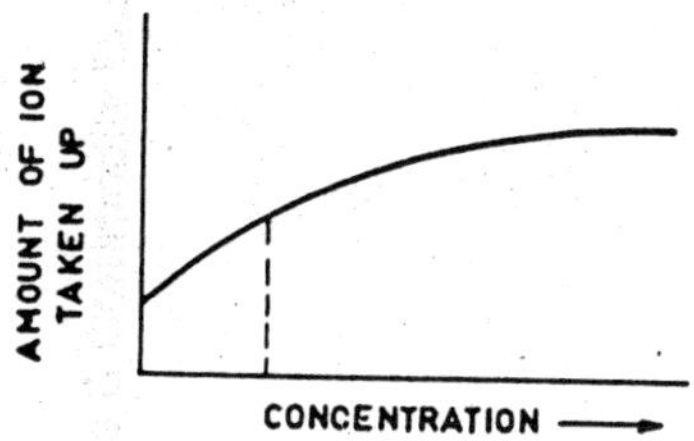

FIG. 3.10. Diagrammatic representation of data presented by Bange (1959) of the relationship between K concentration in solution and the uptake of K by maize.

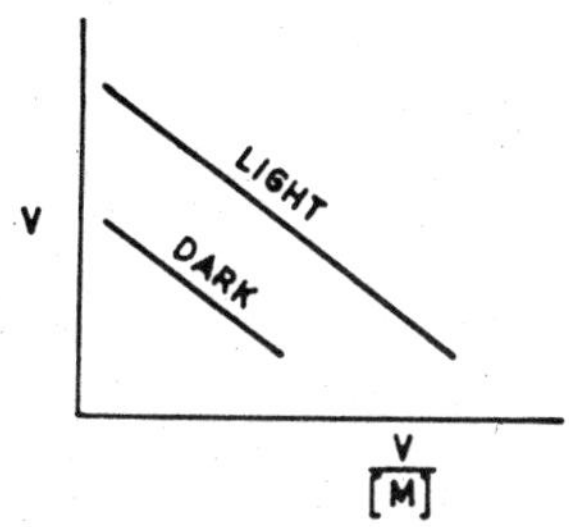

FIG. 3.11. Semidiagrammatic representation of the data presented by Fried *et al.* (1961) on the relationship between Rb concentration and Rb uptake by barley as affected by the presence or absence of light.

elucidating the nature and pathways of the ion accumulation mechanism itself.

A plot of amount of ion taken up vs. concentration of ion in solution invariably gives a curve of the type shown in Fig. 3.10, taken from Bange (1959), where the dashed line represents the point of one-half ion saturation. This concentration equals K_M.

Another probably more precise way of obtaining K_M is to plot v vs. $v/[M]$ as given in Eq. (3.13). The slope of the resultant line is K_M as shown in Figs. 3.11 and 3.12. By considering a mathematical transformation of Eq. (3.13)

$$1/v = K_M/V_{\max} = 1/[M] + 1/V_{\max}$$

and plotting $1/v$ vs. $1/[M]$, the slope divided by the intercept gives K_M as shown in Fig. 3.12, taken from Epstein and Hagen (1952).

When uptake of individual ion species by plants are studied over a wide concentration range, there appears to be more than one site of uptake (where site experimentally refers to a rate limitation) whether uptake of individual cations or anions is being studied. The results of Noggle and Fried (1960) are typical and are shown in Fig. 3.13. Similar results have been reported for Rb, K, Cs, Na, Sr, P, and Br and, in

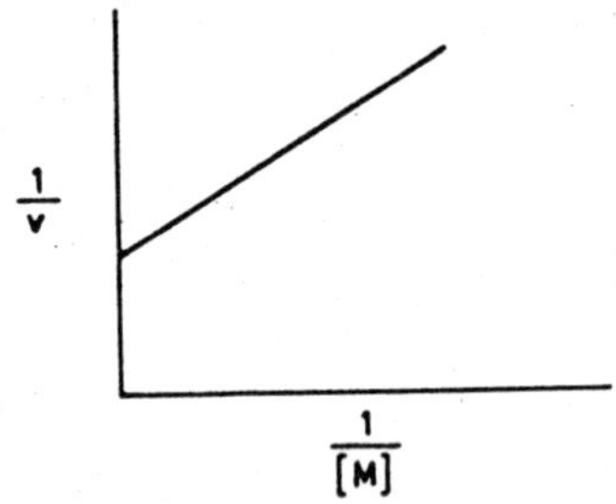

FIG. 3.12. Semidiagrammatic representation of the data of Epstein and Hagen (1952) on the relationship between Rb concentration in solution and Rb uptake by barley.

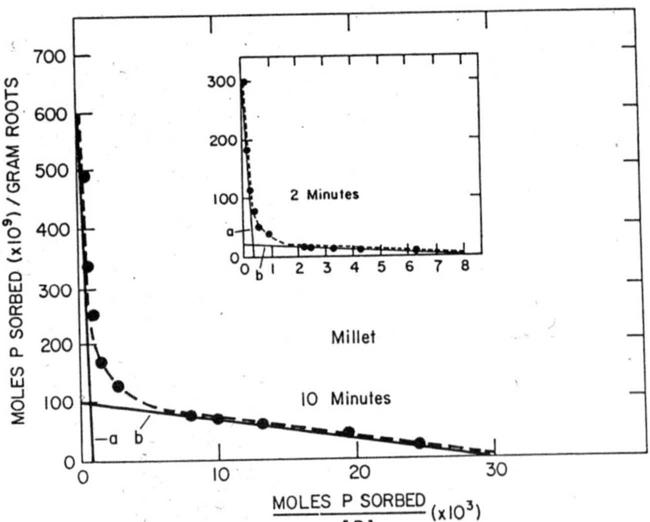

(A)

FIG. 3.13. The effect of P concentration upon the uptake of P by 1 [illegible] roots of (A) millet, (B) barley, and (C) alfalfa in 2 and 10 min[illegible] reactions are separated into lines "a" and "b". From Noggle and F[illegible]

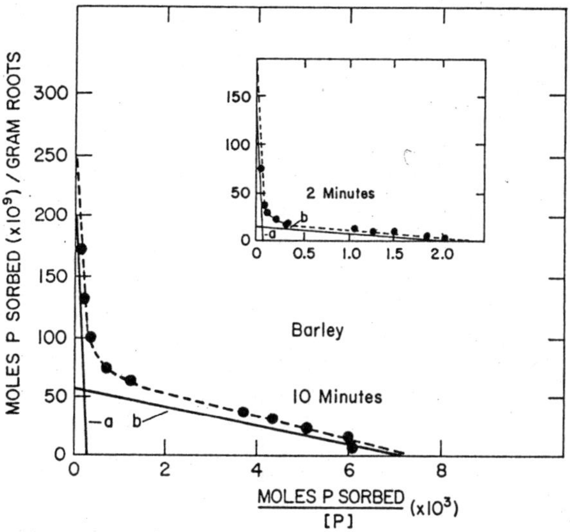

(B)

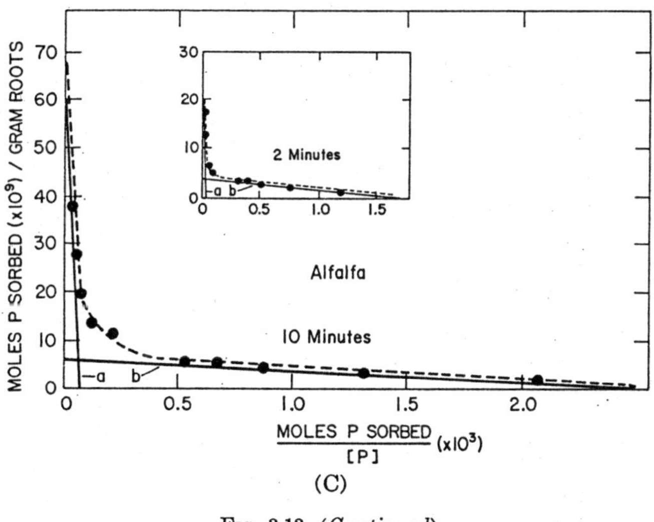

(C)

FIG. 3.13. (*Continued*)

fact, when this is the method of approach, no exceptions to this generalization have been found. In the phosphate system the metabolic products of phosphate uptake from each site can be determined and have been demonstrated to be different (Jackson and Hagen, 1960), thus clearly indicating that at least two sites of phosphorylation are present in barley roots, one predominant at relatively high phosphate concentrations and

TABLE 3.2

MICHAELIS–MENTEN CONSTANTS OF CATION–CARRIER COMPLEXES IN PLANT ROOTS

Ion species	Plant species	K_M At relatively high concentrations	K_M At relatively low concentrations	Ref.
Rb	Barley	4×10^{-3}	2×10^{-5}	*c*
Rb	Barley	4×10^{-3}	—	*d*
Rb	Barley	2×10^{-3}	4×10^{-6}	*e*
Rb	Barley	—	12×10^{-6}	*f*
Rb	Barley	—	4×10^{-5}	*g*
Rb	Barley	16×10^{-3}	$16\text{–}18 \times 10^{-6}$	*h,i*
K	Barley	1×10^{-3}	3×10^{-6}	*e*
K	Barley	4×10^{-3}	3×10^{-5}	*c*
K	Corn	0.3×10^{-3}	—	*j*
K	Corn	0.2×10^{-3}	2×10^{-6}	*k*
K	Barley	11×10^{-3}	21×10^{-6}	*h*
K	Barley	5×10^{-3}[a]	—	*l*
K	Barley	—	9×10^{-6}	*f*
K	Soybeans	0.5×10^{-3}	—	*m*
K	Wheat	—	35×10^{-6}	*n*
Cs	Barley	—	25×10^{-6}	*f*
Cs	Barley	5×10^{-3}[a]	8×10^{-6}	*o*
Cs	Barley	—	2×10^{-5}	*p*
Na	Corn	3×10^{-3}	2×10^{-5}[b]	*q*
Na	Soybeans	2×10^{-3}	—	*q*
Na	Radish	1×10^{-3}	—	*q*
Na	Barley	4×10^{-3}	10×10^{-6}	*e*
Na	Barley	7×10^{-3}	55×10^{-6}	*c*
Na	Wheat	—	11×10^{-5}	*n*
NH_4	Wheat	—	7×10^{-6}	*n*
NH_4	Rice	3×10^{-3}	2×10^{-5}	*g*
NH_4	Maize	—	9×10^{-5}	*r*
NH_4	Maize	—	1×10^{-5}	*s*
NH_4	Ryegrass	—	4×10^{-5}	*t*
NH^4	Barley	3×10^{-1}	2×10^{-5}	*o*
Sr	Barley	—	60×10^{-6}	*u*
Sr	Barley	3×10^{-3}	3×10^{-6}	*e*
Li	Barley	6×10^{-3}	—	*v*

[a] Estimated for the data.
[b] See Bange and van Gemerden (1963).
[c] Jackson and Adams (1963).
[d] Epstein and Hagen (1952).
[e] Fried and Noggle (1958).
[f] Boszormenyi and Bange (personal communication).
[g] Fried *et al.* (1961).
[h] Epstein *et al.* (1963).
[i] Rains *et al.* (1964).
[j] Hanson and Kahn (1957).
[k] Bange (1959).
[l] Overstreet *et al.* (1952).
[m] Kahn and Hanson (1957).
[n] Tromp (1962).
[o] Bange and Overstreet (1960).
[p] Bange and van Gemerden (1963).
[q] Huffaker and Wallace (1958).
[r] Van den Honert and Hooymans (1961).
[s] Becking (1956).
[t] Lycklama (1963).
[u] Epstein and Leggett (1954).
[v] Epstein (1960).

one prevailing at relatively low phosphate concentrations. Tables 3.2 and 3.3. give the values for K_M reported for different plant and ion species. These can be compared to the concentration of ions in the soil solution (Table 2.9).

The results in these tables show how the soil solution concentration and plant uptake are related. Thus at 10^{-6} *M* concentration of phosphate [10^{-9} *M* concentration of HPO_4^{2-}, if the interpretation of Hagen and

TABLE 3.3
MICHAELIS–MENTEN CONSTANTS OF ANION–CARRIER COMPLEXES IN PLANT ROOTS

Ion species	Plant species	K_M At relatively high concentrations	K_M At relatively low concentrations[a]	Ref.
Phosphate	Barley	0.4×10^{-4}	1×10^{-6}	*c*
Phosphate	Barley	3×10^{-4}	3×10^{-6}	*d*
Phosphate	Barley	9×10^{-4}	3×10^{-6}	*e*
Phosphate	Alfalfa	9×10^{-4}	3×10^{-6}	*e*
Phosphate	Millet	10×10^{-4}	4×10^{-6}	*e*
Phosphate	Sugar cane	—	1×10^{-6}[b]	*f*
Phosphate	Barley	5×10^{-4}	5×10^{-6}	*l*
Phosphate	Alfalfa	8×10^{-4}	4×10^{-6}	*l*
Phosphate	*S. humilis*	3×10^{-4}	6×10^{-6}	*l*
Phosphate	*P. lathyroides*	5×10^{-4}	3×10^{-6}	*l*
Phosphate	*D. uncinatum*	4×10^{-4}	4×10^{-6}	*l*
Sulfate	Barley	—	10×10^{-6}	*g*
Bromide	Barley	13×10^{-3}	—	*h*
Bromide	Barley	10×10^{-3}	20×10^{-6}	*d*
Bromide	Barley	—	$10–19 \times 10^{-6}$	*i*
Bromide	Barley	1×10^{-4}	24×10^{-6}	*n*
Chloride	Barley	—	$1–5 \times 10^{-6}$	*m*
Chloride	Barley	—	16×10^{-6}	*n*
Nitrate	Corn	—	20×10^{-6}	*j*
Nitrate	Rice	6×10^{-4}	—	*o*
Iodide	Wheat	3×10^{-3}[b]	—	*k*

[a] If phosphate ion is HPO_4^{2-}, multiply by 0.00064 to obtain K_M.
[b] Estimated from the data.
[c] Hagen and Hopkins (1955).
[d] Hagen *et al.* (1957).
[e] Noggle and Fried (1960).
[f] van den Honert (1933).
[g] Leggett and Epstein (1956).
[h] Epstein (1953).
[i] Böszörményi and Bange (personal communication).
[j] van den Honert *et al.* (1955).
[k] Böszörményi and Cseh (1958).
[l] Andrew (1966).
[m] Elzam and Epstein (1965).
[n] Böszörményi (1966).
[o] Fried *et al.* (1965).

Hopkins (1955) is correct], the carrier site effective at lower phosphate concentrations is half saturated. Increasing phosphate concentration in the soil solution cannot increase phosphate accumulation more than two-fold until the site effective at higher concentrations of phosphate (H_2PO_4—according to Hagen and Hopkins) becomes effective. Table 2.9 indicates that in acid soils it is not uncommon for the concentration of phosphate to be about as large as the K_M of the site effective at lower concentrations, i.e., 10^{-6} M. The other major nutrient anions, sulfate and nitrate, are less well-defined both as to concentrations in the soil solution and K_M. There is a good indication that at least two uptake sites occur with each anion. The K_M of the two sites for phosphate are shown in Table 3.3. In addition, Hagen *et al.* (personal communication) have determined K_M at both sites for bromide; Böszörményi and Cseh (1958) noted two sites in chloride and iodide uptake by wheat roots; and Legget and Epstein (1956) noted a second concentration function for sulfate on barley roots.

The concentration of cations in the soil solution (Table 2.9) indicates that the site effective at the lower concentration is almost always essentially saturated. However, the site that is effective at higher concentrations is generally at a lower degree of per cent saturation. Thus there is always appreciable uptake from the site with the lower K_M, whereas the contribution of the site with the higher K_M is a function of the soil solution concentration of the individual cation. Fertilization increases the concentration of the ion and consequently the turnover of MR of the site with the higher K_M.

Although K_M is relatively easy to determine by plotting uptake as a function of concentration, obtaining either ΣR or k_2 is rather difficult if not impossible. V_{max} which equals ΣRk_2 can easily be obtained from the same data used in obtaining K_M. However, determining the individual constants making up V_{max} requires additional information Since both ΣR and k_2 are controlled by metabolism, it is very difficult to vary one experimentally without varying the other. Attempts have been made to obtain these constants by Hagen *et al.* (1957), Fried *et al.* (1958), Noggle and Fried (1960), and Andrew (1966) based on the observation that when uptake of an ion was plotted against time at various concentrations of the ion, a series of steady state lines was obtained in which the slope intercept ratio was essentially constant (Fig 3.5). Furthermore, the intercept was specific in relation to the ion used. This suggests that the experimentally observed uptake from one site of the ion M, under experimental conditions could be described by Eq. (3.14)

$$[M] = [MR]k_2t + [MR] \tag{3.14}$$

and that MR was the intercept shown in Fig. 3.1. Therefore, the slope MRk_2 divided by the intercept MR directly equals k_2. Once k_2 is known ΣR can be obtained from V_{max} since $V_{max} = \Sigma Rk_2$. Thus, Noggle and Fried (1960) were able to show in a comparison of three species of plants that only the carrier concentration varied appreciably.

Although the results of Hagen *et al.* (1957) and Fried *et al.* (1958) suggest that MR and k_2 can be measured, objections have been raised particularly by Bange and Van Gemerden (1963). Their objection was based primarily on the difficulty of satisfactorily washing the roots, and in an experiment with Cs they concluded that the intercept of the time curve did not represent MR. It is possible, however, that under their experimental condition the ratio of slope to intercept was so high (i.e., that k_2 was so large relative to MR) that MR was experimentally undetectable.

The very possibility that MR can be measured and thus the concentration of carrier and the reaction rate constant is worth pursuing. Of course, the accuracy of this interpretation is on no firmer footing than the carrier concept itself and the model here presented, to which much exception can be taken. Yet certain evidence is clear. It was demonstrated in Chapter 3, Section 1.2 that ions are adsorbed to both living and dead roots, but that the adsorption phenomena was clearly different in living roots than in dead roots. This, of course, can reasonably be explained by assuming that the method of killing the roots caused differences in the adsorption sites by causing chemical changes in the root tissue. It is much more difficult, however, to explain why the exchange process in dead roots follows the well-known valence law and even the lyotropic series, whereas living roots show a marked amount of specificity. Furthermore, the specificity shown for adsorption is clearly the same as the specificity shown in the ion accumulation process. Thus, strict competition between K and Rb has been demonstrated by growing plants in nutrient solution over long periods of time (Collander, 1941; Menzel and Heald, 1955) and as shown in Fig. 3.8 A–C the same can be demonstrated for the effect of K on adsorbed Rb for experimental conditions in which even a divalent cation had no effect.

Another consistent phenomenon linking the specific adsorption sites to metabolic activity is the ability of noncompetitive inhibitors to affect the amount of adsorption. The effect of mannose and 1% oxygen on K adsorption has already been mentioned. (Fig. 3.8A). The same figure shows that DNP may have the same effect. Figure 3.14 demonstrates a similar phenomenon for the effect of NaN_3, and Fig. 3.15 for the effect of succinate and β-hydroxybutyrate (Hagen *et al.*, 1957) on the adsorption of phosphate (intercept of the time curve). Direct evidence of a linking of the initial specific adsorption phase of ion uptake with metabolism has been obtained with storage tissue notably by Laties (1959, 1959a) and Barber and Russell (1961). The latter authors showed that low temperature and DNP could decrease the amount of Rb adsorbed by carrot discs. The results of Laties go even further by appearing to demonstrate an ability to control the production and destruction of a chloride carrier site at will, i.e., a site that both adsorbed chloride that was linked to chloride uptake and a site the presence of which was controlled by metabolic factors.

Considering the clear linkage of the specific adsorption of ions by living roots to metabolism and the demonstration by Hagen *et al.* (1957)

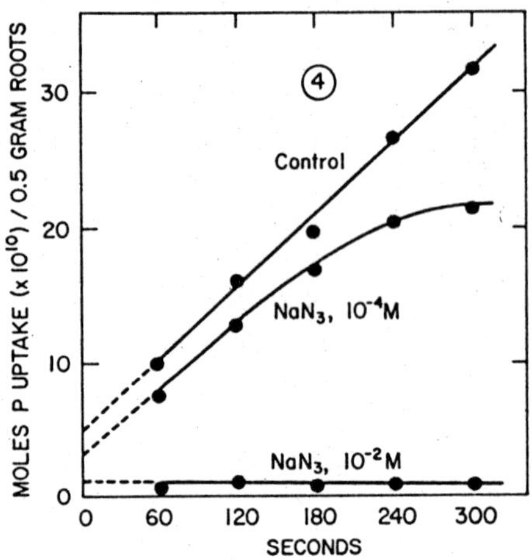

FIG. 3.14. The uptake of orthophosphate by 0.5 g of excised barley roots from 2×10^{-6} *M* orthophosphate in the presence and absence of azide. From Hagen *et al.* (1957).

and Fried *et al.* (1958) of the proportionality of adsorption with the slope of the steady state uptake time curve for phosphate and cations, respectively, it is not at all surprising that both groups of authors concluded that this adsorption by living roots may represent a specific adsorption to carrier sites, the concentration of which could be measured. This permits the additional calculation of the specific reaction rate constant and the total carrier concentration.

The general picture that thus emerges is that ion uptake by plant roots involves an active metabolic process that is specific. It is not enhanced by any nonspecific adsorption of ions that may take place on root surfaces. However, the adsorption of the ion to the carrier site is a first step in ion uptake which may be confused with either nonspecific adsorption which occurs generally in dead or damaged roots (and may occur to some limited extent on living roots) or even Donnan effects which can occur both in living and dead roots. The process can be characterized by three constants: the concentration of carrier, ΣR, the specific reaction rate constant of the rate-limiting step, k_2, and K_M, a constant, functional on the dissociation constant of the ion carrier combination.

Although Eqs. (3.1) and (3.2) are the simplest representation of carrier uptake theory, they do not take into account the necessity that in order to maintain electrical neutrality in the external solution either equivalent

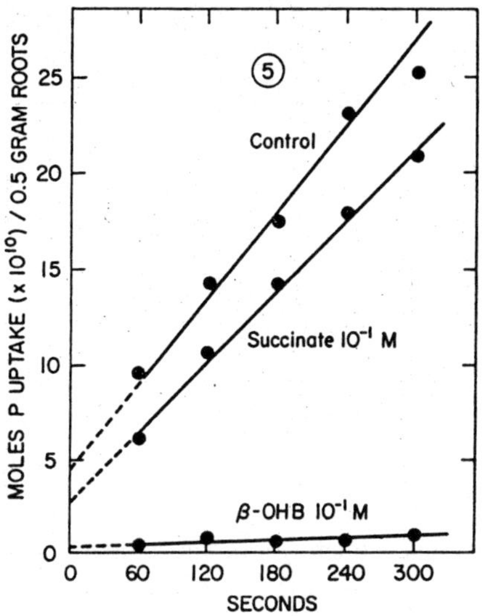

FIG. 3.15. The uptake of orthophosphate from 0.5 g of excised barley roots from 1×10^{-6} *M* orthophosphate in the presence and absence of succinate or β-hydroxybutyrate. From Hagen *et al.* (1957).

numbers of cations and anions must be taken up by the plant or each time a cation or anion is taken up the plant gives off to the solution an ion or ions, with the same total number of charges of the same sign. Any combination of these two processes, properly balanced, would also satisfactorily maintain electrical neutrality.

There is increasing evidence that at least part of this balance is maintained by hydrogen ion exchange during cation uptake. In 1950 Jacobson *et al.* suggested that uptake of K could be represented by the equation,

$$HR + K \rightleftharpoons KR + H^+ \tag{3.15}$$

where HR represented a metabolically produced cationic binding substance and KR a labile organic complex of K. This is equivalent to

Eq. (3.1) and would be referred to as MR. The kinetic consequences of this representation are identical to those of Eqs. (3.1) and (3.2), assuming that Eq. (3.15) would again have to be followed by Eq. (3.2). It has the further consequence, however, that H ion concentration must then affect the uptake of cations. The authors and others (Arnon *et al.*, 1942; Hoagland and Broyer, 1940; Jacobson *et al.*, 1957; Fried and Noggle, 1958; Becking, 1956; Jackson and Adams, 1963) have clearly demonstrated this. Becking (1956) went even further and attempted to measure the amount of H given off during the uptake of ammonium ion by maize roots. He found almost an equivalent exchange ratio, and where deviations occurred they could be explained by the limitations of his technique. Jackson and Adams (1963) in a more direct measurement of H ion release by barley roots concluded "that potassium and sodium ions are absorbed independently of the anions of the absorption solution in exchange for H^+." If this is the mechanism of ion uptake, then H should be fully competitive with all of the cations in the full sense as outlined in Chapter 4. This indeed was found by Fried and Noggle (1958) and recently confirmed for Rb by Rains *et al.* (1964). The concept of the biological constants describing the system remains unaltered, and hydrogen ion competitive effects are subject to the mathematical treatment for competing ions.

2.3. Ion Interactions

Interactions between one ion and another must be either a direct effect on one of these biological constants or must operate through the medium of these constants. Once these are quantitatively evaluated with respect to the biological constants, interaction effects can be quantitatively predicted. An ion may interact with another owing to either of three causes or a combination of either of the three: (1) an interaction with the carrier that does not change the total amount of carrier, but does change the amount of carrier that will combine with the other ion; (2) an interaction with the carrier-producing system such that either more or less carrier is produced; and (3) an interaction with the turnover system such that the biological rate constant of the rate-limiting step changes.

The first type of interaction affecting the amount of carrier that will combine with the other ion can have many causes which are mathematically definable. Equivalent examples have been worked out for enzyme systems (Dixon and Webb, 1958). From the standpoint of ion uptake, one type of interaction is particularly important. This is competitive inhibition or competition.

Competitive inhibition or *competition* refers only to the situation in which the foreign ion reversibly competes with the indigent ion for the same place on the carrier site, i.e., the resultant action of the foreign ion is at the same locus. V_{max} would thus remain the same and from the physicochemical standpoint the inhibition can be overcome by increasing the concentration of the indigent ion. Thus if one writes the equation for the competing ion combining with the carrier as shown in Eq. (3.16)

$$I + R \rightleftharpoons IR \tag{3.16}$$

where R in Eqs. (3.16) and (3.1) refer to the same carrier, I (the competing ion) and IR (a complex of the competing ion and carrier) and considers the equations for the equilibrium constants of both the indigent and competitive ion combining with the carrier plus the conservation equation,

$$[\Sigma R] = [R] + [IR] + [MR] \tag{3.17}$$

a mass action equation can be derived relating the concentration of the ions and the apparent dissociation constants to the amount of active intermediates.

$$MR = \Sigma R - (K_M + K_M[I]/K_i)[MR]/[M] \tag{3.18}$$

where K_i is a constant that is functional on the association of the inhibiting ion to the carrier (i.e., competing ion with carrier) and completely analogous to K_M except that it refers to the competing ion. By substituting for MR and ΣR the respective relationships obtained from kinetic equations of steady state [Eqs. (3.3) and (3.4)] Eq. (3.19) is obtained

$$v = V_{max} - (K_M + K_M[I]/K_i) \cdot v/[M] \tag{3.19}$$

The characteristics of competition are that ΣR and k_2 are unchanged and as a result V_{max} is unchanged. Furthermore, at a given concentration of competing ion, measurements of v as a function of concentration of indigent ion enable one to obtain a constant which includes the constant K_i (i.e., the K_M of the competing ion); and if either K_M of the indigent ion is known or if the experiment is done at two or more concentrations of competing ion, the K_M of the competing ion, K_i, can be calculated.

This has been done for many ions in the ion uptake system and some of the results obtained are given in Table 3.4. Thus, competitive inhibition has been demonstrated between K^+, Cs^+, and Rb^+ (Bange and Overstreet, 1960; Collander, 1941; Epstein and Hagen, 1952; Menzel and Heald, 1955; Fried *et al.*, 1961); Ca^{2+}, Sr^{2+}, and Ba^{2+} appear competitive (Leggett and Epstein, 1956); Cl^-, Br^-, and possibly I^- appear competitive (Böszörményi and Cseh, 1958; Epstein, 1953) as do phosphate and arsenate (Hagen and Hopkins, 1955). Hydrogen appears to compete for all sites that bind cations (Fried and Noggle, 1958; Mortland, 1955) and hydroxyl may compete for all sites involving anions (Hagen and Hopkins, 1955; Jacobson, 1950; Jackson and Adams, Rains *et al.*, 1964). Sodium competition is not clear. There is some indication that Na^+ may compete with one K^+ site and not the other (Bange, 1959). No two nutrient ions have yet been shown to utilize the same

TABLE 3.4

COMPETITIVE INHIBITION CONSTANTS, K_i, FOR ION UPTAKE BY BARLEY ROOTS[a]

Sorbed species	Ion K_m	Competing species	Ion K_i	Ref.
Na	1×10^{-5}	H	2×10^{-5}	*h*
K	3×10^{-6}	H	1×10^{-5}	*b*
Rb	4×10^{-6}	H	1×10^{-5}	*b*
Rb	16×10^{-6}	H	3.8×10^{-5}	*c*
Sr	3×10^{-6}	H	1×10^{-5}	*b*
Rb	4×10^{-6}	K	4×10^{-6}	*b*
Rb	3×10^{-3}	K	4×10^{-3}	*d*
Rb	5×10^{-3}	Cs	9×10^{-3}	*d*
Sr	60×10^{-6}	Ca	30×10^{-6}	*e*
Sr	60×10^{-6}	Ba	80×10^{-6}	*e*
Br	13×10^{-3}	Cl	13×10^{-3}	*f*
Cl	1×10^{-6}	Br	4×10^{-6}	*i*
SO_4	10×10^{-6}	SeO_4	7×10^{-6}	*g*
H_2PO_4	4×10^{-5}	OH	2.5×10^{-8}	*h*
HPO_4	5.5×10^{-10}	OH	2×10^{-11}	*h*

[a] From Fried and Shapiro (1961).
[b] Fried and Noggle (1958).
[c] Rains *et al.* (1964).
[d] Epstein and Hagen (1952).
[e] Epstein and Leggett (1954).
[f] Epstein (1953).
[g] Leggett and Epstein (1956).
[h] Hagen and Hopkins (1955).
[i] Elzam and Epstein (1965).

binding site of ion uptake when the strict definition of competitive inhibition is maintained. This means that no interactions between nutrient ions are at present explainable through the mechanism of competitive inhibition. However, the necessary combinations of nutrient ions at different concentrations have not by any means been exhaustively tested. The suggestion by Epstein (1960) that Ca and Li were competitive was shown to involve an extreme extrapolation and, in fact, could be invalidated by the author's own data (Fried and Shapiro, 1961). Waisel (1962), who also suggested competition between Ca and Li, retracted this in a later publication (Waisel, 1962). The possibility that NH_4

might compete with K since there was a slight suggestion that NH_4 competes with Cs is highly unlikely since Tromp (1962) showed that the NH_4-K interaction was not competitive, and Lisanti and Marckwordt (1963) showed the same for the NH_4-Rb interaction.

Noncompetitive and Uncompetitive Inhibition. There are various ways other than by competition that an inhibitor could interfere with the binding of the nutrient ion to the carrier as indicated by at least two analogous interferences recognized in enzyme kinetics. One of these is noncompetitive inhibition which is assumed to involve a reversible reaction of an inhibiting ion with the carrier in a region other than the active center. The combination of nutrient ion and carrier is unaffected, but the breakdown of the ion carrier complex, MR, is partially or wholly prevented, depending upon the concentration and K_i of the inhibitor. The result is a change in V_{max} but not K_M. A second is so-called uncompetitive inhibition which is assumed to involve only the combination of inhibitor with the ion carrier complex, MR, forming MRI. The result is a reduction in both V_{max} and K_M. Thus, these other types of inhibition are clearly defined mathematically (Dixon and Webb, 1958). In general, attempts have not been made to evaluate whether the interaction of ions during the ion uptake process by plants can be interpreted by such mechanisms. Bange (1962) has suggested that competition occurs between carrier ion combinations for the enzyme required in the breakdown process and the necessary mathematical calculations are included in his report.

The second type of interaction (i.e., the effect on the amount of carrier produced) and the third type of interaction (i.e., the affect on the reaction rate constant) are difficult if not impossible to distinguish from each other at this stage of our knowledge, although attempts have been made to do so. Indeed, if the conclusions of Hagen *et al.* (1957) and Fried *et al.* (1958) with regard to the intercept of the time uptake curve are valid many more such determinations could be made. Noggle and Fried (1960) have compared species of plants using this thesis and have found that the difference among the species was essentially a difference in carrier concentration only. Hagen *et al.* (1957) interpreted the action of various substrates and inhibitors based on the same conclusions and thus located the sites of accumulation of phosphate with respect to the respiratory chain.

The scientific literature is replete with reports of the effect of one ion on the uptake of another without experimental evidence of mechanism. From the standpoint of determining mechanism, the shorter time experiments are more likely to be enlightening. Even the short time experiments, however, cannot always be compared with each other because of the differences in technique. Many investigations, using isotopes, measure the uptake of the ion and the interaction directly. Others measure the net change of the ion within the plant and still others, the net change in solution. Since for every ion taken up one of the opposite charge must be taken up or one of the same charge must be given off by the root, the net change or change in solution concentration may be affected by this maintenance of electrical neutrality, particularly in relation to the pretreatment of the roots. Although this exchange is real, it can at least to some extent confuse the mechanistic picture unless both the uptake and net change are measured or the experimental conditions are set up in such a manner that either uptake or interaction or both are directly measurable.

Tables 3.5 and 3.6 and the summary in Table 3.7 are a partial compilation from the literature of these short-time effects, with the exception of competitive interactions which are listed in Table 3.4.

As much as one can generalize from observations from so many authors with different experimental conditions, different biological material with respect to species, age, and nutritional condition, the following conclusions may be drawn.

(1) Lithium accumulation is inhibited by the presence of such multivalent cations as Mg, Ca, Sr, Mn, Al, and La.

(2) Sodium accumulation is inhibited by the presence of all cations tested at equivalent concentrations to Na. These ions are Li, K, NH_4, and Ca.

(3) Potassium accumulation has been reported to be stimulated by Ca, but has also been reported to be inhibited by Ca. Sodium and ammonium appear to inhibit K accumulation, whereas the effect of other divalent cations such as Mg, Ba, Mn, and Al has varied with the experimental material.

(4) Ammonium accumulation appears to be inhibited by Na, Ca, and La with slight indications of inhibition by Li and K.

(5) Rubidium accumulation usually is stimulated by the presence of Ca and has been reported to be both stimulated and inhibited by Li. Both Na and Mg have been reported to inhibit Rb accumulation as have Ca and ammonium.

(6) Cesium accumulation has been inhibited by all ions tested, i.e., Ca, NH_4, Li, and Na.

(7) Calcium accumulation is inhibited by the presence of Rb, K, Na, Al, and Mg, and Sr is inhibited by Mg.

(8) Bromine accumulation is reported to be stimulated by the presence of divalent cations, but inhibited by the presence of nitrate. The

TABLE 3.5

EFFECT OF COMPLEMENTARY ION ON THE ACCUMULATION OF SPECIFIC CATION SPECIES BY PLANTS IN SHORT-TIME EXPERIMENTS

Plant species	Ion accumulated	Ion or salt producing effect	Effect	Ref.
Maize	NH_4	Ca	Slight inhibition	*a*
Wheat	NH_4	K	None	*b*
Wheat	NH_4	Na	None, inhibition	*b*
Rice	NH_4	Na, Li, Co	None	*c*
Rice	NH_4	K, Rb, Ca	Inhibition	*c*
Barley	Ca	Rb	Inhibition	*d*
Barley	Ca	Rb	None	*e*
Plantain	Ca	Rb	Inhibition	*e*
Barley	Ca	Na	Inhibition	*f*
Wheat	Ca	Al, Mg, K	Inhibition	*g*
Barley	Cs	Ca	Inhibition	*h*
Barley	Cs	Na, Li	None	*i,j*
Barley	Cs	NH_4, Rb, Li, Cs, K, Na, Ca	Inhibition	*i,j*
Barley	Cs	Ca	None to slight stimulation	*j*
Barley	Cs	Ca, Li	Inhibition	*h*
Barley	Cs	NH_4, Na, Mg	None to slight inhibition	*h,j*
Barley	Li	Si, Ca	Inhibition	*k*
Barley	Li	Mg, Ca, Si, La, Mn, Al	Inhibition	*l*
Corn, pea, squash, sunflower	Li	Ca	Inhibition	*l*
Corn, pea, onion, barley, squash, sunflower	K	Ca (10^{-2})	Stimulation	*l*
Barley	K	Ca (10^{-4})	None	*l*
Barley	K	Mg, Ca, Sr, La, Mn, Al (10^{-3})	Slight Stimulation	*l*
Barley	K	SO_4, NO_3, PO_4, HSO_4, HCO_3	None	*m*
Corn	K	Ca	Inhibition	*n*
Wheat	K	NH_4, Na	Inhibition	*o*
Barley	K	Ca, Mg, Si, Al	Stimulation	*o*
Barley	K	Ba, Na	None	*o*
Corn	K	Na	Inhibition	*p*
Barley	K	Ca	Stimulation	*f*
Barley	K	Na	Inhibition	*f*
Barley	K	Ca, La	Stimulation	*q*
Barley	K	Ce, Li, NH_4, Na	Inhibition	*q*
Barley	Rb	Li	Slight Stimulation	*k*

TABLE 3.5 (*Continued*)

Plant species	Ion accumulated	Ion or salt producing effect	Effect	Ref.
Barley	Rb	Li, Na	Inhibition	*r*
Barley	Rb	Li	None to stimulation	*r*
Barley	Rb	Mg, Na, Ba	None	*s*
Barley	Rb	Ca	Stimulation	*t*
Barley	Rb	Ca	Inhibition	*u*
Barley	Rb	Ca	None to stimulation	*k*
Barley	Rb	Na, Mg	Inhibition	*u*
Tobacco	Rb	NH_4	Inhibition	*v*
Mung bean	Rb	Ca	Stimulation	*u*
Barley	Rb	Ca, Mg	Stimulation	*w*
Barley	Rb	NH_4	Inhibition	*w*
Plantain	Rb	Ca	Inhibition	*e*
Barley	Na	Ca	Inhibition	*x*
Barley, maize, onion, pea, squash, sunflower	Na	Ca	Inhibition	*l*
Corn, barley	Na	K, Ca, Li	None	*f*
Corn, soybeans	Na	$K(10^{-4} - 10^{-3})$	None to stimulation	*y*
Corn, soybeans, radish	Na	$K(10^{-3} - 10^{-2})$	Inhibition	*y*
Wheat	Na	K, NH_4	Inhibition	*b*
Barley	Sr	Mg	Inhibition	*z*

[a] Van den Honert and Hooymans (1961).
[b] Tromp (1962).
[c] Fried *et al.* (1965).
[d] Marckwordt (1963); Welte and Marckwordt (1963); Lagerwerff and Peech (1961).
[e] Noggle *et al.* (1964).
[f] Helmy *et al.* (1963).
[g] Johnson and Jackson (1964).
[h] Bange and Van Gemerden (1963).
[i] Handley and Overstreet (1961).
[j] Bange and Overstreet (1960).
[k] Waisel (1962).
[l] Jacobson *et al.* (1961).
[m] Jackson and Adams (1963).
[n] Hanson and Kahn (1957); Kahn and Hanson (1957).
[o] Viets (1944).
[p] Bange and Van Vliet (1961).
[q] Fawzy *et al.* (1954).
[r] Epstein and Hagen (1952).
[s] Fried *et al.* (1958).
[t] Tanada (1956); Tanada (1955); Lagerwerff and Peech (1961); Welte and Marckwordt (1963); Marckwordt (1963).
[u] Tanada (1963).
[v] Skogley and McCants (1963).
[w] Lisanti and Marckwordt (1963).
[x] Heytler (1963); Marschner (1964).
[y] Huffaker and Wallace (1958).
[z] Epstein and Leggett (1954).

chloride which is competitive with Br has been reported in one experiment to be stimulated by fluoride.

(9) Phosphate accumulation is reportedly stimulated by th of calcium nitrate, and sulfate accumulation is stimulated chloride.

TABLE 3.6

EFFECT OF COMPLEMENTARY ION ON THE UPTAKE OF SPECIFIC ANIONS SPECIES BY PLANTS IN SHORT-TIME EXPERIMENTS

Plant species	Ion accumulated	Ion or salt producing effect	Effect	Ref.
Barley	Br	NO_3	Inhibition	*a*
Barley	Br	$CaSO_4$	Stimulation	*b*
Barley	Br	Ca, Mg, Sr, Ba	Stimulation	*b*
Wheat	Cl	F	Stimulation	*c*
Barley	Phosphate	$Ca(NO_3)_2$	Stimulation	*d*
Pea	Phosphate	Ca	None	*e*
Barley	Phosphate	$Ca(NO_3)_2$	Stimulation	*f*
Barley	Sulfate	NO_3, H_2PO_4	None	*g*
Barley	Sulfate	$CaCl_2$	Stimulation	*g*
Barley, alfalfa, soybeans	Cl	Ca	No stimulation	*h*

[a] Epstein (1953).
[b] Viets (1944).
[c] Boszormenyi and Cseh (1964).
[d] Tanada (1956); Tanada (1955).
[e] Savioja and Miettinen (1960).
[f] Russell and Martin (1953).
[g] Williams and Saunders (1956).
[h] Elgabaly (1962).

TABLE 3.7

EFFECT OF COMPLEMENTARY IONS ON THE ACCUMULATION OF A PARTICULAR ION SPECIES

Species of ion accumulated	Effect and species of complementary ions: Inhibition	Stimulation
Li	Mg, Ca, Sr, Mn, Al, La	—
Na	Li, K, NH_4, Ca	—
K	Na, NH_4, Ca, Ce, Li	Ca, La, Mg, Si, Al
NH_4	Na, Ca, K, Rb	—
Rb	Li, Na, Mg, Ca, NH_4	Na, Ca, Li, Mg
Cs	Li, Na, NH_4, Ca	Ca?
Ca	Rb, Na, Al, Mg, K	—
Sr	Mg	—
Br	NO_3	Mg, Ca, Sr, Ba
H_2PO_4	—	$Ca(NO_3)_2$
Cl	—	F
SO_4	—	$CaCl_2$

One set of data omitted from Tables 3.5 and 3.6 are those of Vose and Fried* which are presented in Table 3.8. They report a large number of comparisons under identical laboratory conditions which thus repre-

* Personal communication to be published in the Third Rept. of IAEA Lab Activities, IAEA, Vienna, Austria.

sents a unique situation for comparing the effects of nine different cations on the uptake of a tenth. In all cases the complementary cation was present at the same concentration as the ion under study. A further comparison also included the effect of addition of an equivalent amount of the ion under study. The results are presented in Table 3.8.

The results indicate that the monovalent cations NH_4, Rb, and K are always inhibitory, usually more so than the competition of the ion by itself. Lithium never had any effect and a slight inhibition by Na only showed up once. Cesium, on the other hand, was inhibitory, except for Ca(10^{-5}) and Rb(10^{-5}), the magnitude of this inhibition varying with the cation under study. It markedly inhibited Na uptake and had a lesser effect on the two divalent cations, Ca and Sr, but only at the higher concentration. At the lower concentration of Sr(10^{-5}) and the higher concentration of Rb(10^{-3}) only a slight inhibition was evident.

A final summary of this compilation would suggest that even in short-time experiments complementary ions usually inhibit or have no effect on the uptake of other ions. Although stimulation is not the common effect, an apparent Viets effect is demonstrable with divalent cations. Monovalent cations invariably either have no effect or inhibit. Even the Viets effect is only evident for K and Rb and not invariably present.

Many investigators (Drake *et al.*, 1951; Crooke, 1958; Wiersum and Bakoma, 1959; Crooke *et al.*, 1960; Heintze, 1961; Keller and Deuel, 1957; Smith and Wallace, 1956; Smith and Wallace, 1956; Mouat and Walker, 1959; Crooke and Knight, 1962; Peterburgskii and Nelubova, 1963), have determined the exchange capacity of roots on the assumption that interionic effects between cations can be explained by competition for exchange sites on the roots similar to the competition that exists for the net negative charges on a soil clay surface. This adsorbed cation then must be further assumed to be a step in the ion uptake process. Crooke and Knight (1962) have shown that a positive correlation does exist between cation-exchange capacity of roots and total cations, ash content, excess base, and total trace-element content. These are only observed correlations, however, and they do contain many exceptions. Because a correlation exists does not mean that there is a direct causal relationship. Indeed, there is much evidence that cation-exchange capacity of roots is not causally related to the ion uptake mechanism. The specificity of ion uptake has been amply demonstrated (Epstein and Hagen, 1952; Laties, 1959; Fried and Shapiro, 1961; Fried *et al.*, 1958) and competition between ions of similar chemistry has been shown repeatedly in marked contrast to nonspecific electrovalent effects. The same can be said for the so-called "Donnan" effects which presumably must occur in the neighborhood of a negatively charged surface such

TABLE 3.8

EFFECT OF COMPLEMENTARY CATIONS ON THE UPTAKE OF VARIOUS CATIONS

Cation	Concentration	Effect on uptake: Inhibition			No effect	Enhancement
		$>$Na	$\cong$Na	$<$Na		
Na	10^{-3}	NH_4, Rb, Cs, K		Mg	Li, Ca, Sr	—
Na	10^{-5}	NH_4, Rb, Cs, K		Mg	Li, Ca, Sr	—
		$>$Rb	$\cong$Rb	$<$Rb		
Rb	10^{-3}	NH_4	K	Cs	Na, Li, Mg	Ca(slight)
Rb	10^{-5}	—	NH_4, K	—	Cs, Na, Li, Mg, Ca, Sr	Sr(slight)
		$>$Cs	$\cong$Cs	$<$Cs		
Cs	10^{-3}	K, Rb, NH_4	—	—	Mg, Li, Na, Ca, Sr	Sr(slight)
Cs	10^{-5}	K, Rb, NH_4	—	—		
		$>$Ca	$\cong$Ca	$<$Ca		
Ca	10^{-3}	—	Ba, Mg, Sr, NH_4, K, Rb, Cs	—	Li, Na, Rb, K	—
Ca	10^{-5}	—	Ba, Sr	Mg	NH_4, Cs	—
		$>$Sr	$\cong$Sr	$<$Sr		
Sr	10^{-3}	—	Ba, Ca, Mg, NH_4, Rb, Cs, K	Na	Li	—
Sr	10^{-5}	—	Ba, Ca	Mg, Cs, K, Rb, Na, NH_4, Cs	Li	—

as a plant root or in the so-called "apparent free space," but which has yet to be demonstrated as affecting the ion uptake process as related to the effective concentration of the external solution.

3. Translocation to the Shoot

Nutrient ions accumulate in plant tops as well as in plant roots. The ions normally pass through the roots and are translocated to the tops. The relationship between accumulation in the roots and translocation to the tops is not at all clear. Since the work of Stout and Hoagland (1939) it is generally accepted that the ions move to the shoots primarily in the xylem; however, there is a difference of opinion as to whether a prior active step is a necessary part of this transport system and how a prior step, if necessary, is related to active accumulation in the root.

In its simplest form passive uptake of ions by plants grown in soil is simply a continuation of the soil solution throughout the plant in the so-called "free" space of the plant in both roots and shoots. Much experimental work has been interpreted as at least partially if not wholly supporting this hypothesis (Hylmö, 1953, 1955, 1958; Epstein, 1955, 1956, 1960; Kramer, 1957; Kylin and Hylmö, 1957; Kihlman-Falk, 1961; Allerup and Nielsen, 1962; Jensen, 1962) and the hypothesis has certain attractions, e.g., it obviates the necessity of explaining how ions move from cell to cell, for it essentially suggests that all cells are bathed by this outside solution and only accumulate ions as needed. It further helps explain why plants obtain almost every ion that occurs in their environment. In spite of its attraction, however, the hypothesis of passive movement is less able to explain other observed phenomena, including the marked difference in composition between the plant and its environment, the relatively small difference in composition of the plant with marked concentration changes in the outside environment, the specificity or discriminating ability of the plant with regard to different ion species, the marked effect of respiratory inhibitors and uncouplers of phosphorylation on ion translocation, the ion competition occurring in translocations as well as in root accumulation, the marked effect of temperature, etc., (Welte and Marckwordt, 1963; Lisanti and Marckwordt, 1963; Breuwer, 1956; Mitsui and Ishizuka, 1960; Russell and Barber, 1960; Helder, 1958; Zuev and Golubeva, 1962; Barber and Koontz, 1963; Fried *et al.*, 1961; Skogley and McCants, 1963; Russell and Shorrocks, 1959; Oertli, 1963; Marschner, 1964; Epstein, 1966). As Crafts (1961) has stated in his discussion of the concept that each cell accumulates ions individually from the free space by a process analogous to salt uptake by unicellular algae: "Such a view would visualize the root as being like a polyethelene bag full of yeast cells. In so doing it discounts over 100 years of work by plant anatomists on the structural relations of roots and it fails completely to explain the active pumping of xylem sap reported by Crafts and Broyer (1938) and many others."

Assuming carrier-mediated active transport as a necessary preliminary to the movement of nutrient ions from the solution bathing the roots to the tops of the plant, Fried *et al.* (1961) and Fried and Shapiro (1961) suggested that the process must therefore consist of a sequence of reactions that is describable by equations which could be characterized by the biological constants of summation carrier concentration, ΣR, the half-saturation constant, K_m, and the specific reaction rate constant of the rate-limiting step, k_2. Their two suggested chemical descriptions of the process, consistent with their data are given below:

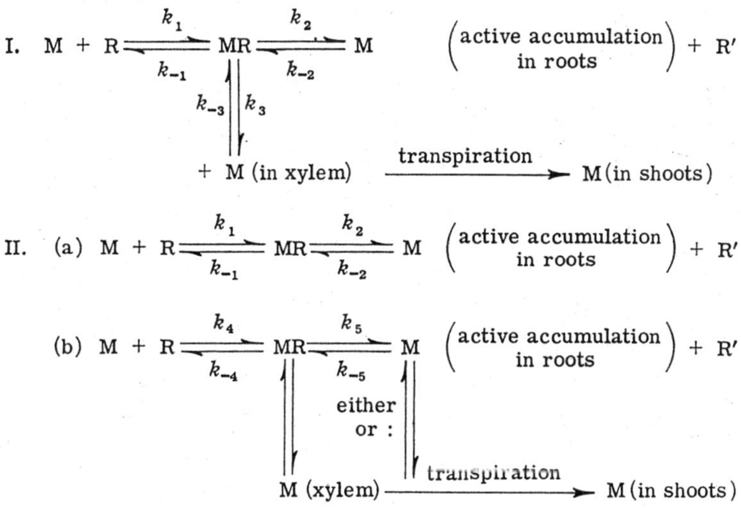

(3.20)

(3.21)

Description I represents two active metabolic steps competing for the same carrier-ion intermediate, one resulting in root accumulation and the other in shoot accumulation.

Description II represents two entirely separate active processes, one leading wholly to root accumulation and the other primarily to shoot accumulation. Fried *et al.* state that "since description I places more restrictions on the system, it is more useful until proven inconsistent with further experimental data." Skogley and McCants (1963) also concluded that their data was consistent with description I, and Jacoby (1964) found that the roots of bean plants prevented the transport of Na to the tops as a result of competition by specific binding sites in the roots with Na transport to the stele.

Interestingly enough, neither one of two suggestions proposes that the ion must first be accumulated in the root before it can be transferred to the xylem from where it moves to the plant tops either by mass flow with the transpiration stream or, as suggested by Bell and Biddulph (1963), by a process of exchange in the case of Ca, except that this possibility of prior root accumulation is not excluded in either suggestion. Both suggestions include the mechanism by which ions accumulated by roots can be retransferred to shoots. This lack of necessity for the ion to be first accumulated by the root before transference to the xylem is consistent with the data of at least two separate experiments by two authors, one involving an anion, sulfate (Epstein, 1956), and the other a cation, Rb (Fried *et al.*, 1961).

The results are illustrated in a combined Table 3.9. It is clear from Table 3.9 that the net uptake during the light period was much higher for Rb when the roots were drawing directly from the solution (treatment 3) than when drawing from the ion previously absorbed by the roots (treatment 2). The sulfate came almost exclusively from the outside solution. This was despite the fact that during the dark period the sulfate roots had already accumulated 0.18 μeq of SO_4 and the Rb roots, 1.8×10^{-7} moles. A similar trend can be noted in the data of Helder (1958) for absorption of Rb from 1×10^{-3} N $RbNO_3$ by barley plants. More recently Lopushinsky (1964) studying the effect of water movement on ion movement into the xylem, found that under certain circumstances much of the salt transported from the bathing solution into the xylem by-passed equilibration with root tissue. Broyer (1950) had done an experiment earlier to test the same hypothesis and concluded "that these close correspondences in magnitude would obtain if the inorganic solute (accompanied by water, osmotically) or influx moves rather proportionately into lumena of both the vacuoles of low-salt root cells and the xylem." Although this data did not take into account the possible isotopic dilution with previously absorbed bromide, it was consistent with either of the two chemical descriptions of the translocation process referred to above.

Descriptions I and II are, of course, in a way oversimplifications, for many intermediate steps may be involved which are not rate-limiting and whose constants become part of the rate constant describing the over-all reaction. A change in environment could, of course, then theoretically even change whatever step in the steady-state system becomes rate-limiting. They are also oversimplified in the sense that finite distances are involved in moving an ion from the outside solution. A more comprehensive picture concerning the physical distances involved was more recently suggested by Oertli (1963) that is essentially a diagram-

TABLE 3.9

EFFECT OF PREVIOUSLY ROOT-ABSORBED ION ON THE SUBSEQUENT AMOUNT OF ION TRANSLOCATED TO THE SHOOT[a]

			Labeled Ion in the Shoots			
			Sulfate μeq/18 plants		Rubidium moles (× 10)/g	
Treatment	In dark	In light	Total	Net	Total	Net
1	60-minute labeled ion + 60-minute H_2O	Harvested	0.06	—	0.12	—
2	60-minute labeled ion + 60 minute H_2O	60-minute (180 for SO_4^{2-}) unlabeled ion + 60-minute H_2O	0.07	0.01	0.20	0.08
3	60-minute unlabeled ion + 60-minute H_2O	60-minute (180 for SO_4^{2-}) labeled ion + 60-minute H_2O	0.32	0.32	0.31	0.31

[a] Ion concentration of Rb in solution was 1×10^{-7} *M*; SO_4^{2-} ion concentration in solution was 1×10^{-3} *N*.

matic physical picture of description II including some free space ion reaching the internal barrier (presumably the endodermis) before actually going through active accumulation (Fig. 3.16). Interestingly enough, Oertli was actually trying to explain guttation.

Examination of bleeding sap from castor oil plant as affected by pressure or absence of K in the culture solution has also indicated that K is transported to the stele without going through a root accumulation process (Bowling and Weatherley, 1964). The concentration duality of ion uptake described in Section 2.2 has its analogous counterpart in

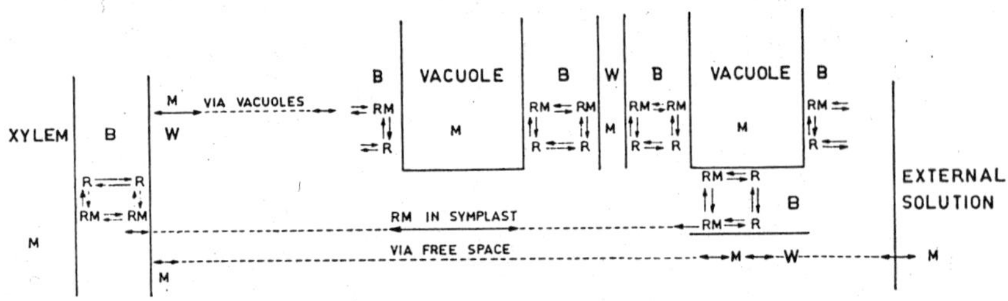

FIG. 3.16. A model of nutrient transfer in roots. *M*, Nutrient; R, carrier; RM, carrier-nutrient complex; B, barrier; and W, cell wall.

translocation Epstein, 1966; Toiri and Laties, 1966; Lüttge and Laties, 1966, 1967). However, this duality does not necessarily mean that the rate limitations for translocation are identical with those for accumulation in cells of the root. Indeed, if as indicated, the translocated ion does not have to go through the prior accumulation step, the rate limitations, even if the same mechanistically, would occur at a different location. A difference in mechanism has been suggested particularly at higher concentrations of the order of $1–50 \times 10^{-3}$ *M* of the ion under study. Movement into the symplasm was considered passive and long distance transport reflected the consequences, (Lüttge and Laties, 1966, 1967). This latter mechanism is not reflected in Eq. (3.20) and (3.21) which only describe metabolically active systems. If the symplast is included in free space, however, then this possible passive movement is visualized in Fig. 3.16.

CHAPTER 4

The Soil-Plant Continuum

Equation (1.1) described the uptake of nutrient ions by the plant as a sequence of reactions which could be described in the quantitative terms of physical chemistry. The first term in this sequence was M(soil) which was shown to refer primarily to adsorbed ion insofar as the next term in the sequence M(solution) was concerned. It was also pointed out, however, that over a long time period minerals containing nutrient ions could be a further source of supply as could organic matter if it was decreasing in amount over the years. In the absence of plant roots the concentration of a nutrient was shown to depend on the nature of this reaction between the ion in solution and the ion on the solid phase and the quantitative descriptions available for describing this relationship were elucidated. They involved a capacity factor and an apparent dissociation constant. The nutrient ion after it is in solution, however, must traverse a phsyical distance in order to reach the ion-accumulating mechanism in the plant. This movement must take place by diffusion or bulk movement with the water phase. Under ordinary transpiration conditions bulk movement can account for the movement of most nutrient ions. Should this bulk movement actually result in more ions reaching the root than are taken up by the plant, then diffusion gradients are set up away from the plant. Also since water is not always moving, all ions at times move to the root under the driving force of diffusion. It would appear that for most nutrient ions the concentration of the ion in the soil solution is the concentration most common to the root surface and ion uptake rates should be predictable on this basis. However, phosphate and, under low transpiration conditions or very low K concentration in the soil solution, K may not reach the root surface at the concentration of the soil solution but at some lower concentration. The prediction and possible modification of this concentration is not at present possible owing to a lack of sufficient experimental information. Even with phosphate and under the special conditions mentioned, K, bulk movement and diffusion determine the concentration that reaches the root, but the relative importance of these two mechanisms is not known. Thus the relative importance of soil solution concentration as compared to diffusivity is not known, since both mechanisms are affected by the same factors. The concentration in the soil solution affects the maximum concentration gradient on the one hand and the amount of ion moved by bulk movement on the other hand and the capacity factor determines the distance of movement necessary by either mechanism. There are certain experimental difficulties that have not yet been resolved. Once the ion reaches the ion-accumulating surface of the root, it is either accumulated by the root or passes into the xylem through a step requiring metabolism from where it is translocated to the tops. This active ion accumulation process can be described in the normal physicochemical sense by three constants, ΣR, the total amount of a specific carrier, k_2, a specific reaction rate constant, and K_m, a constant which relates the concentration of the solution bathing the root to the amount of ion which would be attached to a given amount of total carrier. The nutrient ion which reaches the xylem has probably gone through a prior active step, but is not normally the ion which has previously been accumulated by the root, although it can be. This ion is then translocated to the top of the plant where it is then removed by the cells and utilized in the metabolic system.

This over-all picture of ion uptake in the soil-plant system may be summarized by the accompanying equation where M is the ion and R refers to carrier.

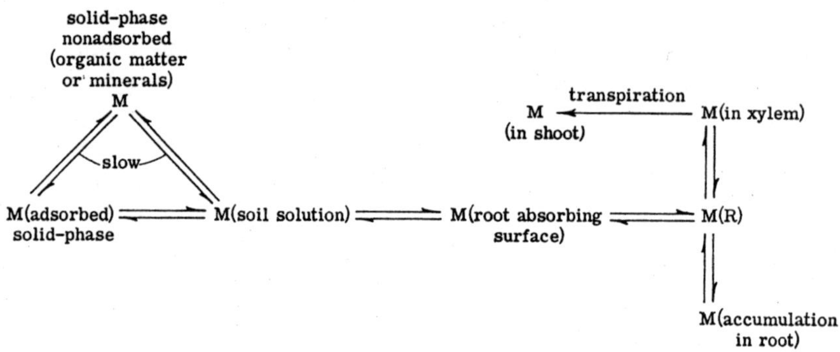

In this continuum the rate-limiting step under steady state conditions for most ions is undoubtedly the active accumulation step that occurs in the root of the plant. If this conclusion is valid, then all the reactions previous to this step are going on at approximately the same rate and are essentially at a pesudo-equilibrium. Since this is a metabolic process occurring in the root, ion uptake should be directly responsive to the concentration of the ion at the ion-accumulating surface and should be markedly affected by temperature changes in the soil. Of course, the continuum may be broken by a low enough moisture content to prevent any ions from reaching the root, but in general terms the factors most important for affecting ion uptake are those which affect this metabolic process. This includes, in addition to the concentration and temperature functions already mentioned, any internal plant factors that result in higher carrier concentrations and higher turnover rates.

PART III

Supplying Nutrients to Plants

Of the 16 known essential elements for plant growth, the soil must supply all the plant needs for 10 of these elements (P, K, Ca, Mg, Cu, Zn, B, Fe, Mn, Mo) and except for leguminous plants all of the need for an eleventh (N). The other five elements (C, H, O, Cl, S) are at least partially and sometimes primarily supplied by the atmosphere. The total supply of the soil-borne nutrients per acre of land is limited, and only a small fraction of this total is in immediate contact with the soil solution. The continual growth of crops on, and removal of products from, cultivated land therefore must eventually require supplemental addition of these nutrient elements. The question is not whether fertilization will be required but rather whether it is economical to initiate fertilization and when, where, what kind, and how much fertilizer should be applied.

Fertilizers are used to supplement the nutrient supply that is naturally present in the soil. One of the factors that determine when fertilization must be initiated and what is a sound fertilizer policy is the actual quantity of nutrients that are removed by the crop. Justus von Liebig's proposals in the middle of the nineteenth century included the concept that the soil had to be supplied with the amount of nutrients that had been removed by the previous crop. It was found by experience, however, that there was not necessarily any relationship between the quantity of nutrients removed by the crop and the fertilizer requirements of the crop. The quantitative effects of removal of nutrients from the immediate environment of the plant root by leaching, gaseous losses, exchange with the adsorption complex, and incorporation in organic matter were not known. Strangely enough, despite the fact that our knowledge of the mechanisms of nutrient transformation, losses, and additions in the soil has greatly increased since Liebig's time, the basis for fertilizer recommendation is still highly empirical.

Many of the factors that affect the nutrient uptake by the plant have been discussed in Part II. Some of the soil factors have been appreciated only recently, particularly the chemical composition of the soil solution in close proximity to the active root surface and the ability of the soil to maintain this composition. Experimentally, it is still difficult to determine the effect of such processes as diffusion, ion exchange, and biological activity on the composition of the soil solution at the root surface.

It is, therefore, not surprising that up to the present time fertilizer recommendations have been based on the results of fertilizer trials, pot experiments, and on the chemical analysis of empirical soil extracts. Although initially the extracting agents were chosen with a view to simulating the extracting power of plant roots, all such approaches failed to give consistent reliable information on fertilizer requirements.

This does not mean that the present methods for the evaluation of the soil fertility status are without any value. The opposite is true. During the past 50 years an enormous quantity of data have accumulated and by means of statistical analysis of the results of field, greenhouse, and laboratory experiments, a basis for fertilizer recommendations can be worked out for many soil types. In some instances, particularly in those countries where the economic situation is such that farmers can follow up the fertilizer recommendations, applications have even been too high and have consequently led to an inefficient use of fertilizers. This has occurred in countries with a very intensive agricultural system, such as the Netherlands, Belgium, and Japan. In the greater part of the world, particularly in the tropical regions of Asia, Africa, and Latin America, little fertilizer is being applied and the economic situation is such that it seems unlikely that levels of application comparable to the temperate regions will be made in the near future.

The plant nutrient in the soil system required for maximum economic yield of crops can be supplied and maintained by the use of fertilizers. Conceptually, this is not difficult. Liebig's original idea that the soil had to be supplied with the amount of nutrients removed by the previous crop needs only a slight upgrading. First, the soil nutrient supply must be brought up to a level where it will produce the maximum economic yield, and then it is this soil nutrient supply that must be maintained. Second, the soil must receive a supplemental fertilizer application sufficient to make up for all losses not only by crop removal but also by gaseous losses, immobilization, and losses from leaching, taking into account all natural additions. This could be called "feeding" the soil, for it represents the concept of bringing the soil up to a certain plant nutritional level and keeping it there. This is analogous to utilizing the soil as the physical support for the plant in a sand culture or nutrient solution and determining the nutrient solution necessary for adequate growth. From an economic standpoint this approach leaves much to be desired; although if the cost of the additional fertilizer is negligible as compared to the value of the crop, this approach has certain merits.

While conceptually the problem of adequate plant nutrient supply is relatively simple, in actual practice it is so complicated that the bases for fertilizer recommendations are still highly empirical. First, it has been noted that there are large differences in the quantity of nutrient required or utilized by different plant species. In extreme cases an adequate level of nutrient for one plant species is actually toxic to another. Second, it has been found that just because a certain quantity of an element is taken up by the plant this does not mean that the element is either essential or is required in that amount. Third, it has become clear that a certain balance of nutrient elements is necessary. Fourth, it has become evident that the concentration of a given nutrient necessary to supply the needs of the plant differs for different plant species even when the total utilization is the same. Fifth, experimental results have shown that the nutrient supply to the crop, when maintained by fertilizer additions, can be varied in amount with the same end result depending upon kind, placement, and time of fertilizer application.

Considering these findings and the necessary associated economic considerations, the approach to adequate plant nutrition in the soil-plant system has become one of feeding the crop. This approach of feeding the crop, while economically more favorable, has made the technical task infinitely more difficult. It necessitates considering the maximum efficient use of the nutrient already present in the soil with its consequent economic implications, realizing that no two soils are identical. It also necessitates making maximum economic use of the fertilizer additions, realizing the large number of fertilizer materials available and the number of combinations of nutrients possible and realizing that all react differently when placed in contact with the same or different soils

Perhaps the development of the modern computer has given us a tool to deal with these complexities and put to practical use the myriad of relationships that have already and will be found. Although this computer approach can be purely empirical, so-called correlation science, it is bound to break down because there are so many exceptions. Nevertheless, at our present state of knowledge this is the only approach possible. However, we should strive for a more fundamental and less empirical approach to the information that will be fed into the computer. This could be an approach which defines as many relationships as possible in terms of the actual concentrations and reaction rate constants necessary to supply a given amount of nutrient to the plant metabolic system at a given rate. Thus, the factors of interest, whether they are a given plant species, a particular fertilizer, a particular placement, or a particular time of application, etc., should be related to a version of Eq. (3.21) which in its simplest form is

$$\text{M(solid)} \rightleftharpoons \text{M(solution)} \rightleftharpoons \text{M(plant root)} \rightleftharpoons \text{M(plant top)}$$

where M is a nutrient ion.

The following chapters are devoted to the various types of information now known or being gathered to aid in making decisions on the problem of supplying nutrients to plants in the soil-plant system. Chapter 5 is devoted to the so-called balance sheet, i.e., the utilization of nutrients by the plant as well as the natural additions and losses; Chapter 6, to the determination of the level of nutrient already present in the soil; and Chapter 7, to the factors necessary for consideration in the final decision concerning the supplementary addition of nutrients in the form of fertilizer.

CHAPTER 5

Plant Nutrient Removals and Mobility

It is customary to divide those elements required by plants, other than C, H, and O, into the macro- and micronutrients. The macroelements, such as N, P, S, K, Ca, and Mg, are required in larger quantities than the micro (or trace) elements, such as Cu, Zn, B, Cl, Mo, Mn, and Fe. Of the micronutrients Mn and Fe are generally required in slightly larger amounts than Cu, Zn, B, Cl, and Mo. By definition all elements are equally essential, and the lack of any one of them will cause plant failure. Considerations of plant growth and nutrition require consideration of the total amount of nutrient taken up by the plant, including nonharvested parts. Only minor amounts are absorbed by the leaf, and enough nutrient must reach the plant root at the right time and rate to furnish an adequate supply. On the other hand, the amount of nutrient actually lost from the soil system due to plants is that removed by the harvested portion of the plant. It is seldom that the whole plant, roots and all are harvested. This is only approached by some tuber crops such as beets and carrots. In fact, when considering tree crops, the amount of nutrient taken up and immobilized is manyfold that removed in the harvested portion. Not only are substantial amounts of nutrients normally left in the soil in roots but also in straw, leaves, and branches.

1. Plant Species

The quantities of nutrients in various crops and the amounts removed by the harvested portions have been extensively reviewed (Aslander, 1958; Rester and Shelton, 1949; Miller *et al.*, 1958; Ochse *et al.*, 1961; Reinwald, 1948; Remy, 1938; Becker-Dillinger, 1937; Bessno, 1941; Goodall and Gregory, 1947; Tisdale and Nelson, 1956; Stiles, 1961; Janes, 1951; and Frear, 1951). The actual quantity of element removal per acre depends not only on the content of the element in the part of the crop that is harvested but also on the total weight of dry matter removed. Since both plant composition and yield vary widely, different amounts of the various nutrients with different plant species are bo

TABLE 5.1
NUTRIENTS REMOVED BY GRAIN CROPS[a]

Crop	N	P	S	B	K	Ca	Na	Mg	Fe	Mn	Zn	Cu	Ref.
Grain crops	60–70	10–15	—	—	60–70	—	—	—	—	—	—	—	*b*
	45	8	—	—	29	6	—	3	—	—	—	—	*c*
Barley													
Grain	41	9.6	4	0.005	13	2	6	4	0.24	0.04	0.26	0.04	*d*
Straw	19	3.5	5	—	39	11	—	3	—	—	—	—	*d*
Total	60	13	9	—	52	13	—	7	—	—	—	—	*d*
Barley (grain and straw)	45	10	—	—	41	10	—	—	—	—	—	—	*e*
Barley	78	16	—	—	64	—	—	—	—	—	—	—	*f*
Buckwheat													
Grain	15	5	3	—	24	26	—	6	0.03	0.02	—	0.001	*d*
Straw	63	10	—	—	86	—	—	—	0.50	—	—	—	*d*
Total	78	15	—	—	110	—	—	—	0.53	—	—	—	*d*
Buckwheat (grain and straw)	60	13	—	—	41	21	—	—	—	—	—	—	*e*
Oats													
Grain	38	6.1	4	—	7	2	—	3	0.15	—	—	0.021	*d*
Straw	36	6.6	15	1.50	76	13	4	14	1.21	—	—	0.066	*d*
Total	74	13	19	—	83	15	—	17	1.36	—	—	0.087	*d*
Oats (grain and straw)	53	10	—	—	52	10	—	—	—	—	—	—	*e*
Rice													
Grain	38	9.6	8	—	7	2	2	5	0.22	0.08	0.08	0.008	*d*
Straw	—	5.2	—	—	55	12	—	4	—	—	—	—	*d*
Total	—	15	—	—	62	14	—	9	—	—	—	—	*d*
Rice													
Grain	21	4.7	—	—	9	1	—	2	1	—	—	—	*g*
Straw	20	4.4	—	—	38	9	—	3	26	—	—	—	*g*
Total	41	9	—	—	47	10	—	5	27	—	—	—	*g*
Total rice	82	6	—	—	53	8	—	6	—	—	—	—	*h*

Wheat													
Grain	52	9.6	4	0.1	10	2	1	3	0.19	0.14	0.12	0.017	*d*
Straw	24	3.9	7	0.2	32	10	—	5	0.50	0.32	0.08	0.014	*d*
Total	76	14	11	0.3	42	12	—	8	0.69	0.46	0.20	0.031	*d*
Wheat (grain and straw)	62	12	—	—	37	8	—	—	—	—	—	—	*e*
Corn and millet	100–120	20–30	—	—	125–150	—	—	—	—	—	—	—	*b*
Corn													
Grain	95	16	8	0.2	15	2	7	9	0.17	0.45	0.28	0.017	*d*
Stalks	60	9	10	0.2	37	19	—	14	1.20	0.96	0.36	0.042	*d*
Total	155	25	18	0.4	52	21	—	23	1.37	1.41	0.64	0.059	*d*
Corn (grain and stalks)	86	17	—	—	118	—	—	—	—	—	—	—	*e*

[a] Data in lb/acre.
[b] Vageler (1933).
[c] Millar *et al.* (1958).
[d] Hester and Shelton (1949).
[e] Rheinwald (1948).
[f] Remy (1938).
[g] Ochse *et al.* (1961).
[h] Gerton (1954).

expected and found. For N, the removal may be about 45 to 320 lb/acre. The removal of P is generally much lower and amounts to 2 to 45 lb/acre. Potassium and calcium vary over a wide range, which may be as low as 5 but also as high as 320 lb/acre. The removal of Mg and S is much lower and generally does not exceed 15 to 20 lb/acre although certain species may remove appreciably higher amounts particularly of S. The removal of Fe and Mn is usually less than 1 lb/acre, whereas the removal of microelements such as Cu, Zn, and Mo is about fractions of an ounce to a few ounces per acre.

1.1. Grain Crops

The following data in Tables 5.1–5.12 are illustrative and therefore only a semiquantitative guide for the removal of nutrients. The examples given are not representative of all the many different conditions under which crops can be produced.

Table 5.1 gives examples of the amounts of nutrients removed by various species of *grain*. Although the yields of plant material are not listed, most of the original sources contained the yield data and the reader is referred to these for such information.

TABLE 5.2

Ratio of Macroelements Removed by Tuber and Root Crops for N = 1[a]

Crop	N	P	S	K	Ca	Mg
Barley	1	0.22	0.16	0.87	0.19	0.13
Buckwheat	1	0.20	—	1.1	0.28	—
Oats	1	0.18	0.29	1.0	0.19	0.26
Rice	1	0.22	—	1.2	0.24	0.13
Wheat	1	0.18	0.17	0.56	0.13	0.12
Corn	1	0.18	0.15	0.71	0.17	0.19

[a] Derived from Table 5.1.

The quantity of N removed is generally about 55 to 80 lb/acre. Lower figures have been reported for rice, about 45 lb/acre. Corn removes substantially more N than the other grain crops, and this removal can be as high as 160 to 170 lb/acre or more. Next to N, K is removed in the greatest quantities, varying between 45 and 140 lb/acre for the different grain crops. Wheat shows a slightly lower removal of K in corn and buckwheat this can be as high as 110 to 140 l ne quantity of P removed amounts to 9 to 25 lb/acre, with mar a

TABLE 5.3

NUTRIENTS REMOVED BY TUBER AND ROOT CROPS[a]

Crop	N	P	S	B	K	Ca	Na	Mg	Fe	Mn	Cu	Ref.
Tubers	50–70	10–15	—	—	100–150	—	—	—	—	—	—	*b*
Beets												
Roots	50	4	2.0	—	27	3	5	2	0.20	0.20	0.040	*c*
Tops	10	4	5.3	—	49	24	29	15	0.98	0.34	0.028	*c*
Total	60	8	7.3	—	76	27	34	17	1.18	0.54	0.068	*c*
Sugar beets	130	23	—	—	160	42	—	—	—	—	—	—
	130	23	—	—	130	76	—	—	—	—	—	*d*
Fodder beets	100	20	—	—	180	51	—	—	—	—	—	*d*
Carrots												
Roots	92	12	6.4	0.20	70	16	14	7	0.24	0.024	0.03	*c*
Tops	100	4	—	0.075	170	150	2	6	1.55	—	0.015	*c*
Total	192	16	—	0.275	240	166	16	13	1.79	—	0.045	*c*
Carrots(roots)	85	16	—	—	110	76	—	—	—	—	—	*e*
Turnips												
Roots	50	7	8.6	0.49	46	10	—	4.8	0.10	0.008	0.02	*c*
Tops	80	4	3.2	—	30	39	0.36	9.6	0.70	0.28	0.01	*c*
Total	130	11	11.8	—	76	49	—	14.4	0.80	0.29	0.03	*c*
Turnips(roots and tops)	89	31	—	—	120	38	—	—	—	—	—	*e*
Potatoes												
Tuber	80	5	2.3	—	100	3	—	2.5	0.21	0.024	0.015	*c*
Tops	120	3	8.1	—	120	49	0.12	13.8	—	—	0.013	*c*
Total	200	8	10.4	—	220	52	—	16.3	—	—	0.028	*c*
Potatoes(tubers and tops)	80	16	—	—	120	31	—	—	—	—	—	*d*
Potatoes(sweet)	5	2	2.4	—	20	2	2	3.2	0.16	0.03	0.008	*c*
Cassava(manioc)	190	38	—	—	430	140	—	—	—	—	—	*c*

[a] Data in lb/acre.
[b] Vogeler (1933).
[c] Hester and Shelton (1949).
[d] Rheinwald (1948).
[e] Becker-Dillinger (1937).

for S and Ca. Magnesium is generally removed at a rate of 5 to 25 lb/acre.

Corn and buckwheat generally remove more micronutrients than the other grain crops. The amount of micronutrients removed is fractions of an ounce to a pound per acre. Corn, buckwheat, and rice remove more Fe than the other grain crops, mainly due to the relatively high Fe content of the straw. The removal of Fe for these crops is about 1.4 lb/acre.

Table 5.2 compares the removal of macroelements by various grain crops relative to N, where N is fixed at value of 1. Compared to the removal of N, only one-fifth as much P is removed by most of the grain crops. The relative removal of S, Ca, and Mg is of similar magnitude to that of P. The removal of K may be similar to N, but is lower for wheat.

1.2. Tuber and Root Crops

In Table 5.3 the removal of nutrients by different kinds of *tuber* and *root* crops is compared. Botanically, the crops listed in Table 5.3

TABLE 5.4

Ratio of Macroelements Removed by Tuber and Root Crops for N = 1[a]

Crop	N	P	S	K	Ca	Mg
Beets	1	0.12	0.12	1.3	0.45	0.28
Fodder beets	1	0.20	—	1.8	0.52	—
Sugar beets	1	0.16	—	1.1	0.44	—
Carrots	1	0.14	—	1.2	0.82	0.06
Turnips	1	0.08	0.09	0.6	0.38	0.11
Turnips(kohlrabi)	1	0.35	—	1.3	0.43	—
Potatoes	1	0.12	0.05	1.3	0.32	0.14
Sweet potatoes	1	0.50	0.45	3.7	0.33	0.60
Cassava	1	0.20	—	2.3	0.74	—

[a] Derived from Table 5.3.

are much less similar than, for example, the grain crops. This explains why such different quantities of the same nutrient are removed by different crops.

With the exception of sweet potatoes most of the tuber and root crops are heavy N and K consumers; K often exceeding the N removal. Beets can also take up considerable quantities of Na which, although not considered an essential elements, is known to influence the growth of this crop. From Table 5.4, where the ratios of macroelements removed for

N equals one have been listed, it is evident that tuber and root crops remove relatively low amounts of P and high amounts of K.

1.3. Legumes

In Table 5.5 the removal of nutrients by *legume crops* is listed. Some crops (alfalfa, red clover, and horse beans) remove relatively large quantities of N, K, and Ca, about 75 to 200 lb N/acre. Leguminous crops, however, may fix large amounts of atmospheric N and under normal

TABLE 5.5
Nutrients Removed by Leguminous Crops[a]

Crop	N	P	S	B	K	Ca	Na	Mg	Fe	Mn	Zn	Cu	Ref.
Alfalfa	155	16	16	0.6	110	120	3	15	0.90	0.48	0.30	0.06	*b*
	220	21	—	—	110	150	—	—	—	—	—	—	*c*
Red clover	126	13	9	0.22	80	92	4	22	1.02	0.6	0.25	0.1	*b*
	120	12	—	—	81	76	—	—	—	—	—	—	*c*
Sweet clover	110	13	—	—	83	—	2	—	0.52	0.20	0.20	0.04	*b*
Yellow lupines	120	12	—	—	52	19	—	—	—	—	—	—	*c*
Legume hay	88	10	—	—	64	39	—	11	—	—	—	—	*d*
Peanuts	140	11	—	—	62	62	—	—	—	—	—	—	*b*
Soybeans	113	18	10	0.05	48	26	—	16	0.63	0.23	—	0.033	*b*
	99	14	—	—	40	—	—	—	—	—	—	—	*d*
Lima beans	95	10	—	0.67	94	61	—	7.6	—	—	—	—	*b*
Snap beans	200	3	3.9	—	46	22	—	2.9	—	—	—	—	*b*
Garden peas	126	6	3.8	—	47	21	0.41	6	—	—	—	—	*b*
Horse beans	310	23	—	—	95	160	—	—	—	—	—	—	*e*
Summer beans	100	10	—	—	62	82	—	—	—	—	—	—	*e*

[a] Data given in lb/acre.
[b] Hester and Shelton (1949).
[c] Rheinwald (1948).
[d] Millar *et al.* (1958).
[e] Becker-Dillinger (1937).

conditions the figures listed do not represent a net loss of N. The P removal is about 5 to 20 lb/acre; K varies between 40 to 110 lb/acre. The Ca removal can be high, for alfalfa, horse beans, and red clover, where it amounts to 75 to 160 lb/acre. Many other crops remove much less Ca (19 to 55 lb/acre), while the removal of Mg is about 13 to 22 lb/acre.

Table 5.6 gives the ratio of macronutrients removed by legume crops where N is given a value of 1. The removal of K is apparently relatively less for legume crops than for tuber crops and to some extent for grain crops.

TABLE 5.6

RATIO OF MACRONUTRIENTS REMOVED BY LEGUME CROPS FOR N = 1[a]

Crop	N	P	S	K	Ca	Mg
Alfalfa	1	0.10	0.11	0.61	0.74	0.12
Red clover	1	0.10	0.07	0.67	0.70	0.18
Sweet clover	1	0.12	—	0.76	—	—
Yellow lupines	1	0.1	—	0.45	0.16	—
Legume hay	1	0.12	—	0.70	0.43	0.12
Peanuts(grain and straw)	1	0.08	—	0.45	0.44	—
Soybeans(grain and straw)	1	0.16	0.1	0.43	0.24	0.14
Lima beans	1	0.11	—	0.98	0.63	0.80
Snap beans	1	0.02	0.02	0.23	0.11	0.02
Garden peas	1	0.05	0.03	0.35	0.15	0.04
Horse beans	1	0.07	—	0.31	0.50	—
Summer beans	1	0.10	—	0.64	0.84	—

[a] Derived from Table 5.5.

1.4. VEGETABLE CROPS

The *vegetables* represent a category of crops that are of heterogeneous botanical origin. This helps explain the rather wide variation in nutrient utilization. Generally, large quantities of N, K, and Ca are taken up (Table 5.7). Relative to N, many vegetable crops are high K consumers.

TABLE 5.7

NUTRIENTS REMOVED BY VEGETABLE CROPS[a]

Crops	N	P	S	B	K	Ca	Na	Mg	Fe	Mn	Zn	Cu	Ref.
Tomatoes	161	22	20	0.09	250	136	3	23	0.7	0.98	—	0.12	*b*
	100	10	—	—	110	82	—	—	—	—	—	—	*c*
Garlic	180	38	—	—	170	130	—	16	—	—	—	—	*d*
Asparagus	16	2	2	0.05	8	1	0.16	—	0.22	0.004	0.006	—	*b*
Onions	76	8	—	—	71	—	—	—	0.20	—	—	—	*b*
	71	15	—	—	63	37	—	—	—	—	—	—	*c*
Cabbage	160	24	—	—	150	110	—	—	—	—	—	—	*e*
Spinach	89	13	—	—	60	25	—	—	—	—	—	—	*c*
Green cabbage	180	39	—	—	330	290	—	—	—	—	—	—	*c*
Brussels sprouts	180	23	—	—	140	110	—	—	—	—	—	—	*c*
Cauliflower	180	31	—	—	190	38	—	—	—	—	—	—	*c*
Endives	71	12	—	—	120	29	—	—	—	—	—	—	*c*
Salad	49	10	—	—	81	22	—	—	—	—	—	—	*c*

[a] Data given in lb/acre.
[b] Hester and Shelton (1949).
[c] Becker-Dillinger (1937).
[d] Zink (1963).
[e] Rheinwald (1948).

TABLE 5.8

RATIO OF MACRONUTRIENTS REMOVED BY VEGETABLE CROPS FOR N = 1[a]

Crop	N	P	S	K	Ca	Mg
Tomatoes	1	0.14	0.12	1.5	0.85	1.4
Garlic	1	0.21	0.12	0.9	0.7	0.08
Asparagus	1	0.12	—	0.5	0.05	—
Onions	1	0.09	—	0.93	0.50	—
Cabbage	1	0.15	—	0.89	0.69	—
Spinach	1	0.15	—	0.67	0.28	—
Green cabbage	1	0.22	—	1.85	1.60	—
Brussels sprouts	1	0.13	—	0.80	0.60	—
Cauliflower	1	0.17	—	1.05	0.21	—
Endives	1	0.16	—	1.6	0.40	—
Salad	1	0.20	—	1.6	0.45	—

[a] Derived from Table 5.7.

The P removal amounts to approximately one-tenth to one-fifth of the N consumption (Table 5.8).

1.5. MISCELLANEOUS CASH CROPS

Tables 5.9 and 5.10 illustrate the removal of macronutrients from a number of *miscellaneous cash crops* of which several are perennial. Potassium is frequently removed at a higher rate than N. Sisal is known

TABLE 5.9

NUTRIENT REMOVED BY MISCELLANEOUS CASH CROPS[a]

Crop	N	P	K	Ca	Mg	Ref.
Tobacco	109	22	180	69	—	*b*
Pineapple	134	47	440	100	53	*c*
Pineapple	140–170	15–30	250–300	—	—	*d*
Cotton	134	37	120	—	—	*e*
Sugar cane	130	35	190	110	97	*c*
Sisal	118	19	140	290	40	*c*
Banana	55	7	150	10	14	*c*
Flax	42	13	67	31	—	*f*
Hemp	62	13	52	95	—	*g*

[a] Data given in lb/acre.
[b] Schmid (1951).
[c] Uexkull (1963).
[d] Vogeler (1933).
[e] Becker-Dillinger (1937).
[f] Anon (1938).
[g] Anon (1951).

TABLE 5.10
RATIO OF NUTRIENTS REMOVED BY MISCELLANEOUS CASH CROPS FOR N = 1[a]

Crop	N	P	K	Ca	Mg
Tobacco	1	0.20	1.6	0.6	—
Pineapple	1	0.35	3.3	0.7	0.40
Cotton	1	0.28	0.9	—	—
Sugar cane	1	0.27	1.5	1.0	0.75
Sisal	1	0.16	1.2	2.4	0.34
Banana	1	0.12	2.7	0.2	0.26
Flax	1	0.32	1.6	0.7	—
Hemp	1	0.21	0.8	1.5	—

[a] Derived from Table 5.9.

to be a high Ca consumer. The nutrient removal is particularly high for crops such as pineapple, sugar cane, and sisal.

1.6. Fruit Trees

Among the *fruit trees* (Table 5.11) the citrus species remove considerable quantities of N, but relatively little is removed by coffee, cacao,

TABLE 5.11
NUTRIENTS REMOVED BY TREE CROPS[a]

Crop	N	P	K	Ca	Ref.
Fruit trees (general)	89	20	110	120	*b*
Apple	110	15	120	—	*b*
Apple	99	16	130	150	*c*
Pears	140	24	180	—	*b*
Peach	76	7	61	78	*b*
Date palms	27	8	36	—	*d*
Coffee	27	2	36	—	*e*
Cacao	18	4	12	3	*b*
Coconut palms	78	17	140	20	*f*
Oil palm	80	8	100	20	*g*
Olives	25	5	36	30	*h*
Oranges	220	21	150	200	*i*
Lemon	160	21	140	179	*i*
Tea	57	6	25	—	—

[a] Data given in lb/acre.
[b] Becker-Dillinger (1937).
[c] Batjer and Rogers (1952).
[d] Uexkull (1963).
[e] Malavolta (1956).
[f] Georgi and Leih (1932).
[g] Blommendal (1937).
[h] Pantanelli (1950).
[i] Oppenheim (1932).

olives, and date palms. The P removal varies greatly and amounts to one-tenth to one-third of the N consumption (Table 5.12). With the exception of cacao, oranges, and tea, the tree crops remove large quantities of K. Also, the Ca removal is considerable except in the case of coconut and oil palms. As previously mentioned, however, the quantity of nutrients immobilized in trees is considerable and usually exceeds several times that amount removed annually.

TABLE 5.12
RATIO OF NUTRIENTS REMOVED BY TREE CROPS FOR N = 1[a]

Crop	N	P	K	Ca
Fruit trees	1	0.22	1.2	1.4
Apple	1	0.14	1.2	—
Pears	1	0.17	1.3	1.0
Date palms	1	0.27	1.2	—
Coffee	1	0.33	1.2	—
Cacao	1	0.21	0.65	1.8
Coconut palms	1	0.17	1.4	0.19
Oil palm	1	0.10	1.2	0.21
Olives	1	0.21	1.4	1.2
Oranges	1	0.10	0.68	0.98
Tea	1	0.11	0.44	—

[a] Derived from Table 5.11.

1.7. GENERALIZATIONS BY ELEMENTS

From the data in Tables 5.1–5.12 certain generalizations can be made about the removal of each of the nutrients elements by harvested crops.

Nitrogen is removed in large quantities by tuber and root crops such as potatoes or cassava. The removal often exceeds 180 lb N/acre. Although legume crops such as alfalfa can remove more than 180 lb N/acre, it is difficult to estimate the net removal because of atmospheric N fixation by symbiotic N-fixing organisms. Vegetables show a wide variation of N removal; asparagus, onions, endives, and salad remove less than 90 lb/acre. On the other hand, the removal of N by most cabbage varieties exceeds 180 lb N/acre. Most of the miscellaneous cash crops are medium N consumers, but flax and hemp may be considered low users. The trees form a heterogeneous group of which citrus trees often remove more than 180 lb N/acre. Coconut palms, oil palms, date palms, coffee, cacao, olives, and tea consume less than 90 lb N/acre.

Potassium is removed in large quantities, more than 180 lb K/acre, by tuber and root crops such as sugar beets, potatoes, and cassava.

Also some vegetables such as tomatoes, cauliflower, and cabbage species are known for their high demands of K. Among the agronomic crops, tobacco, pineapple, and sugar can remove more than 180 lb K/acre.

Phosphorus is removed at much lower rates than N and K. For grain crops the P removal amounts to approximately one-fifth of the N removal. Among the high P consumers, corn, turnips, cassava, cabbage, pineapple, cotton, and sugar cane may remove P at a rate equal to or above 27 lb P/acre.

Sulfur is removed in amounts similar to P, but certain crops are known as high S consumers. These include sugar beet, cabbage, alfalfa, clover, onions, and cotton, which commonly remove more than 15 lb/acre and may remove as much as 40 lb/acre. Cereals, grasses, and potatoes remove lesser amounts, generally in the neighborhood of 10 lb/acre.

Calcium is consumed in appreciable amounts, more than 90 lb Ca/acre, by carrots and cassava. Also the legume crops such as alfalfa, clover, and horse beans are known for their high Ca consumption. Among the vegetables, tomatoes and cabbage remove large quantities of Ca. Extremely high quantities of Ca are required by sisal, and high but lesser amounts by pineapple and sugar cane. Among the tree crops, citrus species are known as high Ca consumers.

Magnesium in general is removed less than either N, K, or Ca but often more than S or P in crop plants. Thus, at one extreme, as much or even more than 100 lb of Mg may be removed per acre by a crop such as sugar cane or, at the other extreme, as little as 2 lb/acre for a crop such as rice. Sugar cane, sisal, pineapple, root crops (particularly the tops), tomato, cotton, buckwheat, and tobacco are known as high users of Mg.

2. Nutrient Removal in Relation to Fertilizer Applications

When feeding the crop is the basis for fertilizer recommendations; the quantity of nutrients removed by a crop can only be a very rough guide. Apart from the nature of the chemical interaction between soil and fertilizer, which is entirely different for each of the nutrient elements, the nature of the root system is a major factor in assessing the level of fertilizer application. A small root system will deplete the soil in the vicinity of the roots to a much greater extent than a well-developed root system, which has a relatively large soil volume at its disposal.

Total N removal by the crop gives a reasonable guide to the quantity of N to be supplied as a fertilizer. Nitrification of NH_4 fertilizers and the relative lack of interaction of nitrate with the solid phase means that the N will appear in the soil solution, and the high mobility of the N ensures that a high proportion will reach the plant roots. Thus, the intensity factor (i.e., the nitrate concentration in the soil solution) is high, and rather rapid uptake by the plant can be expected.

The total P removal is of little value for recommendations of the quantity of P to be supplied as fertilizer. A removal of 26 lb P/acre (60 lb P_2O_5/acre) is extremely high. In contrast to N, the interaction of phosphate fertilizer with soil is very great and the intensity factor soon reaches a very low level. The mobility of P is very limited in most soils and only a portion of the applied phosphate ever reaches the vicinity of the root. Thus, in practice, applications of P have to be much higher than removals because only a fraction of the applied dose is utilized by the growing crop. However, continuous use of supplemental fertilizers in a larger quantity than that being removed results in a gradual increase in the capacity factor P (soil).

Potassium removal and application as a fertilizer are more closely related because the K interaction with the soil is intermediate between N and P. Although the bulk of the K applied is usually adsorbed, the K level in the soil solution does not become so low that it severely limits the mobility of K. Thus, a high proportion of the applied K will reach the vicinity of the root and the intensity factor will be high enough to result in a rapid rate of uptake of the applied K. The actual level of K in solution will be governed by its ratio to other cations present and to the total salt concentration. Thus, owing to the valency effect (see Chapter 2, Section 3.3) the ratio of K to bivalent cations such as Ca and Mg will increase upon dilution of the soil solution. Although under dry conditions the ratio of K to bivalent ions in the soil solution will decrease, the actual K concentration will tend to increase at lower soil-moisture contents. In other words, the soil acts as a buffer for the concentration of K in the soil solution, thus ensuring that the bulk of the K applied as a fertilizer will reach the plant root at a reasonably high concentration in solution.

The removal of Ca can be considerable, about 45 to 90 lb Ca/acre. The reserves of Ca on the adsorption complex and in the soil solution, however, are usually considerable. Moreover, Ca is often concomitantly supplied with other fertilizers or as lime. The need for Ca application as a plant nutrient is, therefore, relatively rare, although its specific need, as in the case of such crops as peanuts, has been clearly established (Colwell and Brady, 1945; Bolhuis and Stubbs, 1955). Magnesium, on the other hand, can become a limiting factor despite the fact that the plant removal is generally less than 45 lb/acre. Although the concentration of Mg in the soil solution is often of the same order of magnitude as Ca, the amount adsorbed, i.e., the capacity factor, is usually appreciably less and Mg is seldom inadvertently added in P or mixed fertilizers. Substantial amounts of Mg are often added to soils as dolomitic limestone. When Mg deficiencies do occur (Schachtschabel, 1956; Hester, 1958), more Mg must be added than removed by the crop owing to the interaction of Mg with the solid phase.

It may be concluded, therefore, that in the case of N and to some extent K the quantities removed together with consideration of the losses or natural additions can give some indication of the quantity to be applied as a fertilizer. For Mg there is less of a relationship and for P, very little.

With regard to other elements such as Mn, Fe, Zn, B, Cu, and Mo, the quantities removed are extremely small as compared with the soil contents, and the supply of these elements as fertilizers is not related to the removal. Apart from the special case where these elements are almost absent in the soil, the deficiency of the microelements is more often related to soil pH and oxidation-reduction status. This explains why deficiencies of these elements are often alleviated by means other than the application of the deficient element. The amounts added are usually so small that economics is not the prime consideration in choosing amounts but rather in choosing the form to be added, and care must be taken to add such an amount or form that will be adequate but not toxic.

3. Nutrient Mobility

All nutrients move in the soil. This mobility is responsible for both leaching losses and gaseous losses. When considering these losses in relation to the amounts of nutrients necessary to supply the requirement of the plant for nutrients, the natural additions, particularly from precipitation and N fixation, must also be considered. The principles governing the mobility of the nutrients insofar as movement in the water system is concerned have been discussed in detail in Chapter 2, Section 4 in connection with the discussion of the movement of ions to the root surface.

Bulk movement of nutrients in the water stream under the influence of moisture gradients owing to water utilization by plants moves nutrients to the plant root. Water moving under the forces of gravity and capillarity also carries ions with it, tending to remove them from the root zone. In addition, surplus water moving through the soil always contains the amount of nutrient ions normally present in the soil solution. Movement of ions by diffusion occurs only over very short distances and need not be considered in relation to nutrient losses.

Since nutrient losses owing to water movement are strictly a function of the amount of water moving and the concentration of the ion in the soil solution, a glance at Table 2.9 gives such information concerning the potential loss of nutrients. The main difference among the major nutrient ions is again P. Its low concentration in the soil solution ensures very little leaching loss.

Some ions fluctuate in concentration in the soil solution to a greater extent than others and, of course, this will affect the amount of these ions that move. The ions that fluctuate the most are those which are not adsorbed by the soil to any appreciable quantity and those most affected by microbiological activity. In this respect the concentration of N in the soil solution fluctuates the most. The fluctuations are related primarily to climatic conditions and the nature of the crop. Generally, after the winter period nitrate will accumulate in the spring but rapidly drop to a minimum value when the crop ripens (Gasser, 1961, 1962, 1962a; Baumann and Maasz, 1957; Daragan-Sushohova, 1960; Shaw, 1962; Stornier, 1962). In Africa, Hagenzieker (1957) observed an increase in N content of the topsoil at the start of rainy seasons in the absence of growing crops. There is a great similarity in the NH_4 and the NO_3 fluctuation. Periods of N accumulation in the tropics are short, only 4 to 6 weeks, and are followed by a rapid drop to a constant low level. When a crop has grown, no fluctuations are observed.

Under grassland Richardson, 1938 the concentration of N fluctuates slightly, usually in the range of 0 to 5 ppm. In the autumn and winter season a slight increase in ammonium and nitrate is observed, presumably owing to the decay of plant parts. In dry years the N content is higher than in wet years. This was also observed for exchangeable K. The N content under grass is lower and fluctuates less than in cultivated fallow soils. The ammonium content is usually slightly higher than nitrate. Simpson (1962) found that the period of drying between consecutive wetting had considerable influence on the extent to which nitrate accumulates under improved pasture. After a heavy rain the nitrate that is produced during summer and early autumn can disappear from the topsoil, resulting in a N poor spring. Rapidly growing crops remove N at a faster rate than it is produced, and when virgin soils are cultivated, a gradual decrease in total N content has been observed (Jung, 1933; Black, 1957). Cultivation under grass, on the other hand, may lead to a gradual build-up of the N level of the topsoil (Richardson, 1938).

The principal factors governing the movement of N in soils as summarized by Harmsen and van Schreven (1955) are as follows:

(1) In fallow soil the mineral N content is lowest during the winter, rises rapidly in spring and the first part of the summer season, maintains

itself at a rather high level throughout the summer, and then drops rapidly to the low winter level with the onset of the rain in autumn.

(2) In cropped land a second minimum is observed in midsummer during maximal growth of the plants, followed by a second maximum after harvest. Accumulation of N is, therefore, suppressed by every crop. Short periods of relatively high concentration of N in the soil are then in late spring and early fall.

(3) The winter minimum can be ascribed to heavy leaching in humid climates and reduced mineralization of organic N at low temperature. The rapid rise in the spring results from the partial sterilization effect of frost. The subsequent drop in mineral N content of the soil is the result of N uptake by the crop and fixation by microorganisms in the vicinity of the roots.

(4) Under perennial crops, especially under a grass cover, the mineral N content remains very low during the whole year and even high applications of N fertilizer are absorbed.

The above factors do not necessarily apply where fall rainfall is low and frost starts early in the fall and remains till late spring.

One of the difficulties in assessing fertilizer recommendations is the fact that the concentration of the ion in the soil solution fluctuates during the year, depending on temperature and rainfall conditions. In addition, a periodic fluctuation in soil fertility from year to year has been observed (van der Pauw, 1962), and these fluctuations may be responsible for large differences in fertilizer response. Since the rate of uptake of a nutrient ion is a function of the concentration of the ion in the soil solution, changes in the chemical composition of the soil solution induced by fluctuating temperature and rainfall conditions may well be responsible for some of the variation in crop yields and response to fertilizers observed from year to year.

Van der Pauw (1962) was able to relate the cyclic periods in rainfall in the Netherlands with the changes in water-soluble P and exchangeable K in the soil. The P content gradually rises during dry years and falls in wet years. This was also observed for exchangeable K which increased in dry and decreased in wet periods. The drop in exchangeable K after prolonged rainfall periods may appreciably surpass 150 to 180 lb/acre with an accompanying decrease in the K ion concentration in the soil solution.

The fluctuations in chemical composition of the soil solution after alternating periods of rainfall and dry seasons can be responsible for a marked variation in yield. For example, the average yields of wheat and rye in the Netherlands after some dry years amount to 1.5 times those obtained after a succession of wet years. For peas these fluctuations in yield are even greater and can be three times higher after dry than after wet years. A low fertility status tends to increase the effect of periodic fluctuations in rainfall (van der Pauw, 1962). Often the periodicity in rainfall is well known for a particular area and, therefore, it would be theoretically possible to counterbalance the unfavorable effects with appropriate fertilizer applications.

4. Losses

4.1. Leaching Losses

Generally, anions such as nitrate, which are neither adsorbed to the soil nor form sparingly soluble compounds, will leach to the subsoil after periods of heavy rainfall. This may also occur to a limited extent with phosphate in sandy soils or peats, although phosphate is generally known to move very little in most soils. Cations such as K, the bulk of which is adsorbed, may still leach out in appreciable quantities depending on the extent to which the adsorption complex is saturated with K and the salt concentration of the soil solution (Black, 1957).

From the analysis of drainage water, lysimeter studies, pot experiments, and experiments involving the use of isotopes as tracers, estimates have been made of the amounts of elements that may have leached out of the topsoil under a wide variety of conditions. Reviews on the amounts of N lost by leaching have been given by several authors (Martin and Skyring, 1962; Millar, 1955; Black, 1957). Nitrogen is generally leached in the form of nitrate (Shaw, 1962), although in some instances leaching losses of NH_4 could be observed (Wilshaw, 1934; Morgan *et al.*, 19·2).

In Table 5.13 the amounts of N lost by leaching have been listed for a number of crops or fallow conditions. It is evident that under continuous cropping the losses of N due to leaching are relatively small if no N applications are made. A period of fallow contributes to a considerable N loss, particularly when N fertilizer is applied. Under grass, leaching losses are almost insignificant, whereas for other crops the leaching losses are often much smaller than the quantity of N which is removed annually by the crop. It is not possible to make any general statement about the expected losses of N owing to leaching because its magnitude depends on a number of factors which have to be taken into consideration for each set of soil, climate, and crop conditions.

Each time water moves down the profile after a rain, the nitrate

TABLE 5.13
LOSSES OF NITROGEN DUE TO LEACHING

Location	Crop	Fertilizer (lb/acre/year)	Leaching loss (lb/acre/year)	Ref.
Ithaca, N.Y.	Vegetable crops and rye cover	143(Ammonium sulfate)	42	*a,b*
		143(Sodium nitrate)	46	*a*
	Continuous timothy	93(Sodium nitrate)	2	*a*
		124	1	*a*
		155	2	*a*
		213	1	*a*
Windsor, Conn.	Continuous tobacco	200(Nitrate)	123	*c*
		200(Ammonium)	108	*c*
		200(Urea)	88	*c*
		200(Organic)	82	*c*
	Fallow	200(Calurea)	225	*c*
	Tobacco	200(Calurea)	97	*c*
	Tobacco and oat cover	200(Calurea)	48	*c*
	Grass	200(Calurea)	28	*c*
	Tobacco	Nil	29	*c*
	Grass	Nil	6	*c*
Riverside, Calif.	Sudan grass	Nil	51	*d*
		100(Calcium nitrate and straw)	73	*d*
		200(Calcium nitrate and straw)	113	*d*
		Nil	18	*d*
		100(Calcium nitrate)	31	*d*
		200(Calcium nitrate)	40	*d*
Rothamsted, U.K.	Fallow	Nil	27	*e*
India, Cawnpore	Fallow	Nil	107	*f*
India, Pusa	Fallow	Nil	70	*f*
	Crops	Nil	13	*f*
Ceylon	Fallow	Nil	329–511	*g*
	Various crops	Nil	3–156	*g*
India	Black cotton soil	Nil	80	*h*
		65 ($NaNO_3$)	102	*h*
New York	Fallow	10 tons manure/10 years	82	*i*
	Rotation with legumes	10 tons manure/10 years	8.3	*i*
	Grass	10 tons manure/10 years	0.3	*i*
	Fallow	Nil	68	*j*
	Crop	Nil	5	*j*

[a] Bizzell (1943).
[b] Bizzell (1944).
[c] Jacobson *et al.* (1948).
[d] Chapman *et al.* (1949).
[e] Russell and Richards (1920).
[f] Leather (1911).
[g] Joachim (1930).
[h] Annet *et al.* (1928).
[i] Lyon and Bizzell (1934).
[j] Lyon *et al.* (1930).

in the soil solution will move downward. This has been illustrated by Soubiès *et al.* (1952). Each rain corresponds to a definite distance of migration of nitrate down the profile (Fig. 5.1).

The coarser the soil, the lower the water-holding capacity and the greater the downward movement of nitrate (Bates and Tisdale, 1957).

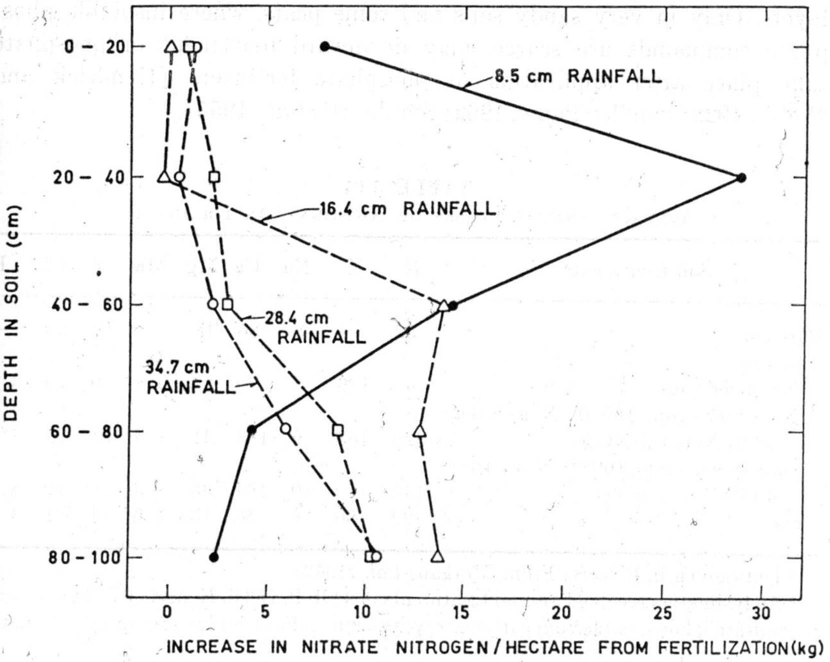

FIG. 5.1. Increase in content of nitrate nitrogen at different depths in a fallow sandy loam soil as a result of fertilization with calcium nitrate at a rate equivalent to 53 lb N/acre. (The measurements recorded for cumulative totals of 2.2, 6.5, 11.2, and 13.7 inches of rainfall after fertilization were obtained 34, 66, 107, and 153 days after fertilization, respectively. The total recovery of added N at the respective times of measurement amounted to 52, 37, 22, and 19 lb/acre.) From Soubiès *et al.* (1952).

When the structural units of the soil consist of dense aggregates, drainage water will move through the larger channels, carrying some of the nitrate with it. That nitrate in the water within the dense aggregate, however, will move more slowly. This explains why the movement of nitrate through the profile sometimes seems to be slower than that of the percolating water, which is restricted to the larger channels (Webster and Gasser, 1959; Moore, 1960).

Losses of P due to leaching are generally insignificant. Even after heavy fertilization of superphosphate the quantity of P that was found to be leached to deeper horizons by lysimeter studies is extremely low (Morgan and Jacobson, 1942). The downward movement of rain water is essentially in equilibrium with the soil and contains approximately 10^{-5} N phosphorus. This means that 25–30 inches of rainfall would correspond to a maximum of approximately 2 lb P/acre leached to deeper layers. Only in very sandy soils and some peats, where insoluble phosphate compounds are scarce, may downward movement of phosphate take place after application of phosphate fertilizers (Hendrick and Welsh, 1928; van der Pauw, 1962; Stanberry *et al.*, 1955).

TABLE 5.14
AVERAGE ANNUAL LOSSES OF NUTRIENTS IN DRAINAGE[a]

Soil treatment[b]	N	K	Na	Ca	Mg	Mn	S	Cl	I
Bare soil	52	94	8	80	18	—	62	20	—
Tobacco									
No cover crop	47	109	8	70	20	0.8	59	20	3.
No cover crop, 160 lb N as urea, 40 lb N as $Ca(NO_3)_2$	225	163	9	183	41	8.9	56	22	7.
Oats cover crop, 160 lb N as urea, 40 lb N as $Ca(NO_3)_2$	142	150	10	154	38	8.3	56	19	8.
Grass sod	22	86	9	55	16	1.0	61	18	4.

[a] Data given in lb/acre. From Morgan *et al.* (1942).
[b] All lysimeters received annual treatments of 44 lb P, 152 lb K, and 30 lb Mg/acre a precipitated, bone, potassium carbonate, potassium sulfate, and magnesium carbonat

Much greater losses of S than P have been reported from lysimete studies in Illinois where the average number of pounds per acre of in the drainage water during a 10-year period varied from 54.9 fro cultivated uncropped land to 35.4 from soil growing alfalfa continuously the soils had only received an amount of 21.5 lb of S annually, containe in superphosphate. Under various cropping systems the S loss amounte to 42–53 lb, (Millar, 1955). Rainfall additions of S in this area, howeve may be appreciable, (Eaton and Eaton, 1926).

Although K and Ca losses due to leaching are small compared the total quantity of these elements in the topsoil, these losses ma be appreciable when compared to other elements. This is illustrated Table 5.14 taken from lysimeter study at the Connecticut Experiment Station, (Morgan *et al.*, 1942). The losses of cations are considerab reduced when the soil is cropped as is illustrated in Table 5.15 tak from a study of the New York Experiment Station at Ithaca, (Lyon et al., 1930).

Leaching losses in lysimeters have been measured very precisely. Yet the figures in terms of the quantities involved have very little meaning except to illustrate those ions that are the most mobile and those that are least mobile and the effect of cultural practices in increasing or decreasing these losses. The reason the actual quantities recorded are not accurate can be grouped under three headings. First, the loss by leaching in the surface soil is not normally the same as the losses in

TABLE 5.15

EFFECT OF CROPPING ON THE LOSSES IN DRAINAGE FROM THE SOIL[a]

Soil	K	Ca	Mg	S	NO_3
Planted	45.9	173.3	37.0	38.2	5.4
Bare	61.0	367.4	65.2	48.5	92.8

[a] Data given in lb/acre/year.

TABLE 5.16

MOVEMENT OF NUTRIENTS IN DRAINAGE WATER AT DIFFERENT DEPTHS IN THE SOIL[a]

Soil	Pressure (inches)	Precipitation (inches)	N	P	K	Ca
Pisonia coral (pH 7.1–8.6)	$1\frac{1}{2}$	12	130	16.1	90	190
	6	12	180	2.8	79	270
	17	12	110	0.6	91	250
Recent deposit coral (pH 8.9–9.2)	3	11	2.0	0.18	1.3	64
	18	11	0.7	0.17	1.0	45
	36	11	0.2	0.09	2.9	40
Glacial till (pH 5.3)	$\frac{1}{2}$	25	1.8	0.15	2.4	8
	6	25	1.2	0.08	1.7	7
	28	25	0.7	0.06	0.4	3

[a] Suction (pressure) controlled with an alumdum filter. Data given in lb/acre. From Cole et al. (1961).

the subsoil yet it is deep in the subsoil that sampling usually takes place. Two reasons for this are that the normal soil is heterogeneous and that the chromatographic effect would normally take place even if the soil was initially homogeneous. An example of actual measurements of leachates from three different shallow depths is given in Table 5.16.

If depths usually used in lysimeters were chosen, the difference would have undoubtedly been even greater.

A second general reason for the inaccuracy of the actual quantities lost is that the moisture regime is seldom the same as would be present under field conditions. In general, where drainage occurs there is an air–water interface, meaning that the soil must be saturated at this point. It is as if the water table were kept at that level without the possibility of moving up or down. Cole *et al.* (1961) used tension plates to get around this difficulty and the results in Table 5.16 do not suffer from this difficulty. Cone-shaped lysimeters inserted at a particular depth also tend to set up forces resulting in abnormal moisture movement.

The third general reason for inaccuracy is that most lysimeters do not meet the requirement of having uniform conditions around the lysimeter, i.e., an appreciable area around the lysimeter must be planted, fertilized, watered, and managed the same way as the lysimeter in order to minimize errors. The prevention of lateral movement of roots or water is not as serious, but the rim or border needs to be as small as possible to prevent temperature and expansion and contraction effects.

4.2. Gaseous Losses

Nitrogen is the main soil-borne plant nutrient that is lost in the gaseous form from soils, although losses of S under reducing conditions are also possible. From a comparison of lysimeter studies carried out in the United States, Allison (1955) arrived at an average figure of 15% net N loss assumed to be through volatilization. The various mechanisms by which N may be lost through volatilization can be summarized as follows

(1) In soils of pH higher than 7 ammonia may be lost, particularly if the surface dries out temporarily. The losses will be greatest if the adsorption capacity of the soil is low and the temperature high. Local decomposition of organic material may be responsible for a rapid rise in pH through formation of NH_3 which is then lost by volatilization. In terms of applied NH_4 fertilizer, the losses of ammonia may be as high as 25% (Allison, 1955, 1966; Harmsen and van Schreven, 1955; Martin *et al.*, 1962). Losses may actually be much higher under some conditions.

(2) Nitrite which is an intermediate in biological nitrification and denitrification reactions will decompose to NO and NO_2 in an acid medium (pH 4–5). Above pH 5.5 nitrous acid is fairly stable. Part of the losses observed in lysimeter studies with acid soils might be accounted for by losses of NO and NO_2.

(3) Another postulated mechanism for gaseous losses of N is as follows:

$$RNH_2 + HNO_2 \rightarrow ROH + H_2O + N_2$$
$$RNH_4 + HNO_2 \rightarrow RH + 2H_2O + N_2$$

According to Gerretsen (1949, 1950) the above reactions could take place under conditions of strong nitrification. Ammonium sulfate should be present and the pH would have to be in the range of 4.5–5.5. Gerretsen and de Hoop (1957) demonstrated appreciable losses of soil-applied ammonium sulfate suggesting that this loss could only be accounted for by the above mechanism. Tracer studies with N^{15}, however, revealed that this reaction does not occur under normal soil conditions (Wyler and Delwiche, 1954). Allison (1963) pointed out that "at pH values where the reaction can occur, conditions for nitrous acid formation by both biological oxidation and reduction are seldom favorable."

(4) Allison (1963) points out that ammonium nitrite can be formed by the reaction of ammonia or urea with nitrous acid according to the following equations:

$$NH_3 + HNO_2 \rightarrow N_2 + 2H_2O \qquad \text{(a)}$$

$$(NH_4)_2CO + 2H_2O \rightarrow (NH_4)_2CO_3 \qquad \text{(b1)}$$
$$(NH_4)_2CO_3 + 2HNO_2 \rightarrow 2NH_4NO_2 + CO_2 + H_2O \qquad \text{(b2)}$$
$$2NH_4NO_2 \rightarrow 2N_2 + 4H_2O \qquad \text{(b3)}$$
$$(NH_4)_2CO + 2HNO_2 \rightarrow 2N_2 + CO_2 + 3H_2O \qquad \Sigma(\text{b1} - \text{b3})$$

The rate of decomposition of ammonium nitrite increases with temperature and is higher under acid conditions. Allison goes on to suggest that gaseous losses of N from soils may be greater via ammonium nitrite decomposition than from mechanism (b2) and (b3) above.

(5) Probably the largest N losses result from bacterial denitrification. Under anaerobic conditions, particularly when fresh organic material is present, nitrates are rapidly reduced to nitrites. Depending on the pH, a mixture of N_2, N_2O, and NO gases are formed as end products of denitrification. Above pH 7, N_2O dominates in the mixture, whereas at a lower pH more NO is found (Wÿler and Delwiche, 1954; Arnold, 1954).

Under anaerobic conditions the reduction of nitrates proceeds very fast (Jones, 1951), but even when oxygen is present, denitrification may continue at a low rate (Wÿler and Delwiche, 1954; Broadbent, 1951; Broadbent and Stojanovic, 1952; Allison, 1963, 1966; Carter *et al.*, 1967).

Rice soils under submersed conditions are known to lose considerable amounts of N through bacterial denitrification (Ponnamperuma, 1955).

TABLE 5.17

NITROGEN NOT ACCOUNTED FOR IN LYSIMETER STUDIES IN THE UNITED STATES[a]

Location	Crop	Fertilizer material	Amount applied (lb N/acre/year)	Amount not accounted for (lb N/acre/year)	Loss (%)
Ithaca, N.Y.	Vegetables and rye	Ammonium sulfate	143	44	24
		Sodium nitrate	143	40	20
	Timothy	Sodium nitrate	213	45	21
			155	24	15
			124	12	9
			93	4	4
Geneva, N.Y.	Timothy	None	—	11	17
	Barley	(Low fertility soil)	—	—	—
	Wheat	—	—	—	—
	Timothy	None	—	46	36
	Barley	(High fertility soil)	—	—	—
	Wheat	—	—	—	—
Windsor, Conn.	Tobacco	Nitrate	200	7	3
		NH_4	200	17	8
		Urea and cyanamide	200	18	9
		Organic fertilizers	200	24	12
		None	—	11	0
	Fallow	Calurea	200	45	19
	Tobacco	Calurea	200	45	19
	Tobacco and oats	Calurea	200	64	31
	Grass	Calurea	190	81	42
	Tobacco	None	—	7	13
	Grass	None	—	22	50
Tennessee	None	Calcium nitrate	989	150	17
		Magnesium nitrate	989	164	19
		Ammonium chloride	989	236	20
		Ammonium phosphate	1111	280	25
		Ammonium sulfate	1224	168	14
Riverside, Calif.	Sudan grass,	Straw	32	1	1
		Straw and calcium nitrate	131	22	9
		Straw and calcium nitrate	237	32	10
	Sudan grass and mustard as cover crop	None	—	26	20
		Calcium nitrate	100	18	10
		Calcium nitrate	200	29	11

[a] Derived from Allison (1955).

These losses are dependent on the pH and redox potential and may be very great as revealed by experiments using N^{15} (IAEA, 1964).

Table 5.17 reviews the N losses that could not be accounted for in lysimeter studies in the United States (Allison, 1955). The losses are presumably due to volatilization. As these losses have been calculated from the difference between gains and losses in nutrient balance studies, a relatively large error tends to make the data for unaccounted loss rather unreliable. Nevertheless, the losses are apparently significant.

Generally, the losses due to volatilization of nitrogen as shown in Table 5.17 are affected by the rate of fertilizer application. Losses occur for all materials tested regardless of accompanying ion or whether ammonium or nitrate is the source of N.

5. Natural Additions

The natural additions of plant nutrient other than from plant residues, manure, or fertilizers come from the atmosphere or rain. Nitrogen, S, and Cl are the most important elements in this respect, and the natural additions of these nutrients to the soil are often significant with respect to supplying them to the plant.

TABLE 5.18
CONTRIBUTION OF NITROGEN IN PRECIPITATION

Location	N in precipitation[a]	Ref.
United States(lysimeter studies)	4–9	Allison (1955)
Europe(130-year records)	7.2	Ericksson (1952), Henzell and Norris (1962)
United States	6.7	Ericksson (1952)
Trinidad	2.2	Ericksson (1952)
British Guiana	2.9	Ericksson (1952)
India	3.9	Ericksson (1952)
Congo	4.8	Meyer and Pampfer (1959)
Sumatra	14.9–41.8	Roelofsen (1941)
Saigon	9.8	Vailard-Goudou and Richard (1956)

[a] Data given in lb N/acre/year. From Henzell and Norris (1962).

The additions of N with rainfall were found to be about 4 to 9 lb N/acre/year in lysimeter experiments in the United States (Allison, 1955). Apart from nitrate, nitrite, and ammonia, N may occur in precipitation in the form of organic compounds. The review of Henzell and Norris (1962) summarizes the total N content of precipitation in various regions of the world. Table 5.18 is derived from their summary.

The amount of N added in the rainfall was appreciable only in the one report from Sumatra. Generally, the quantities are too small to be of much agricultural significance except on soils cultivated with nonleguminous crops for centuries without addition of supplementary nitrogen.

The origin of N in precipitation may be due to electrical discharges during thunderstorms by which nitrates are formed from oxygen and nitrogen gas. Another source may consist of the nitric and nitrous oxides produced by bacterial denitrification in the soil.

The quantities of S in rainfall may be considerable as is illustrated by Table 5.19 (Whitehead, 1964). The large amounts of S in the precipitation are primarily a result of the use of fossil fuels, particularly in industrial centers where SO_2 and other gaseous S compounds are continuously released to the atmosphere. This results in a sufficiency of S in these areas. In these areas additions are sometimes so high that public health authorities are worried about air pollution and in some areas manufacturing concerns are forced to remove these compounds from their waste gases. The quantities of S in precipitation in rural areas of the United States and other parts of the world are often not adequate to meet the crop requirements. Additional contributions of S to the atmosphere come from volcanoes, sulfur springs, and bogs.

The atmosphere as a source of plant nutrients is of importance with regard to N. There are two different mechanisms known by which N from the air can be fixed, i.e., by means of symbiotic and nonsymbiotic fixation. As both processes occur simultaneously, an accurate estimate of symbiotic N fixation by legumes is difficult, and the data obtained from field and lysimeter studies should be considered with caution (Henzell and Norris, 1962). The amount of N fixed by *Rhizobium* in the root nodules of legumes is also a function of the amount of readily utilizable soil N and is thus affected by competition with the roots of other plants such as weeds for this utilizable soil N. Generally, less atmospheric N is fixed when immediate N supply to plants is high.

Table 5.20 compares the N amounts fixed by various hay, green manure, and food crops in different parts of the world. The data clearly illustrate that considerable amounts can be fixed by means of symbiotic fixation. There is no indication in the data that the amount fixed in association with legumes differs under different climatic conditions.

A number of nonlegumes have been reported to fix atmospheric N in root nodules. By means of studies involving the use of N^{15} it has been shown that a number of tree species (*Alnus, Myrica, Hippophae*) can obtain atmospheric N in this way (Henzell and Norris, 1962; Coleman, 1939; Bond, 1956; Stevenson, 1958; Harris and Norrison, 1958; Steward and Bond, 1961).

Some investigators have claimed that agricultural crops such as buckwheat can fix atmospheric N (Schanderl, 1947). These results were

TABLE 5.19
SULFUR IN PRECIPITATION[a]

Location	Sampling period	Sulfur (lb acre/year)	Ref.
Goldalming, Surrey, England	1957–1958	28.7	*b*
Portishead, Somerset	1957–1958	14.5	*b*
St. Helens, Lancashire	1957–1958	74.4	*b*
Loch Katrine, Scotland	1957–1958	11.6	*b*
W. Germany (rural)	1957–1958	10–12	*c*
W. Germany (industrial)	1957–1958	80	*c*
Denmark (10 stations)	1957–1961	9.6–13.4	*d*
Sweden (33 stations)	1954–1957	1.7–9.2	*e*
Sweden (industrial)	1954–1957	41.7	*e*
Kikuyu, Kenya	1955–1956	0.1	*f*
Michigan	1957–1959	8.0–12.5	*g*
Indiana	1946–1947	20–33	*h*
Indiana (industrial)	1946–1947	127	*h*
Kentucky	1953–1956	7–16	*i*
Virginia	1953–1956	12.7–33.5	*j*
Florida and Southeast	1953–1956	5.4 (Mean)	*l*
Victoria, Australia (coastal)	1954–1955	3–7	*m*
Victoria, Australia (inland)	1954–1955	<2	*m*
Kojonup, W. Australia	1951	0.1	*n*
Perth, and Nedlands, Western Australia	1957–1959	1.2–6.0	*o*
New Zealand (inland, rural)	1960	0.4–2.8	*p*
Taita, Nr. Wellington, New Zealand	1956–1958	7.5	*p*

[a] From Whitehead (1964).
[b] Dept. of Scientific and Industrial Res. (1960).
[c] Buchner (1958)
[d] Jensen (1963).
[e] Johansson (1959).
[f] Hesse (1957).
[g] Cressman and Davis (1962).
[h] Bertramson *et al.* (1950).
[i] Seay (1957).
[j] Lutz (1957).
[l] Jordan and Bardsley (1958).
[m] Hutton and Leslie (1958).
[n] Rossiter (1952).
[o] Droner (1960).
[p] Miller (1961).

proven erroneous (Roschach, 1960), and up to the present no definite indication exists that such agricultural crops are able to fix N from the atmosphere.

The nonsymbiotic fixation of N from the atmosphere has been proved

TABLE 5.20

SYMBIOTIC FIXATION OF ATMOSPHERIC NITROGEN BY LEGUMES[a]

Crop	Location	Amount of N fixed (lb N/acre/year)	Reference
Hay and pasture			
White clover	New Zealand	539	Sears and Evans (1953)
		100	Walker *et al.* (1959)
Cooksfoot	Belteville	160	Wagner (1954)
Ladino clover	Pennsylvania	100	Washko and Pennington (1956)
White clover	Hurley, U.K.	30	Cowling and Green (1956)
White clover (pure)	Kentucky	148	Karraker *et al.* (1950)
White clover and Kentucky blue grass	Kentucky	95	Karraker *et al.* (1950)
Red clover	Cornell	146	Lyon and Bizzell (1934)
Alsike clover	Cornell	136	Lyon and Bizzell (1934)
Subterranean clover	New South Wales	43	Donald and Williams (1954)
Trifolium species	South Australia	65	Cook (1939)
Lucerne	Arizona	221	Smith (1944)
	Arizona (mohave clay)	733	Smith (1944)
	New York	206	Collison *et al.* (1933)
		260	Collison *et al.* (1933)
		223	Collison *et al.* (1933)
Lucerne and Kentucky grass	New York	162	Collison *et al.* (1933)
Lucerne	Kentucky	297	Lyon and Bizzell (1934)
Tropical pastures (*stylosanthes, heteropogon*)	Queensland	85	Miles (1949)
Tropical pastures (*pueraria, Melinis panicum*)	Puerto Rico	169	Abruña and Figarella (1957)
Tropical pastures	Puerto Rico	134–175	Vincente-Chandler *et al.* (1953)
Lucerne glanca	Mauritius	271	
	Hawaii	462	
Green manure and cover crops			
Green manures (red clover, white clover, vetch)	Iowa, Nebraska	30–100	Fribourg and Johnson (1933); Kroontje and Kehr (1956)
Madrid sweet clover	Iowa, Nebraska	100–200	Fribourg and Johnson (1933); Kroontje and Kehr (1956)

TABLE 5.20 (*Continued*)

Crop	Location	Amount of N fixed (lb N/acre/year)	Reference
Green manure and cover crops (*Cont.*)			
Hairy vetch	New Jersey	133	Sprague (1936)
Various clovers	New Jersey	40–92	Sprague (1936)
Hairy vetch, Austrian peas	Southern United States	16–60	Bailey *et al.* (1930); Coleman (1939); Grissom (1950); Obenshain and Gish (1941)
Hairy vetch	North Carolina	75–95	Kamprath *et al.* (1958)
Purple vetch	California	106	Chapman *et al.* (1949)
Sweet clover	California	118	Chapman *et al.* (1949)
Tropical green manure	Indonesia	57–224	Anon (1963)
Food crops			
Soya beans	United States	26–62	Norman (1943)
		38–56	Bear (1942)
	Cornell	102	Lyon and Bizzell (1933)
Field beans	Cornell	57	Lyon and Bizzell (1933)
Pigeon peas	India	87–133	Sen (1958)

[a] Data derived from Henzell and Norris (1962).

for five different groups of microorganisms: (1) bacteria (*azotobacter, clostridium*), (2) photosynthetic bacteria, (3) actinomycetes, (4) fungi and yeasts, and (5) blue-green algae.

Although direct proof of fixation is readily obtained, particularly with the aid of N^{15} labeling techniques, little quantitative data are available for field conditions that show the actual magnitude of the nonsymbiotic N fixation contribution to plant nutrition. Most investigators consider that the amount of N fixed by the above-mentioned organisms is small and does not exceed 20 lb N/acre/year (Allison, 1955; Jensen, 1950; Norman, 1946) and is more likely only of the order of a few pounds of N per acre per year (Allison, 1955; Henzell and Norris, 1962). Suggestions of higher levels of nonsymbiotic N fixation still continue to appear (Moore, 1963; Karaguishieva, 1963).

Considerable increases in crop yields have been claimed as a result of inoculation of the soil with *azotobacter*. Clear confirmation of these observations, particularly in relation to ascribing any increase to increased N fixation has been lacking (Allison, 1947; Henzell and Norris,

1962), but such reports continue to appear in the literature (Czubchik and Konashevich, 1962; Rakhno and Ryys, 1963). It is difficult to understand how important quantities of N could be fixed by bacteria because considerable amounts of organic matter would be required as an energy source. To fix 18 to 45 lb N/acre, 1 to 2.5 tons of organic matter of the same nutritive value as glucose would be consumed as an energy source in the process, and this can hardly be imagined to occur in arable soils (Jensen, 1950).

Under flooded conditions in rice cultivation, blue-green algae have been reported to fix considerable amounts of atmospheric N. In contrast to bacteria, the energy required for N fixation is provided by photosynthesis. This not a general property, as most cultures of blue-green algae will not fix N. Nitrogen fixers appear to be rare in northern soils but relatively abundant in tropical and semitropical regions (Wilson and Burris, 1947; Cameron and Fuller, 1960).

In India and Japan considerable weight is attached to the fixation of N by algae. In pot experiments, a total of 20 lb N/acre was fixed (Watanabe *et al.*, 1951). The amount of N fixed by blue-green algae in the soils of India was reported to vary from 14 to 63 lb N/acre/year. Fixation of N apparently increases when rice is grown and phosphate and lime are applied (Hernandez, 1956).

6. Conclusions

The balance sheet of nutrient losses and gains in the soil is important for resource planning and even for determining the order of magnitude of such important items as gaseous losses or amounts of atmospheric N fixation. The balance sheets, however, do not contribute appreciably to the practical decision with regard to fertilization. Yet the observations reported previously suggest pertinent conclusions.

The observations very strongly suggest that N fertilization, insofar as possible, should strongly follow the principles of feeding the crop. When applied too far in advance—particularly in the absence of ground cover, there is real danger of loss particularly by leaching or even in the gaseous form. The principle of trying to maintain this by adding back the amount removed is likely to be particularly wasteful.

Additions of P, however, are not likely to result in appreciable losses even if made far in advance or in the absence of ground cover, assuming no erosion losses. Nevertheless, the time of contact of phosphate fertilizer with soil results in certain transformations which may affect the effectiveness of the applied phosphate. This is a chemical reaction, and the nature and extent of the reaction depends upon the chemical nature of the soil and the chemical nature of the fertilizer. The amount of P removed by the crop is not necessarily a good criterion to use in ferilization. To a limited extent, however, it might be possible to raise the whole plow layer up to a high level of P fertility and then base fertilizer additions on crop removal without the danger of incurring severe losses either by volatilization or leaching.

The K balance sheet suggests intermediate considerations between N and P. Appreciable leaching can occur under certain climatic conditions if attempts are made to build up the level of exchangeable soil K. However, this is not as serious as with N. In addition, since the K concentration in the soil solution is not only dependent upon the amount present but on the presence of other exchangeable cations, decisions with regard to K fertilization require knowledge of this chemistry.

The guide to S fertilization may, unlike all the other elements, be determined more by the amount in the precipitation than by any other factor, although the difference in removal by the different plant species is certainly an important factor. Once it is determined that a need exists, the principle of feeding the crop should apply, just as is the case with N.

Considering the amount of the major nutrients contained in a given crop and the amount removed by the harvested portion, it is evident that for most nutrients cultural practices based on getting more out of the soil reservoir are doomed to failure. The soil reserves are not great enough to stand the drain of intensive cultivation over a long period of time without a consequent drop in yield unless supplemental additions, either natural or in the form of artificial fertilizers, are made.

CHAPTER 6

Determination of Soil Nutrient Supply

The soil nutrient supply is a natural resource whose measurement is of interest in national and industrial planning. Measurement of this supply, however, has come about primarily as a need in agricultural practice both as an attempt to understand the effects of soil treatments and to predict the plant growth response to applied nutrients. A discussion of plant nutrients supply inevitably involves the term available nutrient or nutrient availability. In a critical study of the term "amount of available nutrient" Kalthoven (1956) came to the conclusion that many definitions are in use which deviate considerably from each other. None of these definitions seems to cover the term "availability" adequately. One thing is certain—a nutrient is not just *available per se.* It is *available to the plant.* In the last analysis *only the plant can judge whether a nutrient is available or not.*

Soil analyses for nutrient level are made in the *laboratory,* in the *greenhouse,* and the *field.* Although measurements under these three conditions have the same goal, each is uniquely qualified to contribute certain types of information to the over-all measurement of soil nutrient level. Laboratory measurements usually reflect a definite chemical fraction of the nutrient; greenhouse measurements reflect that portion of the nutrient in the soil that is available to plants; and field measurements reflect the volume relationships or size of the nutrient reservoir for a particular plant species grown in a particular soil under the natural plant environment.

1. Laboratory

Laboratory measurements of the supply of soil nutrient may be rapid and precise. In a practical sense, any large-scale testing program can be done only by a rapid, relatively inexpensive procedure. Thus, greenhouse and field techniques for evaluation of soil nutrient supply can only be for research purposes or for the ultimate purpose of determining which laboratory procedure consistently reflects the soil-plant relationship system most effectively.

The laboratory procedures usually involve a somewhat arbitrary extractant in contact with the soil for an arbitrary length of time. Results of these extractions are then correlated with other measurements of soil nutrient supply obtained in the laboratory, greenhouse, and field.

Equation (1.1) is repeated below to indicate which part of the continuum the laboratory test can measure.

$$\text{M(solid)} \rightleftharpoons \text{M(solution)} \rightleftharpoons \text{M(plant root)} \rightleftharpoons \text{M(plant top)} \quad (1.1)$$

It has already been pointed out that M(solid) is a capacity factor, whereas M(solution) is an intensity factor, and their importance in this continuum has been discussed in Chapter 2, Section 3.

Laboratory methods can only involve measurements of M(solid) and M(solution) and their rates of reaction and cannot take into account any plant factors. All laboratory methods remove M(solution) as it occurs in the laboratory, but except where the laboratory method is the extraction of the soil solution itself, this intensity factor is not normally measured. All laboratory methods except extraction of the soil solution extract a part or all of M(solid), in its present meaning, in addition to the M(solution). Unfortunately, the portion of the solid phase extracted it not always that portion which is in active interchange with the ion in the soil solution. Since the laboratory procedure can, at best, measure only M(solution) and the M(solid) in equilibrium with this solution, then the ideal laboratory procedure would measure both of these factors in one operation. In a given situation, however, if either the capacity or intensity dominate to such an extent that the other can be neglected with respect to plant nutrient supply, then the measurement of only the dominant factor is a reasonable approximation of the soil nutrient supply.

Unfortunately, M (solution) is not a constant in the soil–plant system at all times even though the rate of uptake of the ion by the plant is proportional to this concentration as well as to the concentration of the other reactants involved in the ion uptake mechanism (see Chapter 2). The actual concentration at a given time may be affected by (1) the rate at which the crop removes elements from the soil solution; (2) the rate at which these elements transfer from the solid phase into the soil solution; (3) the length of the diffusion path; (4) the diffusivity of the ion; (5) the magnitude of M(solid) and the reaction rate constants; (6) climatic factors (such as *rainfall and temperature* which affect both ionic and biological processes); and (7) the history of the soil sample (involving such things as previous vegetation and how the sample was taken and stored prior to measurement, etc.). Suffice it to say that any laboratory procedure which measures M(solution) must pick up some arbitrary environmental situation and cannot take into account the effects produced by the presence of the plant in the soil-plant system.

The laboratory techniques may be divided into two broad categories: (1) chemical methods and (2) biological methods. Chemical methods refer to those techniques in which the soil is extracted by different chemical extracting agents. Biological methods refer to those techniques in which the soil is extracted by different biological agents such as bacteria, fungi, algae, seedlings, and even entire plants.

1.1. Chemical Methods

Bergman (1958) and more recently Williams (1962) reviewed the extractants that have been used for soil-testing purposes. An attempt will be made to classify the current methods of analysis in relation to their ability to measure intensity, capacity, or rate factors.

1.1.1. *Intensity Factors*

Only a few attempts have been made to determine the concentration of elements in the soil solution of soils at normal moisture contents. This is not surprising because the extraction techniques that are available are still to a large extent unsatisfactory. Not only is the amount of solution involved only a few milliliters or less, but the quantities of nutrient ions contained in them is minute for many elements. Attempts have been made to analyze soil solution expressed by the pressure membrane, hydraulic press, displacement by organic and other liquids etc. but, except for special situations, no method has as yet been developed that can be applied satisfactorily on a routine basis in a soil-testing laboratory. The chemical analysis of fractions of a milliliter of solution with a total salt concentration of 0.01–0.001 *N* offers considerable analytical problems. Activation analysis may have some possibilities in the future in this respect (IAEA Rept., 1963).

The attempts to measure ion concentration of the soil solution can be divided into two categories: (1) those attempts to measure the concentration as it exists at normal moisture contents and (2) those attempts to measure the concentration in a water phase in equilibrium with the solid phase, requiring amounts of water greater than those existing at normal moisture contents.

a. *Normal Moisture Content.* The measurement of the ion concentration in the soil solution at normal moisture contents requires some means of extraction of the actual soil solution. This has taken essentially several forms: compaction, displacement by a nonaqueous liquid such as alcohol, displacement by water, displacement by gas, suction, centrifugation, and equilibrium extraction.

i. *Compaction* is simply the application of pressure on the soil-water-air system (Burgess, 1922; Lipman, 1918) in order to "squeeze" out the soil solution, much as one squeezes a sponge. This is only possible with soils high in clay from which the coarse sand and gravel have been removed. The moisture content of the soil must be close to the water-holding capacity. The method is not feasible either for sandy soils or where the moisture content of the soil is relatively low. The soil must be packed in a suitable strength filter to prevent loss of fine clay particles. The method is of limited applicability insofar as the nature of the soil and its moisture content are concerned, in addition to the effects of negative adsorption and the charge on the filtering membrane.

Negative adsorption is discussed more fully in Chapter 2, Section 3. The charge on the filtering membrane in which the soil has been wrapped and through which the solution must pass can also appreciably affect the composition of the extracted soil solution.

ii. *Displacement by nonagueous liquids* include both liquids that are miscible with water and those that are not. Examples of the latter are paraffin oil and other oils, while the former include the various low molecular weight alcohols and even acetone. The method involves the packing of a previously moistened soil in a column and application of enough of the displacing liquid over the soil column to force suitable amounts of soil solution out the bottom. A closed system may be used and external pressure applied to hasten the displacement. According to Parker (1921), liquids which were nonmiscible with water (such as benzene, kerosene, and ethyl acetate etc.) did not displace the soil solution, although Morgan (1916) considered that his method was successful. Parker considered ethyl alcohol the most successful of the three miscible displacing liquids since it displaced the highest proportion of the contained soil solution.

The displacing liquid initially displaces soil solution which forms a zone of saturation below the displacing liquid. When the saturated zone reaches the bottom of the soil in large enough quantities to overcome the surface tension, clear solution free of alcohol or other displacing liquid drops from the soil. This is obviously a saturation extract and can only represent the soil solution at one moisture content, water saturation.

iii. *Displacement by water* has been far the most common method of obtaining soil solution since soil scientists became interested in soil solution (Anderson *et al.*, 1942; Burd *et al.*, 1931; Burd, 1923; Eaton, 1939; Magistad *et al.*, 1945; Pierre, 1931). In technique, it is identical to that of displacement by nonaqueous liquids and, indeed, the results should be similar. A typical example of a simple apparatus is shown in Fig. 6.1.

Figure 6.1 indicates the possible use of air pressure for hastening sample collections, but this obviously is not necessary and any laboratory can use the water-displacement method without any more equipment than a cylindrical column with a hole underneath.

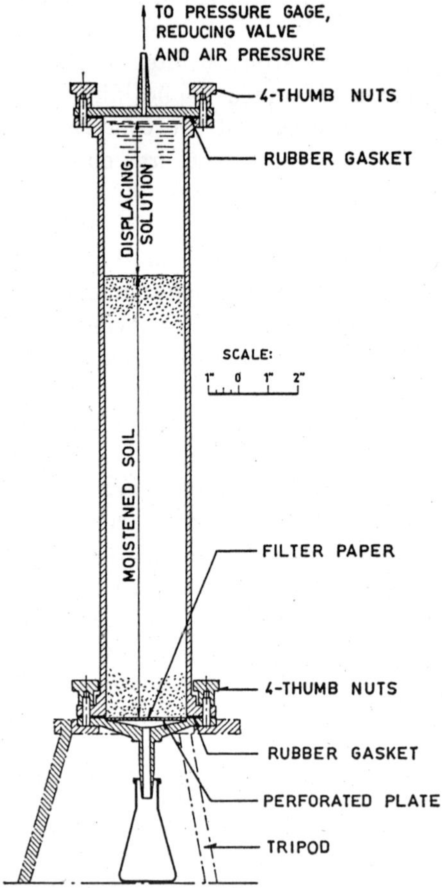

FIG. 6.1. Brass cylinder for displacing soil solutions. From White *et al.* (1936).

Again, this is a saturation water extract and only represents the composition of the soil solution when enough water is present to give zero water tension.

iv. *Displacement by gas* has essentially taken two forms, the use of pressure membrane apparatus and variations of this (Richards, 194‑ Eaton *et al.*, 1960). The use of the centrifuge (Davies and Davi‑ or suction (Dutt, 1962) also involve replacing the water b‑ special conditions which are discussed separately.

The pressure membrane apparatus or various versions thereof is one of the common methods of obtaining samples of the soil solution. Apparatus of this kind can now be purchased commercially. Its development and present use is a result of the practical needs of salinity problems. Obtaining saturation extracts has now become a routine laboratory procedure when dealing with saline soils (L. A. Richards, 1954; Luken, 1962; Reitemeier, 1946; Magistad and Reitemeier, 1943).

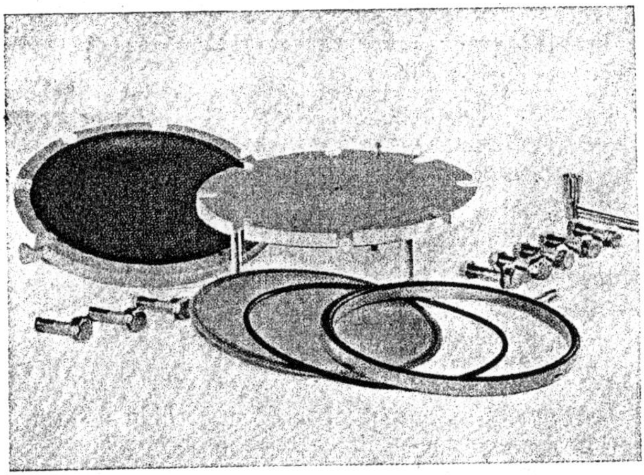

FIG. 6.2. Photographs of disassembled pressure-membrane extraction apparatus. (From Catalog 60 of Soilmoisture Equipment Co., Santa Barbara, Calif.)

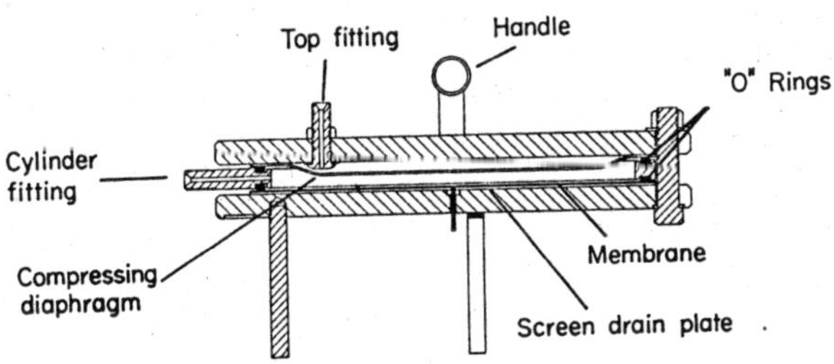

FIG. 6.3. Section view of extractor. (From Catalog 60 of Soilmoisture Equipment Co., Santa Barbara, Calif.)

Figure 6.2 is a photograph of a commonly used pressure membrane apparatus when disassembled and a commercially available apparatus is shown in Fig. 6.3. The principle is relatively simple. A small amount of soil (25 g is usually used in the apparatus shown in Fig. 6.3) is brought to the desired moisture content and placed on a cellophane or similar type membrane over an outlet hole. The container is screwed tight to prevent gas leakage. Various pressures of an inert gas, usually N, can then be applied, forcing the soil moisture that will move under that pressure out of the soil, through the membrane, and out through the outlet tube. Since the pressure can be varied at will, a sample can be collected which presumably represents the concentrations of salts at the various moisture contents (equivalent to the applied pressure) in the normal useful soil moisture region, e.g., between "so-called" field capacity and the wilting coefficient (i.e., between $\frac{1}{3}$ and 15 atm. pressure). The soil solution begins to flow out of the outlet hole after pressure is applied. Unfortunately, each successive increment of solution does not have the same composition. This is due primarily to the phenomenon of "negative adsorption" (see pp. 22, 36). Basically, owing to the negative charge on the clay, the concentration of anions in the solution adjacent to the clay particle is lower than in the soil solution a further distance out. As the soil solution is forced out by the gas, more and more of this water adjacent to the particle is removed, resulting in the appearance of a decreasing concentration of the anions in the solution with each successive extract. Since each anion must be accompanied by an equivalent number of charges in the form of cations, the cation concentration also decreases. The last increments removed may thus be essentially salt-free. If the moisture content of the soil is high enough, such that the ions in this zone of negative adsorption are small in quantity in relation to the total amount of ions in the soil solution, the initial concentration of ions removed by the pressure membrane apparatus may well be representative of the soil solution concentration. Another less serious complication for most ions is the charge on the membrane through which the solution is filtered. This may even result in the retention of some ions. Other ions may be more specifically adsorbed by the membrane, thus giving a faulty picture of the soil solution concentration.

Table 6.1 gives the results of analyses of soil solutions from 17 soils.

These were soils used in an investigation on salinity and, in general, are high pH soils obtained from western United States. All values have been adjusted to the water content corresponding to a pressure of 15 atmospheres (wilting point).

At these low moisture contents and with high pH soils, many of which are calcareous, the results show that the Ca concentration is generally 10^{-2} N or slightly above, and the Mg is close to this figure. The Na concentration is approximately 10^{-3} N except where the soils start becoming saline, and then the values increase sometimes so considerably that one soil solution was even 3 N in Na. The K concentration is generally lower but varies between 10^{-4} and 10^{-2} N. The anions chloride and sulfate generally varied between 10^{-3} and 10^{-1} N, while the HCO_3 was fairly steady at 5×10^{-3} N. The nitrate, which may not be a true figure owing to the effect of storage on nitrate production, was nevertheless slightly over 10^{-3} N. In general, these are rather concentrated solutions comparable in many respects to commonly used nutrient solutions and are the same order of magnitude as obtained by other investigators

TABLE 6.1
COMPOSITION OF SOIL SOLUTIONS[a]

	Ionic concentrations in extracted solutions									
	Cations				Anions					
Soil no.	Ca	Mg	Na	K	CO_3	HCO_3	SO_4	Cl	NO_3[b]	Total salts
75	6.9	11.7	4.6	0.7	0.0	4.7	1.9	2.4	14.3	23.9
66	14.3	7.0	1.4	0.2	0.0	6.7	1.3	3.1	10.2	22.9
83	25.4	3.5	1.1	0.5	0.4	3.7	1.1	1.1	22.5	30.5
65	24.8	8.0	1.2	0.8	0.0	7.6	1.1	2.5	(23)	34.8
77	11.1	4.7	9.8	0.5	1.1	6.5	7.4	5.2	5.7	26.1
183	20.4	6.6	4.3	1.8	0.3	3.4	4.3	5.3	(19)	33.1
84	23.3	6.0	22.7	1.0	0.3	4.1	26.3	14.8	(7)	53.0
56	24.4	19.8	40.9	0.9	0.0	3.1	49.5	27.0	11.3	90.9
79	27.8	14.8	19.2	2.1	0.3	6.7	17.4	10.4	28	63.9
85	45	52	24	9	1	6	53	20	43	130
58	6.4	1.7	100	2.1	5.6	5.6	66.2	28.1	9.0	115
222	25	29	125	2.5	0.2	3.0	128	27	15	182
211	37.4	15.6	25.3	3.9	0.0	2.6	40.7	16.1	(22)	82
86	12.3	210	183	3.3	0.4	12.9	325	31.2	13	409
207	89.9	65.8	224.2	3.5	0.0	2.8	51.0	296	(33)	383
57	787	317	561	2.9	0.0	3.9	19.4	1553	89	1668
62	50	190	2906	97	0.0	4.7	284	2954	18.9	3202

[a] Data given in milliequivalents per liter. From Magistad and Reitemeier (1943).
[b] Nitrate concentrations in parentheses were determined by combining analytical values with appreciable anion deficits because of apparent loss of nitrate prior to the time of analysis.

(Luken, 1962; Eaton *et al.*, 1960) for similar high pH soils. Prominent by its absence is phosphate, which, generally, is at least partially adsorbed by the filter used in the pressure membrane apparatus, making P determinations unreliable for small volumes of solution. This is at least one of the reasons why, although the pressure membrane apparatus has received such extensive use in determination of total soluble salts, it has not been generally used for extracting soil solutions of acid soils.

Barber *et al.* (1963) recently investigated the Ca, Mg, K, and P concentrations in saturation extracts of 135 acid to neutral soils from north central United States. Extraction was at relatively low pressure (50 lb/square inch N_2) using the pressure membrane apparatus with filter paper. The results were published in the form of histograms as shown in Figs. 6.4–6.7. The concentration of salts in the soil solutions of these noncalcareous, nonsaline soils was much lower than that found in solutions extracted from soils from the western United States. The Ca concentration was generally between 0.5 and 2×10^{-3} *N*, whereas the K concentration was 10- to 20-fold lower at a level of 0.5 to 2×10^{-4} *N*. Phosphorus was a hundredfold lower than K, generally falling between 0.7 and 4×10^{-6} *M*.

v. *Suction* as referred to here is probably the most common means of removing water from soil in the laboratory. Since the water is replaced by air, this is really a displacement-by-gas method which is applicable at pressures below 1 atmosphere. The common filtration on a Buchner funnel is one of the best examples of this. The suction method consists essentially of connecting the soil sample with a column of water to which suction is applied (Briggs and McCall, 1938; Kapp, 1938; Krügel *et al.*, 1935; Parker, 1925). The Buchner funnel with filter paper has only limited use because air can get into the pores of the filter paper which breaks the water connection. Ceramic cells or other finely porous material, however, can successfully be used up to a pressure of almost 1 atmosphere. The extract thus obtained represents the concentration of the soil solution at very high moisture contents only. Nevertheless, it is the soil solution as it might be right after a rain and for certain types of investigations is quite adequate and useful as a description of the concentration of the ions in the soil solution.

vi. *The centrifuge* is not commonly used for obtaining the soil solution although ever since Briggs and McLane (1907), followed by Olmstead (1937) and Russell and Richards (1938) introduced centrifugation of soil samples for determining moisture equivalent, soil moisture has been extracted in this way. Davies and Davies (1963) have recently described a method they considered suitable for extracting soil solution using the centrifuge which essentially made use of a glass wool pad to hold back the soil but transmit the water. There seems no *a priori* reason why this method should not be suitable and it is a method that can be performed with the normal apparatus present in most laboratories. It is potentially applicable over a wide moisture range, limited only by the capabilities of developing enough centrifugal force while keeping mois-

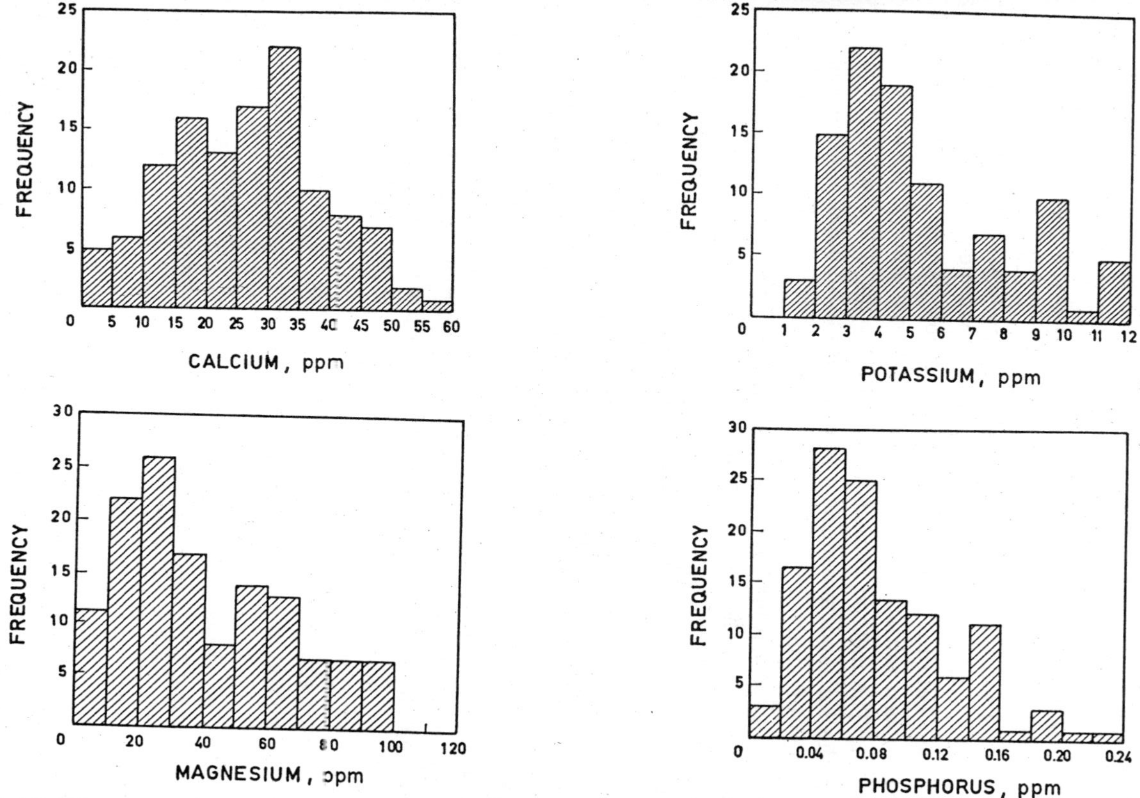

FIGS. 6.4–6.7. Histograms showing frequency of occurrence of Ca, Mg, K, and P contents of saturation extracts from 135 north central U.S. soils. From Barber *et al.* (1963). Reprinted with permission of copyright owner, *Am. Chem. Soc.* [Fig. 6.4 (top left), Ca; Fig. 6.5 (bottom left), Mg; Fig. 6.6 (top right), K; Fig. 6.7 (bottom right), P.]

ture films continuous. It does have the disadvantage of not permitting the ready discarding of the first portion of the soil solution removed. This is often a precautionary measure that is not necessarily a requirement. Evaporation during prolonged centrifugation may also cause some distortion.

vii. *Equilibrium extract* refers simply to placing a water absorbent in contact with the soil, allowing equilibrium to occur, then separating the absorbent from the soil, and analyzing the solution that moved into the absorbent. Actually, little work of this type is reported in the literature, although Gardner *et al.*, (1937) reported on its use, and this type of work has been initiated in at least one other laboratory (IAEA Lab. Rept., 1963). The method consists simply of placing filter paper in contact with the soil and measuring the equilibrium solution. Two serious problems, one of analysis of such small quantities of solution and the other of obtaining a properly charged absorbent, can, in general, be met. One way of meeting the sensitivity problem is to use activation analysis (IAEA, 1963), and one way to meet the charge problem is to coat the paper with suitable material, such as by treatment with $AlCl_3$ when analysing for cations (Bolt, Pers. comm.).

b. *High Moisture Content.* The measurement of nutrient ion concentration at high moisture content has taken essentially two forms: the measurement on leachates from a soil sample and the measurement on equilibrium solutions at various soil solution ratios. Either of these forms may involve salt solutions.

Soil leachates obtained from lysimeters represent leachates at the same moisture content as those obtained by displacement (i.e., at zero moisture tension). They might have been included under water displacement except that they are leachings from heterogeneous soil *in situ* rather than a homogeneous soil in the laboratory. They do not, therefore, represent the soil solution at the surface but rather the soil solution at the point of collection. They are, of course, relatively accurate representations of the losses of nutrients, but not necessarily of the concentration in the rooting zone. The technique is still widely utilized (Allison, *et al.*, 1959; Shaw and Robinson, 1960; Pratt, *et al.*, 1960; Owens, 1960; Cole *et al.*, 1961; Shilova and Korovkina, 1961; Shilova and Korovkina, 1962). It should be realized that even if the soil were homogeneous, a chromatogram would be formed with the concomitant separation of ions that takes place in a chromatographic column and also, owing to soil heterogeneity, the solution coming out of the bottom of the column may bear little resemblance to the soil solution around the roots.

Measurement of *equilibrium solutions* at various soil solution ratios is not normally a technique for determining the concentration of ions in the soil solution. It is often considered, however, that some ions do not vary much in concentration in solution at various soil solution ratios and other ions may vary so regularly that extrapolations can be attempted. Many comparisons have, therefore, been made (Anderson and Keyes, 1942; Eaton, 1939; Hibbard, 1923, Hoagland, *et al.*, 1920; Magistad *et al.*, 1945; Parker, 1921). One of the most revealing is that of Reitemeier (1946) who studied the effect of water dilution on the concentration of ions in a soil-water mixture in which he also determined the concentration in the 15-atmosphere extract. These analyses were made on six high pH soils. The results are reproduced in Figs. 6.8–6.13. Interestingly enough, his results indicate that to some extent one can generalize about many of the ions measured, particularly nitrate and chloride.

All the chloride and nitrate ions in all the soils appear to be in solution. Starting with saturation extract and getting more dilute, the total chloride and nitrate removed appears fairly constant, although an initial reduction was always evident upon higher dilution. Of the six soils, all but soil 65 had solid phase sulfate or carbonate salts present. In these five soils, the divalent cations either remained about the same or increased. The monovalent cations generally appeared to increase with increasing dilution. In soil 65, however, both Mg and Ca showed a marked decrease, whereas Na and the other monovalent cations appeared to increase. This apparent increase in monovalent ions and decrease in divalent is readily explained by the various theories presented in Chapter 2, Section 3.2, showing that ion exchange theory predicts that dilution shifts adsorption in the direction of a greater adsorption of the divalent ion. Even the initial reduction in Cl^- and NO_3^- is readily explained by "negative adsorption" which can be described as follows. Owing to the negative charge of the clay particles, the anions are actually repulsed and in close proximity to the surface of the particle, the concentration of anions is lower than in the outside solution. Thus, at low moisture contents, the total anion calculated from sampling a fraction of the soil solution may be as shown in Fig. 6.14. The lower portion of the curve in Fig. 6.14 reflects filtering effects of the pressure membrane technique (see p. 156). Reitemeier (1946) must have sampled from the region of the shaded area at the low moisture-content samplings. Of course, upon higher dilution the amount of negatively adsorbed ions becomes a neglible part of the total chloride or nitrate present, and the total ion present, as calculated from these samplings, remains constant.

Other observations include those of Moss (1963) who found that the

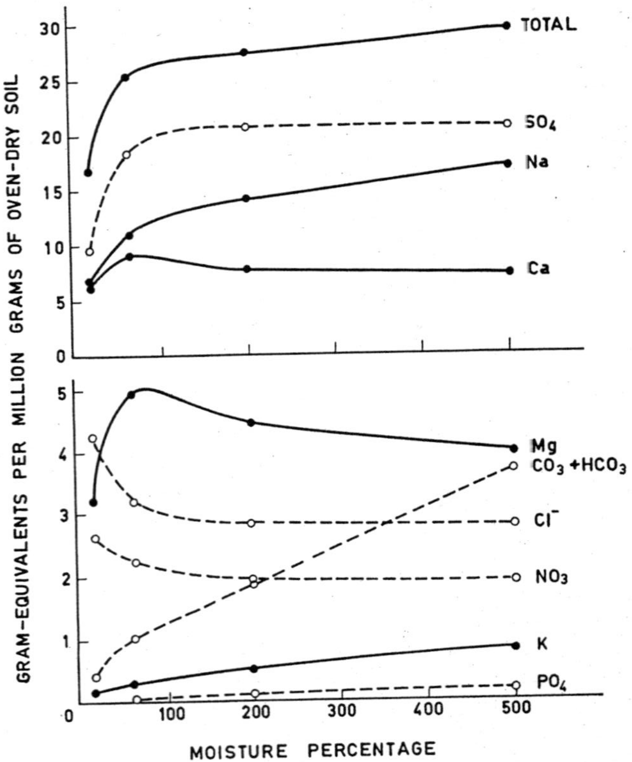

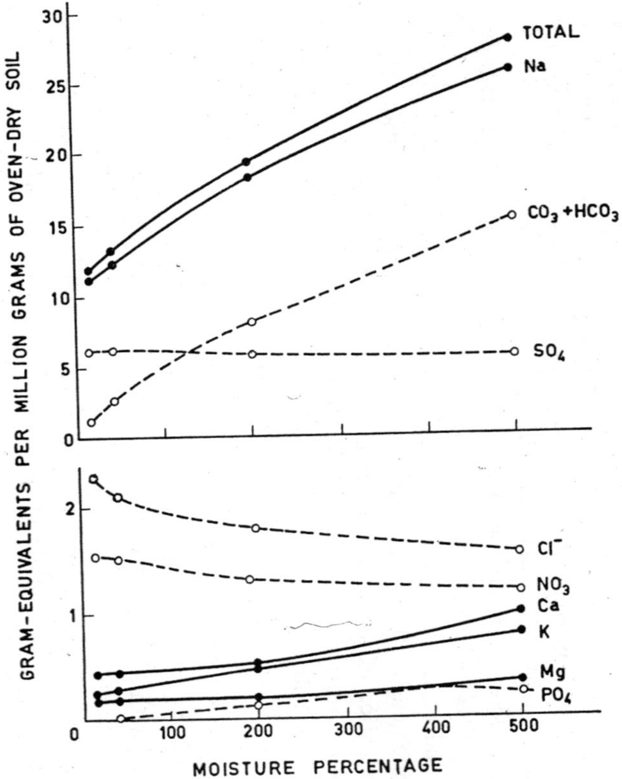

FIG. 6.8 (left). Effect of moisture content on dissolved ions of Imperial clay loam (soil 56). From Reitemeier (1946).

FIG. 6.9 (right). Effect of moisture content on dissolved ions of Indio very fine sandy loam (soil 58). From Reitemeier (1946).

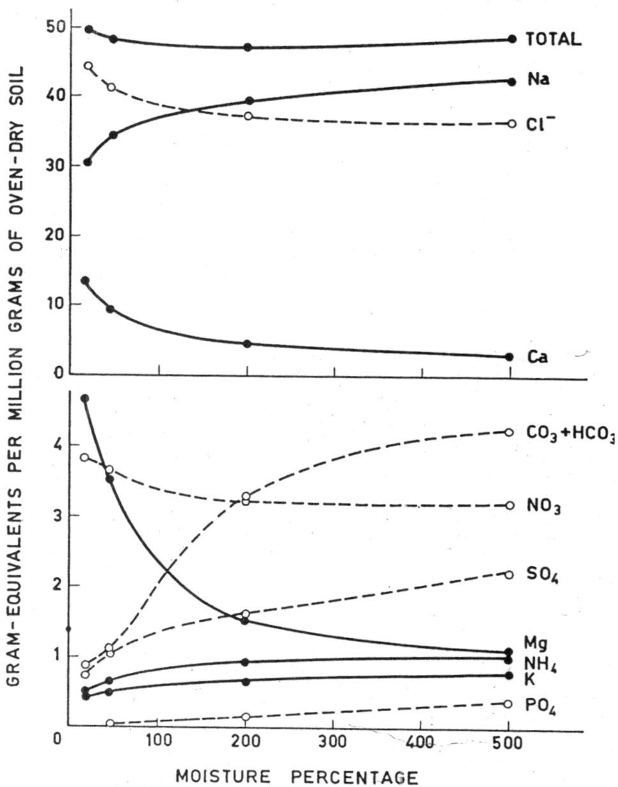

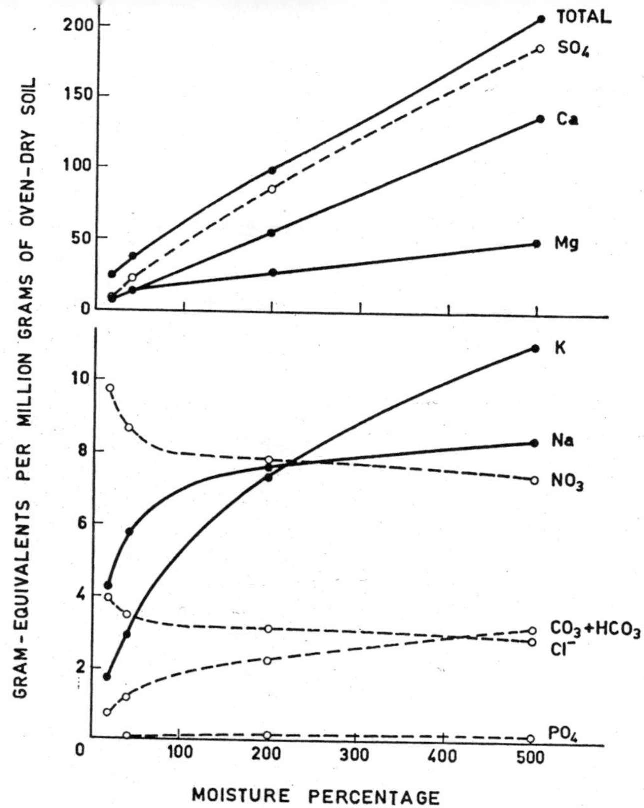

FIG. 6.10 (left). Effect of moisture content on dissolved ions of Palouse loam (soil 65). From Reitemeier (1946).
FIG. 6.11 (right). Effect of moisture content on dissolved ions of Regan loam (soil 85). From Reitemeier (1946).

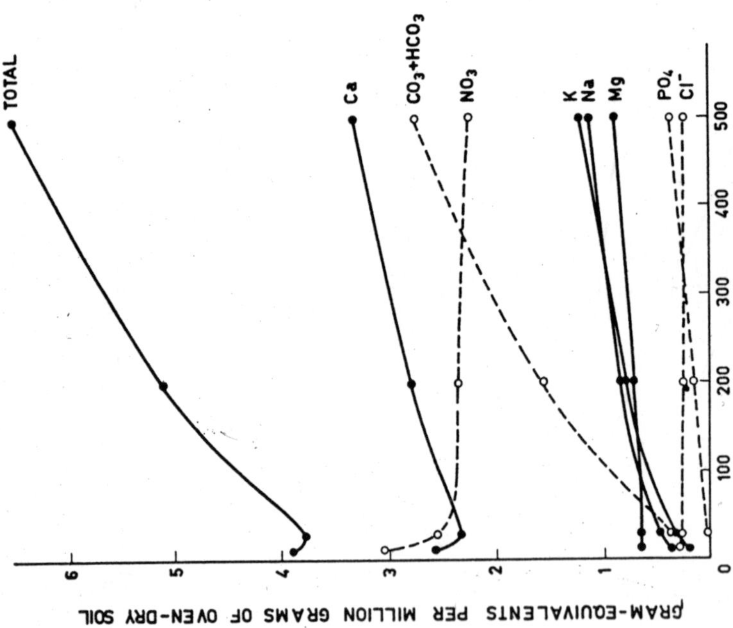

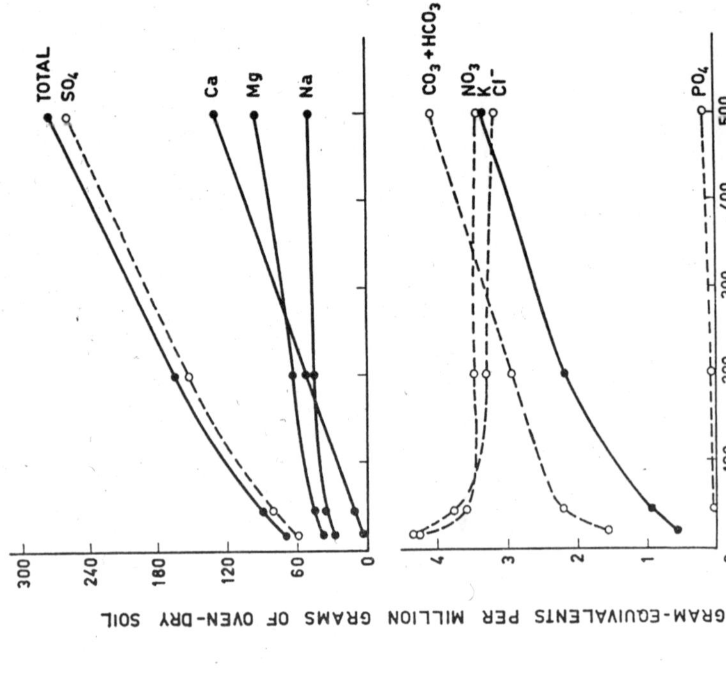

FIG. 6.12 (left). Effect of moisture content on dissolved ions of Fort Collins loam (soil 86). From Reitemeier (1946).
FIG. 6.13 (right). Effect of moisture content on dissolved ions of Hesperia sandy loam (soil 183). From Reitemeier (1946).

total sum of Ca and Mg in solution could be predicted over the field range of moisture for three soils and over a much wider range of dilution for one of the soils by use of the appropriate constants, and those of Shapiro and Fried (1959) who showed that phosphate concentration in solution at widely differing soil–water ratios could be predicted if the appropriate soil constants were known. Wild (1959) was unsuccessful in characterizing the phosphate concentration at various dilutions in terms of the phosphate potential when 0.01 *N* $CaCl_2$ was used as an extractant.

The reliability of the methods of extracting soil solution may be summarized by the statement of Gedroiz in 1906 as quoted by Shilova and Korovkina (1952): "Any attempt at isolating unchanged solution from

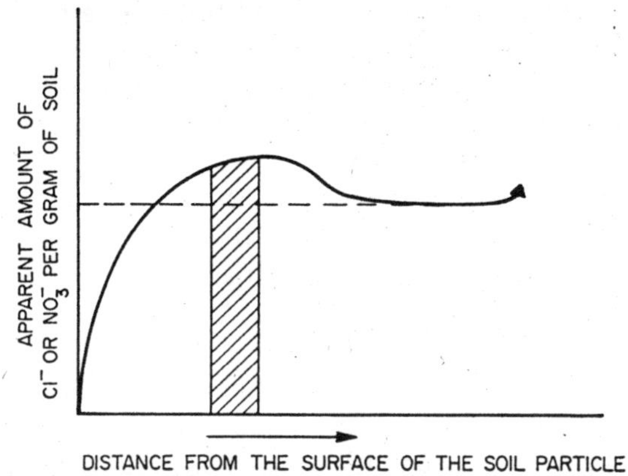

FIG. 6.14. Diagrammatic sketch of the apparent amount of anion per gram of soil as a result of sampling a hypothetical fraction of the soil solution with the pressure membrane apparatus.

soil at a comparatively low moisture content will be futile because it is virtually impossible to do so without the expenditure of considerable effort. Liquid is maintained in soil by molecular forces of considerable magnitude. The very act of isolating a soil solution alters its concentration."

Nevertheless, two methods are promising, namely, gas displacement by use of the pressure membrane, because it utilizes such a thin layer of soil, and equilibration with a thin layer of filter paper, since this is a capillary equilibration with a thin layer. Of course, if the relationship of ions in solution to that in the solid phase is understood well enough, than any extraction procedure can be utilized and the necessary information calculated from it.

The reliability of any of the experimental methods for predicting soil solution P concentrations is, unfortunately, not usually subject to experimental test at normal moisture concentrations. Two difficulties, yet to be overcome successfully are the valid extraction of soil solution without changing the phosphate concentration and the analytical determination of the very low concentrations of phosphate normally present in the soil solution (often in the presence of interfering ions). Both of these problems, however, are amenable to attack, the soil solution extraction by some equilibrium technique using coated filter paper and the sensitivity by the possible use of such new sensitive detection methods as activation analysis.

A fair approximation of the composition of the soil solution can be made by extracting the soil with small amounts of water (Ulrich, 1961). The amount of water should be kept as low as possible because the dilution of normal soil solutions will tend to increase the ratio of monovalent to divalent cations in solution, a manifestation of the valency effect in ion exchange equilibria in soils. Because of a high dilution factor, most of the earlier attempts to obtain the natural soil solution by means of water extractions were theretofore of limited significance (Nemec, 1926; Gedroiz, 1926; Spurway, 1929; Rohde, 1933). The same can bc said for the attempts to extract the soil solution with CO_2-saturated water (Mitscherlich, 1907; Pfeiffer, 1913; Mitscherlich, 1909; Hahn, 1927; Gehring, 1931; Christensen, 1930; Behrens, 1935; Webster, 1959).

Because Ca and, to a lesser extent, Mg are dominant in many natural soil solutions extracting agents containing $Ca(HCO_3)$ or $MgCO_3$ (Behrens, 1935; Aslyng, 1954; Webster, 1959; Marel, 1947; Chaminade, 1940) are likely to introduce smaller changes in the cation ratio in the soil solution than pure water. Schofield (1955) adopted 0.01 *M* calcium chloride solution to obtain an estimate of the chemical potential of monocalcium phosphate in the soil solution. This was essentially a measure of phosphate concentration at a given Ca ion concentration. Ulrich (1961) who also expressed the phosphate concentration in the form of monocalcium phosphate potential, indicated that at about a potential of $pCa + pH_2PO_4 = 7$, the supply of P to the plant becomes critical as indicated by water culture, sand culture, and greenhouse experiments. At higher potentials, definite responses to phosphate were recorded. Below 7 no relationship between phosphate potential and yield was observed. Since the phosphate concentration at a given Ca ion concentration determines the phosphate potential, the conclusion may not be drawn that the uptake of phosphate is governed by the potential of monocalcium phosphate (Wild, 1964).

At about 0.1 ppm P in the soil solution, the supply of P may not be high enough to maintain maximum yields. In practice, it is not possible on the basis of chemical potential alone to forecast whether a particular level of nutrient in the soil solution can be maintained by the solid phase during plant growth.

Similar to phosphate potentials, the lime potential, which is pH-$\frac{1}{2}$pCa (Aslyng, 1954) and the K-Ca potential pK-$\frac{1}{2}$Ca (Woodruff, 1955), have been utilized for the description of the chemical composition of the soil solution in equilibrium with the solid phase.

All these expressions are affected by the so-called valency effect which is predictable from ion exchange constants. These expressions represent soil characteristics, but are unlikely to be of much value in relation to the actual supply of nutrients to the crop. Evidence obtained by Broeshart (1962) and Lagerwerff and Peech (1961) suggests that the uptake of cations by plants from diluted soil solutions in which the reduced ratio of monovalent and divalent cations is kept constant is governed by the concentration of cations and not by their reduced ratio. A K–Ca potential in the soil solution higher than 2.6 has been associated with K-deficient plants (Woodruff, 1955; Ulrich, 1961). These findings should be interpreted with caution. The drop in K–Ca potential accompanies a drop in K concentration which may well be the cause of K-deficient plants. The effect of K on Ca and vice versa, which is often referred to as ion antagonism, need not be directly related to the uptake process, but may involve other plant physiological mechanisms.

1.1.2. *Capacity Factors*

The total quantity of an element that is able to transfer to the soil solution when the latter is depleted by crop removal or leaching may be defined in various ways. If the time factor is not taken into account, essentially all solid phase elements may eventually move into the soil solution in the course of weathering processes. It is obvious that such a definition would be of little importance in terms of plant production. A more realistic definition of the capacity factor is the reserves that can move into the soil solution during plant growth. Appreciable quantities of elements may be brought into the soil solution over longer periods by weathering processes, exchange, and slow diffusion.

The sources of the nutrients in the solid phase that may act as capacity factors are widely different. In the early history of soil fertility research, extracting agents were sought that purported to extract the soil in the same way as the plant root (Bergman, 1958). The nature of the uptake process, the many chemical forms of the nutrients in the soil, and the different kinds of nutrients in the soil system, are now better understood. Consequently, many different soil extractants are in use for the extraction of specific chemical entities.

In contrast to the intensity factor, which is known to be located in one source (i.e., the soil solution) many different sources of plant nutrients may contribute to the supply of nutrients to the soil solution and each at a different rate. Since there are so many different soil types of widely different composition and physicochemical characteristics, a universal method for the analysis of the capacity factor for a given nutrient element is difficult and in specific cases may be impossible. As our knowledge of the soil-plant system improves, new methods of measurement are still being suggested for particular plant nutrients.

a. *Origin in the Solid Phase.* The supply of NO_3 and NH_4 ions to the soil solution is governed by the rate at which organic material is decomposed by microorganisms and essentially is a function of the moisture content and temperature of the soil in addition to the amount and kind of organic matter. There is virtually no relationship between the total N content of the soil and the N supply to plants (Harmsen, 1955; Bergman, 1958).

The estimation of the capacity factor of P is complicated by the presence in one soil of several phosphate sources which may be responsible for the supply of phosphate in the soil. The relative importance of each source depends not only on the source itself but also on the nature of the soil, including the pH and even the oxidation–reduction status under a given set of environmental conditions.

The P in the soil may be divided into four categories with respect to the supply of phosphate to the soil solution: (1) Organic phosphate, which is bound in lecithin, nucleic acids, or phytins; (2) surface phosphate, which is partly held against positively charged surfaces and is ill-defined insofar as molecular species; (3) sparingly soluble phosphates such as the phosphates of Fe, Al, and Ca; and (4) occluded phosphate inside particles with oxide coatings.

The fractionation of soil P into several categories has been attempted with reasonable success (Dean, 1938; Chang and Jackson, 1957); but this information is not likely to contribute much to an estimate of the plant nutrient supply in a given season. Since the major immediate renewal source for the soil solution is surface phosphate, the phosphate extractant should logically remove a part or all of this. Surface phosphate is here defined as the phosphate on the surface, irrespective of the nature of the bond with which it is held. Exchangeable P is an attempt to measure this fraction and can only be defined in terms of the method used. Adsorbed phosphate, on the other hand, has the connotation of phosphate held by ionic bonds only.

The supply of cations such as K, Ca, Mg, Na, etc., in the so[illegible] on is a function of the amounts adsorbed at the negatively charge[illegible] of the clay minerals and organic matter. The organic matter may be the all-important source in sandy soils.

In addition to the exchangeable fraction, salts may contribute to the supply of cations, e.g., Ca may be present as gypsum or calcic or dolomitic limestone and in the form of the many calcium phosphate compounds. Apart from exchangeable Mg, this element also occurs as sulfate, carbonate, or dolomite. When the conditions are favorable for weathering as in the humid tropics, primary minerals such as muscovite and biotite may release appreciable quantities of Mg (Marel, 1947).

Iron and Mn are adsorbed to the negative adsorption complex and are present as hydroxides and oxides as well as in chelated forms. In general, the supply of those elements to the soil solution is related to the pH and oxidation-reduction status of the soil.

The foregoing discussion serves to illustrate the problems which confront the soil chemist as to the source of nutrients making up the capacity factor when suitable extraction techniques have to be selected for the evaluation of the individual capacity factors.

b. *Types of Extractants.* Strong concentrated acids will generally dissolve much more of any kind of nutrient than its capacity for supplying nutrient during plant growth. With the possible exception of the utility of this information for soil classification purposes, the analysis of concentrated acid extracts are rarely used for soil fertility diagnosis (Behrens, 1935; Sprengel, 1837; Hasenbäumer, 1931; Opitz, 1907).

Less concentrated acids have been and are still used for the extraction of cations and phosphate. These acids will generally remove the bulk of adsorbed cations in most soils together with a part of the calcium and magnesium phosphate (Teakle, 1928; Doughty, 1930). Iron and Al are not appreciably removed by these dilute acids. For routine soil fertility testing there is an advantage in using such acid extractants, because the extraction procedure, which usually consists in simply shaking aliquots of soil with excess extractant, generally gives clear filtrates. Examples of such acid extractants are 0.2 N HNO_3 (Fraps, 1902, 1909; Moore, 1902; Frear and Erb, 1918), 0.1 N HCl (Pauw, 1956), 0.2 N HCl (Bray, 1929; Kirssanoff, 1934, 1939) 0,002 N H_2SO_4 plus 0.3% ammonium sulfate (Truog 1930, 1937, 1938), sodium perchlorate in 0.1 N $HClO_4$ (Long and Seatz, 1953), and other acid solutions buffered at a particular pH (Sigmond, 1931; Bondorff and Steenbjerg, 1932; Demolon, 1933) or mixtures of acids and complexing agents such as 0.03 N H_2SO_4 and NH_4F (Miller and Axley, 1956).

For the extraction of exchangeble cations, neutral salt solutions are convenient. Ammonium, sodium, or barium salts are generally used to exchange with the soil cations. The bulk of the adsorbed cations wil' appear in the extract if the soil is either shaken with or leached with an excess of extracting solution. Buffered solutions of NH_4Cl (Kelly *et al.*, 1931; Peech, 1945; Schollenberger, 1945; Joffe and Levine, 1946), NH_4OAc (Martin, 1929, Hoagland and Martin, 1933) NH_4NO_3 (Gedroiz, 1933; Nehring, 1943), and NH_4F in HCl (Bray and Kurtz, 1945) also have been used.

c. *Extraction of Individual Nutrients.*

i. *Nitrogen.* Methods for the analysis of mineral and total N have been developed by several investigators (Koenig, 1924, 1929; Goy, 1928; Horne, 1955; Giesecke, 1931; Hermann, 1949; Prince, 1945) but these methods generally yield highly variable results for successive soil samplings owing to the rapid fluctuation of mineral N in the soil. A much more realistic approach for obtaining data that are related to the N supply in the soil solution is the determination of the rate of mineralization of organic N into mineral N (Harmsen, 1955) as discussed in Chapter 2, Section 1.2.

ii. *Potassium.* The bulk of the K supply to the soil solution is furnished by the release from exchangeable K in contrast to the solubilization process of K-bearing minerals. Some of the K in soil is slowly released because of its position between lattices in clay minerals. The process appears to be at least partially reversible. This slowly released K is often referred to as "fixed" K. The fixation usually decreases at low pH and increases with liming. The depletion of K from the soil solution by plants can result in bringing more of the "fixed" K into the exchangeable form or into the soil solution; thus the fixed K may contribute to the supply of K to the plant (Chaminade, 1940; Standford, 1946; Walsh, 1945; Reitemeier, 1951; Wiklander, 1954; Hauser, 1941; Wiklander, 1950; Karlson, 1952; Gilligan, 1938). The weathering of K minerals may also contribute to the K supply in the exchange positions or the soil solution and decreases in the order biotite > muscovite > orthoclase > microcline (Blanck, 1912; Blanck, 1913; Schachtschabel, 1937).

The amounts of readily exchangeable K extracted by widely different agents usually agree within rather narrow limits. This is illustrated in Table 6.2.

In some clay soils containing expanding lattice-type clays Na may open the lattices, resulting in a release of more K than when ammonium salts are used. Shrinkage or closing of the lattices occurs with ammonium and K salts. Although boiling 1 *N* HNO_3 does not remove all the K in the soil, exchangeable K is only a small fraction of the nitric acid-extractable K (Reitemeier, 1951).

In general, the content of readily exchangeable nutrient cation, K, in particular, shows a positive correlation with the yield of plant material. The methods of extraction with dilute acids or salt solutions yield data of similar magnitude, reflecting a combination of the K intensity and capacity factor.

iii. *Phosphorus.* One of the first extractants used in an attempt to measure soil phosphate level was 0.5–2% citric acid (Dyer, 1894; Gerlach, 1896; Dean and Dean, 1929; Hoffman, 1930; Lesch, 1931; Demolon, 1932; Behrens, 1935; Pauw, 1956). The action of citric acid apparently results in phosphate release owing to a complexing of Fe, while the acid action of the citric acid results in a dissolution of calcium phosphates. Thus the citrate acid method extracts relatively

TABLE 6.2
EFFECT OF METHOD OF EXTRACTION ON THE AMOUNT OF POTASSIUM REMOVED[a]

Soil type:	Clay	Loam	Silt	Sandy
Number of samples:	21	20	12	22
Method	Amount of K removed[b]			
1 N NH_4OAc	13	9.6	6.4	8.3
Monochloracetic acid + calcium monochloracetate (Egner)	11	8.2	6.1	8.6
1 N $NaNO_3$ (Bray)	15	9.8	7.7	9.9
2.4% acetic acid (Williams and Stewart)	11	9.5	6.1	8.2
1 N HNO_3 (Reitemeier)	132	51	55	64

[a] Derived from Semb and Uhlen (1955). [b] Data given in mg K/100 g of soil.

large quantities of phosphate as compared to currently used phosphate extractions and probably dissolves more phosphorus than is involved in the current nutrient supply to the plant.

Other weak acids or their sodium or ammonium salts have been and are used (Behrens, 1935; Solomon and Smith, 1956; Schachtschabel, 1941; Herrmann, 1943; Joret and Hebert, 1955). Acetic acid and oxalic acid are weaker than citric acid and dissolve less Ca-bound phosphate. Their complexing power for Fe and Al is also less than citric acid and, consequently, the quantities of P extracted are less. The Morgan extractant, which is sodium acetate-acetic acid buffered at pH 4.8, has also been commonly used (Morgan, 1935; Morgan, 1941; Knickmann, 1955).

The complexing action of ethylenediamenetetraacetic acid (EDTA) can be used to extract phosphate (Viro, 1955; Wallace *et al.*, 1955), but not enough information has been obtained up to the present to decide whether a reasonable estimate of P supply can be obtained with this reagent. Ammonium lactate-acetic acid methods, originally developed by Egner and Riehm, have become standard for determining readily soluble soil P, particularly in western Europe (Semb and Uhlen, 1955; Egner, 1932; Riehm, 1942, 1948; Lemmermann, 1946; Finck and Schlichting, 1955; Egner *et al.*, 1960; Scheffer *et al.*, 1952; Kertscher, 1955; Riehm 1953, 1956). This extractant has both acid and complexing properties and is buffered to the extent that during extraction the pH is very little changed and the solubility of phosphate during the extraction procedure is hardly altered. This method can also be used in the determination of exchangeable K, i.e., in exchange for ammonium. This duality makes it attractive for routine soil testing.

The combination of complexing and acid extraction has also been applied to phosphorus by Bray and Kurtz (1945) by means of ammonium fluoride-hydrochloric acid mixtures at various pH levels. Depending on the strength of the acid more or less Ca-bound phosphate is removed. The fluoride both exchanges with phosphate and complexes with Fe, thus bringing phosphate into solution. The two proposed extracting solutions have the same strength of ammonium fluoride, namely 0.03 N, but so-called P_1 is in 0.025 N HCl at pH 2.5–3.5, while the so-called P_2 is 0.1 N HCl at a pH of 1.5–1.6. P_1 removes much less Ca-bound P than P_2. Another suggested extracting solution is P_4 which is 0.5 N NH_4F in 0.1 N HCl (Bray and Kurtz, 1945).

The mechanism of action of alkaline-extracting solution such as 1% K_2CO_3, $(NH_4)_2CO_3$, or Na_2CO_3 (Das, 1933; Hockensmith, 1933; Whitney and Gardner, 1936; Kühn, 1935) is a removal of phosphate by means of anion exchange and hydrolysis owing to the high pH of the solutions. The method of Olsen (1954) using 0.5 M $NaHCO_3$ solution buffered at a pH of 8.5, is a relatively recent method that is receiving more attention for the estimation of surface-bound phosphate, especially for calcareous soils. At such a high pH basic Ca precipitates do not dissolve. Extraction with CO_2-saturated water apparently removes part of the exchangeable P but the actual source of the P has not been identified (Rennie, 1959; Arnaud, 1950).

An entirely different approach for extracting readily exchangeable P was worked out by Amer and co-workers (1955). They used anion exchange resins in the chloride form as the extractant. After shaking a soil and resin suspension for approximately 6 hours, the initially rapid dissolution of soil P is virtually completed. Although this method does not give much information concerning the source and nature of the phosphate bonds, good agreement with other conventional extraction methods has been obtained. The method can be modified to obtain a measure of the rate of P release.

A comparison of various phosphate-extracting solutions was made in Taiwan (Chang and Juo, 1962) to determine the source of the P in 26 soils. The soils were classified according to their content of aluminum, iron, and calcium phosphate as determined by the phosphate fractionation procedure of Chang and Jackson (1957). Seven extracting solutions were compared with the phosphate chemistry of the soils. These were 0.5 M $NaHCO_3$ (Olsen), sodium acetate-acetic acid (Peech), 0.03 N NH_4F + 0.025 N HCl (Bray No. 1), 0.03 N NH.F + 0.1 N HCl (Bray No. 2), 0.5 N NH_4F + 0.1 N HCl (Bray No. 4), 0.05 N HCl + 0.025 N H_2SO_4 (North Carolina), 0.002 N H_2SO_4 (Truog). The results of the comparison are shown in Table 6.3.

TABLE 6.3
CORRELATION BETWEEN THE DETERMINED VALUE OF AVAILABLE SOIL PHOSPHORUS AND THE TOTAL AMOUNT OF EACH FORM OF PHOSPHATE

Form of phosphate	Correlation coefficients (r)[a]						
	Olsen	Peech	Bray No. 1	Bray No. 2	Bray No. 4	North Carolina	Truog
			11 Soils dominating in iron phosphate				
Calcium	+0.843[c]	+0.811[c]	+0.819[c]	+0.808[c]	+0.819[c]	+0.825[c]	+0.831[c]
Aluminum	+0.949[c]	+0.773[c]	+0.893[c]	+0.903[c]	+0.839[c]	+0.794[c]	+0.715[c]
Iron	+0.872[c]	+0.726[b]	+0.785[c]	+0.739[c]	+0.910[c]	+0.699[b]	+0.807[c]
			8 Soils dominating in iron and calcium phosphates				
Calcium	+0.384	+0.319	+0.258	+0.614	+0.486	+0.840[c]	+0.825[b]
Aluminum	+0.570	+0.347	+0.704[a]	+0.641[a]	+0.496	+0.237	−0.154
Iron	+0.636[a]	+0.657[a]	+0.344	+0.588	+0.717[b]	+0.533	+0.577
			7 Soils dominating in calcium phosphate				
Calcium	+0.224	+0.137	+0.431	+0.905[c]	+0.709[a]	+0.869[b]	+0.827[b]
Aluminum	+0.948	+0.580	+0.887[c]	+0.644	+0.762[b]	+0.560	+0.651
Iron	+0.241	+0.152	+0.342	−0.098	+0.028	−0.033	+0.047

[a] Significant at 10% level. [b] Significant at 5% level. [c] Significant at 1% level.

In spite of the limitation of drawing conclusions from correlation data alone, the Table 6.3 tends to illustrate that the various techniques extract different portions of the soil P in the various soils. On soils in which iron phosphate predominates all extraction procedures were reasonably well correlated. On soils containing Ca-bound phosphates, the acid methods give similar results. Apparently the bicarbonate and ammonium fluoride-extracting agents determine Al-bound exchangeable phosphate but no Ca-bound phosphate. In the intermediate groups in which both Ca and Fe phosphates predominate, the bicarbonate, sodium acetate, and high fluoride concentration extractions are mutually correlated. Similar results were obtained by Blanchard and Caldwell (1964).

All techniques for the determination of P supply factors in the soil dealt with a measure of readily soluble exchangeable or surface P.

The organic P fraction may be a P source and therefore might be partly included in the capacity factor. The analysis of organic P can be done in various ways, such as alkaline extraction (Dean, 1938; Ghani, 1943, or indirectly as the difference between acid-soluble P before and after ignition of the soil sample (Mehta, 1954; Kaila, 1961, 1963; Damsgaard-Sorensen, 1946). It should be stressed, however, that if cultural practices are designed to maintain organic matter content, it should not be considered as source of P.

Another category which is unlikely to contribute to P supply is the occluded P. Only under reduced conditions such as in submerged rice soils will some of the Fe oxide coatings dissolve, mobilizing some phosphate (Bartholomew, 1931; Bauwin, 1957).

iv. *Sulfur.* With regard to S supply, except under deficiency conditions, a secondary source during the growing season is the organic matter which may have tied up some of the S renewed by the rainfall during the year. Again this presupposes, however, a net loss of organic matter. In arable land there will be a steady release of S, depending on climatic conditions (Whitehead, 1964).

Below pH 6.5 some exchangeable SO_4 is present (Williams and Steinberg, 1962), but the nature of the exchange sites is still not known. Hydrated Al and Fe but also kaolinites apparently have an ability to adsorbe sulfate (Kamprath and Nelson, 1957; Berg and Thomas, 1959; Chao *et al.*, 1962, 1962a; and the mechanism has been suggested (Jackson, 1963; Schell and Jordan, 1959; Chao *et al.*, 1962).

With regard to the determination of the potential S supply in the soil, still little is known and no generally suitable extraction procedures have yet been developed. Exchangeable SO_4 can be extracted with a KH_2PO_4 solution (Ensminger, 1954; or with $CaCO_3$ (Williams and Steinberg, 1962). An entirely different approach is extraction with 0.001 N HCl (Little, 1953, 1957, and 1958). Kilmer and Nearpass (1960), on the other hand, used 0.5 M $NaHCO_3$, at pH 8.5 as the extractant of choice. Interestingly enough, this is the same extractant receiving more attention for extraction of P. Their results showed good correlations with sulfate "A" values obtained in the greenhouse. None of these methods is likely to give results that are related to the over-all SO_4 supply of soils. The supply of SO_4 from the atmosphere may be considerable (See Chapter 5, Section 2).

v. *Micronutrients.* The entities involved in the capacity factors of micronutrients such as Fe, Mn, Zn, Cu, and B are little known, although useable soil-testing methods have been developed for some.

The rate of Fe supply is related to pH and redox potential. The CO_2 partial pressure and moisture conditions may also have a marked influence on the mobility of soil Fe. Under oxidized conditions Fe^{3+} predominates, and the content in the soil solution is extremely low as a result of bonding to the negatively charged clay minerals, the complexing with organic matter, and the low solubilities in the presence of low concentrations of hydroxide ion.

Some organic Fe^{3+} compounds are soluble. Extraction of Fe^{3+} (Wiklander *et al.*, 1949) is based on shaking with excess 0.5 *N* H_2SO_4. Part of the Fe in oxidized soils is adsorbed as $Fe(OH)_2^+$ or $FeOH^{2+}$, and under reduced conditions (e.g., in submerged soils) the concentration of Fe^{2+} in the soil solution increases rapidly. It is difficult therefore to characterize the Fe capacity factor in the soil, because the capacity changes continuously, depending on external conditions, and any method of extraction gives results that are difficult to interpret in terms of Fe supply to plants (Grunes and Jenny, 1960).

Manganese chemistry in soil is similar to Fe to the extent that Mn^{2+} and Mn^{4+} concentrations are governed by pH and redox potential. Under acid conditions and low redox potential, Mn^{2+} will predominate. Various extraction techniques have been used, all of which bear little relation to the source of supply of Mn to the soil solution. Among the extractions for the determination of exchangeable Mn are citric acid (Masoni, 1916), $Ca(NO_3)_2$ (Heintze, 1938), and $Mg(NO_3)_2$ (Steenbjerg, 1933). A more realistic approach is the determination of Mn by reduction with hydroquinone (Jones *et al.*, 1951; Schachtschabel, 1954; Finck, 1954; or peroxidisulfate (Nydahl, 1949), although this too has its limitations as a reflection of the supply of nutrient Mn.

The factors which govern the supply of Cu for plants are incompletely known so that no suitable extractant has yet been found. The supply of Cu seems to depend on the pH, the nature of the organic material, and the climatic conditions. Copper is tightly bound by clay minerals and organic matter, owing to its strong complexing properties. Salt solutions such as NH_4Cl are unable to replace any Cu (Smith-Brun, 1945). In order to remove a substantial amount of Cu, strong complexing agents such as KCN or EDTA have to be used (Smith-Brun, 1945; Kanswar, 1954; Viro, 1955). Also strong acids such as 1 *N* HNO_3 and HCl are able to remove most of the "exchangeable" Cu (Lucas, 1948; Chen & Bray, 1953; Henrikson, 1956). Copper can be adsorbed as Cu^{2+} and also as complex monovalent ions such as $CuOH^+$, $CuCl^+$, or $CuCH_2COO^+$ (Bower and Truog, 1941; Menzel and Jackson, 1950).

Zinc to some extent, behaves like Cu. It is tightly adsorbed to clay minerals and organic matter as Zn^{2+}, $ZnOH^{+}$, and $ZnCl^{+}$. Replacement with neutral salts is very incomplete. At high pH the Zn supply may become limiting in citrus-growing areas and also for corn on some calcareous soils. Extraction of exchangeable Zn can be performed with 0.1 *N* HCl and dithizone with reasonable results (Tucker and Kirk, 1955; Shaw and Dean, 1952; Nearpass, 1956).

The mechanism of B release is not known. Although B is amphoteric, it apparently is adsorbed as a cation. The bulk is contained in minerals, however, such as tourmaline and in silicates such as biotite and muscovite. Boron extraction is usually done with hot water (Berger and Truog, 1940, 1944; Truog, 1945; Baird and Dawson, 1955).

vi. *Summary.* Summarizing the chemical methods available for the determination of soil capacity factors, the extractants currently used are designed to extract the readily exchangeable or soluble nutrient, although the true mechanisms of the supply of elements to the soil solution are not fully understood.

With regard to macronutrients, extraction of total mineral N is meaningless as the supply is mainly governed by the decomposition of organic matter and this, in turn, depends on climatic conditions. This is also true for S, although here other means of supply (the atmosphere) may be important.

The evaluation of the capacity factor for K is usually done by determining the quantity of readily exchangeable K. The results are relatively independent of the nature of the anions that are associated with the exchanging cation. The size of the cation may have a slight effect on the quantity of readily exchangeable K that is removed. This explains why such good agreement is obtained when different extractants are compared. In practice, soil-testing laboratories often choose an appropriate extracting agent for the determination of readily soluble, exchangeable, or surface phosphate and use the same extract for the determination of exchangeable K.

The difficulty in determining the P capacity factor of soils is much greater because so many different phosphate combinations may contribute to the supply of P to plants. The current extraction methods are a compromise between acid solubiltiy for the dissolution of Ca-bound phosphate and complex formation for extraction of Fe- and Al-bound phosphate. Alkaline extractants supposedly exchange surface-bound phosphate from all sources, although the removal of Ca-bound phosphate is very small. These methods generally agree reasonably well with resin extraction methods and isotope exchange methods when soils of similar chemistry are compared.

The extraction of microelements is beset with the difficulty that the nature and mechanism of the supply sources are not yet understood. Iron and Mn extraction methods are of limited value because the supply of these elements is related to environmental factors that change during crop development such as pH, redox potential, and climate. This is particularly true for soils under submerged conditions. Copper and Zn are strongly bound by complexing agents and their removal can only be accomplished by the use of strong complexing agents.

1.1.3. *Rate Measurements*

The rates of release of elements from the soil have been determined by two laboratory chemical techniques: (1) by successive extraction techniques in which the quantity of element removed by repeated extraction may give a measure of the rate of release into the soil solution; and (2) by isotopic techniques in which the rates of equilibration of a tracer in an equilibrium system is studied as a function of time (see Chapter 1, Section 3 for full discussion).

All methods have a common difficulty in that the rates of appearance are studied under conditions that may be quite different from those in natural soils. There is a danger that rate processes are evaluated that would not take place at the same rate in normal soils.

Successive extraction of P was investigated by Fried and Shapiro (1956) who compared continuous leaching with water, both to the successive equilibration of soil and water for 1 hour and to the leaching of soils with water in tall pots in which millet was growing. The three methods gave virtually the same results and indicated that the initial phosphate content of the soil solution was not related to the rate of supply of P obtained in successive fractions. Soils known as "phosphate fixing" released much more P during the extraction than "nonfixing" soils, although the latter initially had a higher P content in the soil solutions. The rate of supply of P for a Bridger, Caribou, and Nibley soil was estimated at 13–15 lb/acre/hour and exceeded the rate of adsorption by plants by a factor of at least 250 (Fried *et al.*, 1957) and probably much greater (Shapiro and Fried, 1959).

Phosphorus-32 equilibration can also be used, and Fried and Shapiro (1960) have shown that such determinations result in rates of release similar to those obtained by continuous leaching. Rate constants can also be calculated in this manner (Scheffer *et al.*, 1960; Wiklander, 1950).

When exchange reactions are involved, the rates of supply of cations are usually fast, as indicated by the rate of Ca exchange as determined by Borland (1950). Several rate constants, however, are involved in K release, depending on the source of the K (Sumner, 1962).

Other chemical techniques for determining nutrient supply actually depend on the rate of release. Even single extractions that are not equilibrium extracts should be interpreted in terms of the rates involved. The results of any leaching or successive extraction are of necessity a rate function. Resin and Soxhlet extraction are two examples of such procedures.

1.2. Biological Methods

Laboratory biological methods use bacteria, fungi, algae, and even higher plants as extracting agents instead of laboratory solutions. Although a definite chemical fraction of the soil nutrient is not extracted, the micro organism is assumed to reflect the same relative level of nutrient as the plant. This assumption is not directly applicable, although correlations may exist. The environment both in the laboratory and in the soil in the laboratory is different from that in the field. The requirements of the organism for the particular nutrient and for that nutrient in relation to other nutrients is also different; it also follows that the rate-limiting step or steps in getting the nutrient ion to the surface of the organism may be entirely different. However, micro organisms have two advantages in certain cases. First, incubation of soil results in microbiological turnover of organic matter and release of N that may correlate with this occurrence in the field. Although this has the same limitations as mentioned above, no chemical extractant has even shown any partially satisfactory correlation. Second, micro-organisms are often sensitive to extremely low concentrations of elements, and for micronutrients, this may temporarily be the only analytical method available.

The two laboratory biological methods that will be discussed are incubation techniques of the soil without inoculation and the microbiological methods involving the use of fungi or algae as nutrient extractants.

1.2.1. *Incubation Techniques*

The determination of the rate of mineralization of N offers a good example of the incubation technique. The procedure is simple in principle. Soil samples are incubated at a suitable temperature for a specific length of time that may vary from 1 week to several months, and the quantities of mineral N produced are measured in water extracts. The final data are expressed in milligrams or parts per million of nitrate per unit time. In practice, this method can yield highly variable results if a number of precautions are not taken and the sampling, incubation, extraction, and analytical techniques are not highly standardized.

The soils should be sampled at comparable periods of the year. It was shown in Chapter 1, Section 2 that the nitrification rate is highly dependent on the amount of organic matter that has accumulated in the soil. This, in turn, depends on the conditions of cultivation, fertilizer application, nature of crop, and climatic factors (Harmsen and Schreven, 1955; Starkey, 1931; Goring and Clark, 1949; Fitts *et al.*, 1953; Richardson, 1938; Drouineau and Lefevre, 1949, 1949a; Duchaufour, 1951). More reproducible results are usually obtained from samples taken in the spring or late winter. At that time, interference from fresh organic residues is smallest and, moreover, in temperate climates, the winter period partially sterilizes the soil, resulting in a subsequent stimulation of N mineralization.

Measurable nitrate is low during the initial nitrification of fresh organic matter with a high C:N ratio because of the incorporation of the N into the microorganisms which are increasing rapidly in population. Often the mineralization-time plot is curvilinear rather than linear if the soil sample contains fresh organic material.

A high C:N ratio of samples can give a completely false picture of the N availability in organic and forest soils. Under natural conditions, it is known that these soils have very unfavorable conditions for the mineralization of organic N, whereas under the conditions of incubation large quantities of N may be released. This is why the incubation method can only be applied to samples from arable land where the organic N is in less stable compounds than in organic and forest soils (Duchaufour, 1951; Jolivet and Elias, 1953; Bel *et al.*, 1951). Two factors compensate in relation to the C:N ratio of samples (Harmsen and Schreven, 1955; Thompson and Black, 1950; Cornfield, 1952). Because of the correlation between C and N contents, soils with a high C content may tend to show a high N mineralization rate. If, however, the ratio of C:N is high, nitrate does not accumulate. This explains why, for example, in the presence of the fresh organic matter in grassland nitrate cannot be detected. Thus, the method is of little use for grassland samples, unless extremely long incubation times are used (Richardson, 1938).

Although relatively constant rates of mineralization are generally obtained, variations in rate may sometimes be found owing to the simultaneous action of microorganisms that assimilate N after it has become available through demineralization (Gerretsen, 1942). Often a decreasing rate of mineralization is found with increasing incubation time. This has been ascribed to the increasing acidity when nitric acid is produced during mineralization (Allison and Sterling, 1949) or to the increasing effect of organic matter synthesis by microorganisms (Harmsen and Lindenbergh, 1949). When the samples are leached with water during the incubation time, straight lines generally are obtained (Jensen, 1950; Stanford and Hanway, 1953; Acharya and Jain, 1954).

From the foregoing, it is clear that the results obtained from incubation experiments are highly empirical and comparisons among samples are only valid if the samples belong to the same soil type and have identical sampling times and pretreatment. There is not much to be gained by adding fertilizers or inoculating the soil with active strains of microorganisms because the characteristics of the soil will be changed. Eventually then only the decomposition of organic matter in a highly artificial medium is studied, with little resemblance to the rate of N mineralization under normal field conditions (Harmsen and Schreven, 1955).

In the extraction procedure of the incubated soil samples, solutions of KCl or NaCl are used if the ammonium ion is of interest. Although the NH_4 that is produced may be adsorbed or even fixed in between the lattices of clay minerals, dilute salt solutions will generally remove all NH_4 which is readily exchangeable (Harmsen and Schreven, 1955; Drouineau and Lefevre, 1949; Barshad, 1948; Allison *et al.*, 1951; Bower, 1951; Nemec and Koppova, 1932; Cornfield, 1952, 1953; Allison *et al.*, 1953). It may be assumed that any NH_4^+ which is fixed will not be able to move rapidly into the soil solution.

The analysis of mineral N in the extracts can either be done by analyzing nitrate, nitrite, and ammonia separately, or by a method that determines the bulk of all N forms. The disadvantage of the latter method is that some dissolved organic compounds containing N, such as amino acids or unknown compounds, are determined as mineral N. This gives rise to exceedingly high mineralization figures for soils that have a low N-supplying capacity under normal conditions. Care should also be taken that any digestion method of the soil extracts transforms all nitrate into ammonia, otherwise the values obtained for mineralization will be too low, particularly when the digestion period is too short.

Another way to overcome the analytical difficulties in the analysis of NH_4^+ and NO_3^- in the soil extract is to incubate the soil under waterlogged conditions. Although this condition differs materially from that of normal field conditions, good results have been obtained (Waring and Bremner, 1964). Under waterlogged conditions only NH_4^+ is produced, and it can be easily determined in the supernatant solution. Further advantages of this technique are that small samples may be used and aeration is not necessary. The incubation time is relatively short, 1 to 2 weeks, because the rate of mineralization under reduced conditions is rapid.

The rate of N mineralization can also be estimated indirectly from the amount of CO_2 liberated during incubation. Usually carbohydrates, such as cellulose, are added to the soil samples as a source of energy for microbiological activity. Although the method has the disadvantage that entirely artificial conditions are created which have little in common with soil under natural conditions, the advantage of the method is that no complications arise due to mineral N accumulation. The extraction procedure and analysis of mineralized N are replaced by a simple determination of CO_2 production as a function of the time (Holben, 1932; White *et al.*, 1949; Bould, 1948; Lees, 1949; Lees and Porteous, 1950).

1.2.2. *Microbiological Techniques*

Specific microorganisms such as *Azotobacter* and *Aspergillus,* which form cells or mycelia rapidly under aerobic conditions, have been used to measure the level of nutrients in the soil. The principle of the method is that the amount of growth is a function of the concentration of the element which is growth-limiting. A review of microbiological methods was compiled by Bergman (1958).

a. *Azotobacter.* The *Azotobacter* method was originally worked out by Winogradsky (1925) and consists of growing *Azotobacter* on soil plaques. The number and size of the colonies is a function of limiting growth factors, particularly Ca, P, and K. Sachet and Steward modified the Winogradsky method for the study of mineral deficiencies in Colorado soils. For a discussion of the method and the technical details the reader is referred to Vandecaveye (1948).

b. *Aspergillus.* The *Aspergillus* method has been used for the determination of K, P, and particularly trace elements such as Zn, Fe, Mo, and Cu. Generally, the mycelium weight or the contents of the element in the mycelium is taken as a criterion. The color of the spores formed by the mycelium is generally taken as an index of Cu level, a darker color of the spores corresponding to a higher Cu content.

A technique frequently used in the *Aspergillus* method is that developed by Niklas and co-workers (Niklas *et al.*, 1930, 1933; Niklas and Toursel, 1940). For the standard series and the soil samples, a 1% citric acid culture solution is used containing sugar, peptone, and nutrient elements. The soil and standard series are inoculated with an *Aspergillus* spore suspension, and after incubation for 4 to 6 days, the mycelium is weighed. The weight of the dry mycelium of the soils is then compared with those obtained from standard solutions of the nutrient. The method actually is a citric acid extraction in which microorganisms are used instead of chemicals to analyze the extract. For K and P, this method is therefore of little significance because it is easier and quicker to determine K and P in citric acid by means of chemical methods. For trace elements, where proper extraction techniques are often not available and chemical analysis of extracts is rather involved, there is some advantages in using a microbiological method (Stapp and Wetter, 1953;

Acock, 1941; Jensen, 1951; Swaby and Passey, 1953; Tucker *et al.*, 1953; Wetter, 1954; Spicher, 1954; Roschach, 1956).

In practice, the results are a function of soil and microbiological effects, making interpretation difficult. This interaction does not occur in the standard solutions where soil is absent. The presence of certain elements or compounds in the soil may be responsible for a stimulation of mycelia production. It is known, for example, that Na may have an effect on mycelia production (Vageler and Alten, 1931) and organic substances such as humic acids may have a specific effect on mycelium growth (Kiessling, 1931; Kiessling and Schmidt, 1932; Eno and Reuzer, 1951; Eno and Reuzer, 1955).

Mutations are frequently induced when the growth medium of the fungus is changed. Such mutations can be responsible for deviations in mycelium growth (Mehlich *et al.*, 1933; Stock, 1933). It is therefore necessary that each time an innoculation is made the same pure strain of the microorganism is used.

Another factor which interferes with the determination of trace elements such as Cu or Mo is that during mycelium growth the medium becomes very acid, which may result in dissolution of Cu which otherwise would have been insoluble (Mulder, 1954; Bortels and Wetter, 1954).

In general, it may be concluded that apart from many interfering factors concerning the techniques of microbiological methods, little information is gained with regard to the supply of plant nutrients in normal soils. Quantities of elements extracted by the fungus are unlikely to correspond with either capacity or intensity factors and, in addition, the microorganisms cannot be equated with higher plants.

c. *Algae.* Recently, a method using soil algae as a test organism for N, P, and S was developed by Tchan and co-workers (Tchan, 1959; Tchan *et al.*, 1961; Tchan *et al.*, 1963). This method eliminates several of the interfering factors intrinsic in the *Azotobacter* and *Aspergillus* techniques. One advantage in using algae is that no alterations in soil pH are involved during incubation. Furthermore, being autotrophic, no additions of organic matter have to be made as a source of energy. The determination of algae production is done by means of ethanol extraction followed by the determinations of chlorophyll content in the extracts by spectrophotometry.

1.3. Tracer Methods

The two laboratory tracer methods that have received attention are the so-called "E" and "L" value. The latter actually involves the growing of plants, but is conceptually the same as an "E" value, except that the plant is utilized as the pipette for sampling the soil solution under normal growing conditions. As mentioned previously, rate constants can also be calculated.

1.3.1. *Phosphorus-32 Exchangeable Soil Phosphorus*

A recent chemical method suggested as a laboratory test for determining a soil P fraction that might be closely related to soil P supply is the P32-exchangeable P referred to as Pe by Wiklander (1950) and Talibudeen (1954), and as the "E" value by Russell *et al.* (1954). It is a direct application of the isotopic dilution principle and is an attempt to measure the amount of phosphate in the soil that is in equilibrium with the phosphate in the soil solution in the same way that Paneth and Vorwerk (1922) measured the specific surface of Pb salts by determining the amount of Pb on the surface in equilibrium with Pb in solution. This technique was extended and utilized by Kolthoff and his associates (Kolthoff, 1936; Kolthoff and Rosenblum, 1933), and noted by Imre (1937) in the 1930's. It has been successfully used in determining the surface properties of various phosphate salts (Caro and Hill, 1956; Olsen, 1952; Ulrich, 1961).

The method is based on the principle that when P^{32} is added to a system containing P^{31}, the P^{32} will distribute itself such that at equilibrium the ratio of P^{32} to P^{31} (the specific activity) will be constant for all the phosphate participating in the equilibrium reaction. The reaction can be illustrated as follows:

$$\text{surface } P^{31} + \text{solution } P^{32} \rightleftharpoons \text{surface } P^{32} + \text{solution } P^{31}$$

at equilibrium

$$\text{surface } P^{31}/\text{surface } P^{32} = \text{solution } P^{31}/\text{solution } P^{32}$$

In 1948 McAuliffe *et al.* suggested that this principle could be applied to soil systems and demonstrated that "the amount of rapidly equilibrating surface phosphate parallels levels of phosphatic fertility measured by anion exchange, modified Truog solubility, or Neubauer growth experiments." Their techniques involved the addition of carrier-free P^{32} to an already equilibrated soil-water system. Various authors, including Russell *et al.* (1954), Gunnarsson and Frederiksson (1951), Olsen (1952), and Fried (1957) suggested modifications of the basic procedure in order to either reduce the analytical problems associated with the method or to obtain additional information with a minimum of effort.

a. Effect of Time of Equilibrium. Although equilibrium is rapidly established between added P^{32} and the surface phosphate in crystalline phosphates, unfortunately it is apparently not rapidly established in the soil system (McAuliffe *et al.*, 1948; Dean and Magistad, 1931; Russel *et al.*, 1954; Ulrich, 1961; Seatz, 1954; Olsen, 1952; Talibudeen, 1957). Typical results from McAuliffe *et al.* (1948) are illustrated in Fig. 6.15.

b. Effect of Added Carrier. Various investigators (Russell *et al.*, 1954; Gunnarson and Frederiksson, 1951; Fried, 1958) have suggested that the addition of appreciable carrier (unlabeled) P to the system would not change the measured value of exchangeable P. Russell *et al.* (1954) showed on one soil that the addition of from 25 to 100 mg of P carrier to 100 g of soil gave different values of exchangeable P after 4 hours but the same value of exchangeable P after 48 hours. Similarly, Fried (1957), also working only with one soil, compared the

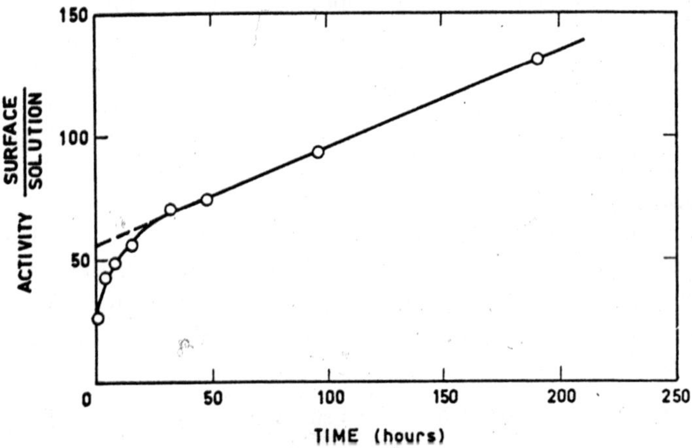

FIG. 6.15. Kinetic exchange of phosphate in solution (P^{32}) with surface phosphate (P^{31}) on a Caribou soil No. 451586. From McAuliffe *et al.* (1948).

exchangeable P obtained after 48 hours upon the addition of 0.5, 20, and 80 mg of carrier phosphate per 100 g of soil. The results are summarized in Table 6.4.

Although both the amount of added P and the percentage of P removed from solution varied markedly, the amount of P^{32}-exchangeable P remained constant. If this observation is valid for soils in general, then one of the technical difficulties of this method can be overcome, i.e., the inability to determine the very low P concentration in a soil-water extract when carrier free P^{32} is used. Not only is this technical difficulty overcome, but information concerning the so-called "fixation" of added phosphate by the soil is obtained. Different results for exchangeable P may be obtained upon addition of carrier when shaking is for short periods of time, 0.5 to 4 hours (Russell *et al.*, 1954 and Amer, 1962).

c. Other Considerations. Talibudeen (1954, 1957) recognizing the requirement that "sufficient phosphorus was present in solution to make accurate analyses possible" suggested the alternative method of using an extractant with the P^{32} which had as its major effect an increase in the proportion of the P^{32}-exchangeable P that remained in solution. Chang (1962) similarly sampled P^{32}-equilibrated soils with common extractants. Talibudeen found that 0.001 *M* ammonium citrate in 0.002 *M* KCl not only increased the proportion of P in solution but gave more rapid attainment of isotopic equilibrium, also resulting in higher amounts of P in solution. This appears to give a measure of an intrinsic reproducible soil property. It gives no information, however, concerning the "fixation" properties of the soil. It has yet to be ascertained whether the information obtained with the added carrier is not the more useful.

TABLE 6.4
SURFACE PHOSPHORUS OF A CARIBOU SOIL AS DETERMINED BY EQUILIBRATION FOR 48 HOURS WITH SOLUTIONS OF P^{32} CONTAINING DIFFERENT AMOUNTS OF P^{31} (mg P/100 g soil)

P^{31} added in solution	P^{31} found in solution after 48 hours	P^{32}-exchangeable phosphorus
0	2	13
5	3	11
20	11	11
80	58	13

For most soils the amount of phosphorus in solution is a negligible part of the total P^{32} exchangeable phosphorus. However, for sandy soils and wide soil-solution ratios or where extracting solutions are used, the phosphorus in solution (after equilibration) contributed by the original soil may be appreciable.

Other factors may affect the measured P^{32}-exchangeable P other than time of shaking with the P^{32}-equilibrating solution. These are the presence of organic anions (Arambarri and Talibudeen, 1959), exchangeable base status (Arambarri and Talibudeen, 1959; Wiklander, 1950; Russell *et al.*, 1954) temperature (Arambarri and Talibudeen, 1959; Wiklander, 1950), time of shaking prior to P^{32} equilibration (Amer, 1962), drying (Wiklander, 1950; Russell *et al.*, 1956), and soil:solution ratio (McAuliffe *et al.*, 1948; Talibudeen, 1957). Any empirical method for determining P^{32}-exchangeable P should take into account these various factors.

d. Evaluation of the Method. Many investigators have assumed that this measurement must be related to soil phosphate supply to the plant. Other have tested this assumption. Thus, Gunnarsson and Frederiksson (1951), Mattingly and Talibudeen (1960), Olsen (1952), Fried (1960), and Russell *et al.* (1957) found a positive relationship between P^{32}-exchangeable soil P and plant response. The plant response was measured either as yield or uptake of P or both. The results of Fried (1960) are reproduced in Table 6.5.

Table 6.5 indicates that under conditions of this measurement soils with a P^{32}-exchangeable P value of less than 180 would almost invariably respond to added phosphate, those with a value above 240 would seldom

TABLE 6.5
RELATIONSHIP BETWEEN ISOTOPE-MEASURED SURFACE PHOSPHORUS IN THE LABORATORY AND THE RESPONSE OF 74 SOILS TO APPLIED PHOSPHORUS IN THE GREENHOUSE

Surface P (lb P_2O_5/A)	Number of soils responding	Number of soils not responding
0–60	2	0
61–120	11	1
121–180	19	2
181–240	18	9
>240	3	19

respond, while those with values between 181 and 240 were intermediate in response. The exceptions are worthy of note. The lack of yield response by soils low in P^{32}-exchangeable P (one soil in class 61–120 and in class 120–180) was probably because yield was limited more severely by factors other than P supply. Those soils responding, despite high values of exchangeable P^{32}, may have responded only because other factors were so ideal that the maximum yield level was higher in these soils. This would result in a higher P requirement to obtain maximum yield. Alternately, consistent with the observation of Russell *et al.* (1957), these soils may have been so different from the others, particularly in their characteristics of phosphate sorption from solution, as not to be part of the same population. This is not a serious difficulty, since the method used to determine P^{32}-exchangeable P included the addition of carrier P at a concentration of 2 ppm in the equilibrating solution. Thus, a measure of laboratory sorption (fixation) was [illegible]ained on each soil and, as Russell *et al.* (1957) have shown with high phosphate-sorbing soils, response is inversely related to sorption. The actual magnitude of this correction requires further evaluation.

The I.D.F. value is a result of similar observations of a possible overestimation in high fixing soils. I.D.F. = $P_{st} - P_o \times (C_t/C_i)$ where $P_0 = P^{31}$ added initially, P_{st} = total concentration of P in solution at time t, and (C_t/C_i) = ratio of P^{32} counts per unit volume at time t and initially. The I.D.F. as stated by the authors is an intensity value which represents "the soils contribution to P contents of equilibrium solutions of soil–solution systems." (McConaghy and Stewart, 1963 and McConaghy *et al.*, 1966.) Since the amount of phosphorus left in solution upon addition of phosphorus is low in so-called high-fixing soils, the I.D.F. value must be relatively low in such soils.

An interesting observation was made by Mattingly and Talibudeen (1960) when they compared the removal of P by cropping with rye grass to the amount of P^{32}-exchangeable P (measured in 0.02 *M* KCl after P^{32} equilibration for 170 hours) before and after cropping. The results were essentially identical.

Although not all evidence suggests that this single measurement can be usefully employed to predict response to added fertilizer (Russell *et al.*, 1957), this test is clearly a good measure of the capacity factor, the amount of M(solid), as discussed in Eq. (1.1). The measurement of a capacity factor for phosphate does not give sufficient information for predictive purposes over a wide range of soils. A measure of intensity, M(solution), is required in addition and, if possible, the effect of applied fertilizer on the intensity factor. By using a carrier concentration of P of a suitable level (e.g., 2–4 ppm), an estimate of the intensity factor is obtained at one level of added P without any additional labor.

The determination of P^{32}-exchangeable P (using added carrier) requires two laboratory determinations as compared to the one required by other soil tests. Both determinations are simple: one is merely a radioactive P^{32} count of the supernatant or filtered solution; the other is the usual P determination necessary in all laboratory tests for a measure of soil P supply. Twice as much information is obtained, and these two pieces of information are probably adequate to evaluate a wide variety of soils with the same test procedure.

The limitation to using P^{32}-exchangeable P as a routine soil test is related to the necessity of handling radioactive materials in a routine soils laboratory. This increases the cost, requires more highly trained personnel, and requires taking the necessary health-hazard precautions, thus almost limiting the testing to centralized laboratories. This means that, other than use at the research level, such a program can only be initiated where facilities for such a routine are available. But it is *the only test that measures both capacity and intensity* in the same sample, and it remains for the research worker to determine whether the increased information is worth the extra effort.

1.3.2. *"L" Value*

Sigurd Larsen was undoubtedly the first to suggest that when labeled phosphate fertilizer was added to a soil at different rates and plants were grown, specific activity analysis of the plant material would give a constant value if isotopic dilution of the added chemical was assumed to take the place in the soil system (Larsen, 1950, 1952). Larsen assumed that the reason for the constant value obtained was that equilibrium had been reached between added phosphate salt and the exchangeable phosphate in the soil. However, this equilibration is usually very slow (Larsen and Gunary, 1965).

At about the same time, Gunnarson and Fredriksson (1951) made similar observations. The quantity measured has since become known as the "L" value. Both Larsen and Gunnarsson and Fredriksson recognized that "E" values and the values they obtained could be equated, and later authors (Schofield, 1955; Russell *et al.*, 1956; Talibudeen, 1957; Newbould and Russell, 1963; Larsen and Sutton, 1963) have referred to the quantity measured as "labile phosphorus" or the "labile pool." The "L" value can thus be considered as an "E" value in which the plant is being used as the means of sampling the soil solution upon equilibration of added P with exchangeable soil P. It has considerable theoretical advantage over the "E" value in that the conditions of measurement are identical to those that confront the plant, when grown in the soil and, furthermore, it will reflect any effect that the plant has on the soil phosphate system. An "E" value and "L" value are conceptually the same, and an "L" value can therefore be defined as the amount of P in the soil and in the soil solution that is exchangeable with orthophosphate ions added to the soil, as measured by a plant growing in the system. It was partially this conceptual equivalence that led Russell *et al.* (1957) to conclude that "E" and "L" values are sufficiently similar for them to be regarded as alternative measures. The choice of the two procedures for this purpose should therefore depend mainly on convenience and economy of labor. On these grounds the "E" value procedure is to be preferred." Actually, Russell was referring to the agronomic use. There is no doubt that from the agronomic standpoint a laboratory method is preferable to *any* plant-growing method, since the laboratory method is normally much cheaper and subject to mass production methods of analysis of soil samples. From the standpoint of obtaining quantitative data under conditions of normal plant growth, however, the "L" value is to be preferred providing the assumption of complete equilibration is essentially obtained (Fried, 1964).

1.3.3. *Rate Constants*

The calculation of rate constants by the isotope technique consists of introducing carrier-free amounts of a radioactive isotope of a nutrient element into an equilibrium suspension. If the element is not available carrier-free, its specific activity should be high enough that the equilibrium conditions in the suspension are not changed.

The isotope will equilibrate with all isotopically exchangeable fractions of the elements in the solid phase. When the decrease in activity of the equilibrium solution is measured as a function of time, graphical analysis of the data enables an estimation of the rate constants of the exchange reactions to be made.

If only one kind of exchange reaction takes place, the exchange equation for element M may be described by

$$\mathrm{M} \underset{-k}{\overset{k_s}{\leftrightarrows}} \mathrm{M_s}$$

with rate constants k and k_s in which $\mathrm{M_s}$ is the amount of isotopically exchangeable element in the solid phase and M the quantity of the element in the equilibrium solution.

If a quantity of $\mathrm{M_0}^*$ of a radioactive tracer of M is introduced into the suspension at $t = 0$, the increase in activity of $\mathrm{M_s}$ due to the reaction $\mathrm{M} \rightarrow \mathrm{M_s}$ at a rate V may be represented by

$$\overrightarrow{V} = k\mathrm{M}^* \tag{6.1}$$

As soon as $\mathrm{M_s}$ becomes labeled, the back reaction $\mathrm{M_s} \rightarrow \mathrm{M}$ will start at a rate $\overleftarrow{V}$ represented by

$$\overleftarrow{V} = k_s\mathrm{M_s}^* \tag{6.2}$$

The net rate of increase in activity of $\mathrm{M_s}$ is given by the algebraic sum

$$d\mathrm{M_s}^*/dt = \overrightarrow{V} - \overleftarrow{V} = k(\mathrm{M}^* - \mathrm{M_s}^*)$$

or, because $k\mathrm{M} = k_s\mathrm{M_s}$, (6.3)

$$d\mathrm{M_s}^*/dt = k\mathrm{M}(\mathrm{M}^*/\mathrm{M} - \mathrm{M_s}^*/\mathrm{M_s})$$

From the equation of conservation of the total amount of activity,

$$\mathrm{M}^* = \mathrm{M_0}^* - \mathrm{M_s}^* \tag{6.4}$$

Substitution of Eq. (6.4) into Eq. (6.3), integration, and rearrangement will then give

$$\ln[1 - M_s^*(M + M_s)/M_0^*M_s] = -kt(M + M_s)/M_s \qquad (6.5)$$

When isotopic equilibrium has been attained at t_∞, the specific activities of isotopically exchangeable M_s and M in solution will be identical, and for t_∞ and $M_{s\infty}^*$

$$M_0^*/(M + M_s) = M_{s\infty}^*/M_s \qquad (6.6)$$

Substitution of Eq. (6.6) into Eq. (6.5) finally gives

$$\ln(1 - M_s^*/M_{s\infty}^*) = -k[(M + M_s)/M_s]t \qquad (6.7)$$

When $\ln(1 - M_s^*/M_{s\infty}^*)$ is plotted against t, a straight line will be obtained with a slope of

$$-k(M + M_s)/M_s = -kM_0^*/(M_0^* - M_\infty^*) \qquad (6.8)$$

from which k can be calculated.

When a chemical determination of M is made, M_s can be calculated from the isotope dilution law because at equilibrium:

$$M_\infty^*/M = M_0^*/(M + M_s) \qquad (6.9)$$

As in practice, M^* is measured instead of M_s^* and $M_0^* = M^* + M_s^*$ at any time, Eq. (6.7) can be written as

$$\ln(M^* - M_\infty^*) = -k[(M + M_s)/M_s]t \qquad (6.10)$$

Working with soil, clay or fertilizer suspension, a plot of the function of Eq. (6.7) or Eq. (6.10) vs. t will generally be curvilinear presumably because several exchange processes M_1, M_2, . . . , M_n with rates k_1, k_2, . . . , k_n will take place simultaneously. This assumes diffusion is not rate-limiting.

If reaction M_1 is virtually completed and M_2 has not yet proceeded to an appreciable extent, a graphical analysis of the composite curve can be made. This is done by a process of continuous subtraction, i.e., in a plot of

$$\ln(M^* - M_\infty^*) \qquad \text{against } t, \text{ or} \quad \ln(1 - M_s^*/M_{s\infty}^*)$$

against t, M_n is extrapolated and deducted from the composite curve. The logarithm of the differences are again plotted against t and the M_{n-1} line is extrapolated and deducted from the composite curve. This process is repeated until the original curve has been separated into a number of straight lines. For each straight line, the corresponding k, M, and M_s can then be calculated. In practice, such a procedure will introduce large errors which tend to accumulate in the processes that proceed at the relatively higher rates.

The observations are valid and easily reproducible. However, the assumption that diffusion is not rate-limiting is only an assumption. Indeed, the observations on soils can all be equally well explained on the basis of differences in the diffusion rates of ions located differently in the solid phase. The results often depend on the experimentally chosen soil solution ratio (Ulrich, 1961). Extrapolation of apparent rate constants to field conditions is not advisable.

2. Greenhouse

Greenhouse measurements of plant nutrient levels are neither rapid nor as precise as laboratory techniques. They are, however, uniquely capable of describing or measuring that portion of the nutrient in the soil that is available to plants. Although the measurements do not reflect the volume relationships between soil and plant as they occur in the field, they do reflect the quantity of plant-available nutrient in a unit weight or volume of soil. Although less rapid and precise than laboratory measurements, greenhouse measurements are more rapid and more precise than field determinations.

Many routine methods have been suggested involving the growth of plant under controlled conditions. Among these are the Neubauer and Mitscherlich techniques and their modifications and other specialized techniques such as Colwell's (1946) age value determination of B deficiency, Dean's (1954) extrapolated yield of P method, and Stanford and De Ment's (1957, 1959, 1959a) short-term nutrient absorption technique.

Although these techniques are subject to various errors, they may be useful under special conditions. The common difficulty in all of these methods is their dependence to a greater or lesser extent on the amount of plant growth. Since amount of growth varies with environmental conditions, including the level of other nutrients, light intensity, temperaures, season, and moisture, growth differences may not be due to differences in the ability of the soil to supply the nutrient under investigation. Tracer methods which focus their attention on the specific nutrient and give results independent of yield of plant material represent a potential advance in determining soil nutrient supply for those elements with suitable isotopes.

2.1. Yield Response Relationships

The general experience in growing plants in the field or the greenhouse is that increasing increments of nutrient results in a decreasing rate of yield response, i.e., if all other factors remain the same, as fertilizer nutrients are added in units of equal size, each increment in yield response is smaller than the preceding increment.

Although this generalization is valid in the usual growth-response curve, many investigators have found that where extreme deficiencies occur, the first unit of nutrient added may result in very little yield response. The addition of further increments will give an increasing rate of response, finally reaching an inflection point after which the rate of response goes through a maximum and starts to decrease, thus giving an over-all sigmoid-shaped yield-response curve. The inflection point of maximum rate of response, however, is always at a very low yield level, never exceeding 50% of the yield level at which measured response ceases and more often at a level below 10% of this maximum yield. This inflection point is not normally observable in the field. Examples of these two types of response are given in Figs. 6.16 (Terman *et al.*, 1962) and 6.17 (Steenbjerg and Jacobson, 1963).

These general observations have led to various attempts to describe yield response within a rigid mathematical framework. The first and most famous of these was that of Mitscherlich who developed the basis of his theory in 1906. He assumed that for a nutrient not present in adequate amounts, a certain maximum yield was obtainable by the addition of this nutrient. The amount of yield response due to the application of an increment of the nutrient was presumed proportional to the proximity to this maximum yield. This presumption does not allow for an inflection point. Mitscherlich (1923) expressed this mathematically in the following form:

$$dy/dx = (A - y)c \tag{6.11}$$

where dy is the increment change in yield when dx of the nutrient is added, c is the proportionality constant; A, the maximum yield; and y is the yield obtained when x is the amount of the nutrient added.

Integration of the equation results in

$$\log(A - y) = K - cx \tag{6.12}$$

when x, the amount of nutrient added, equals 0, the yield may be defined as a, and k equals $\log(\mathrm{A} - \mathrm{a})$. Therefore

$$\log(A - y) = \log(A - a) - cx \tag{6.13}$$

If b is defined as the amount of x already in the soil, then substituting b for x when $y = a$

$$\log(A - y) = \log A - cb \tag{6.14}$$

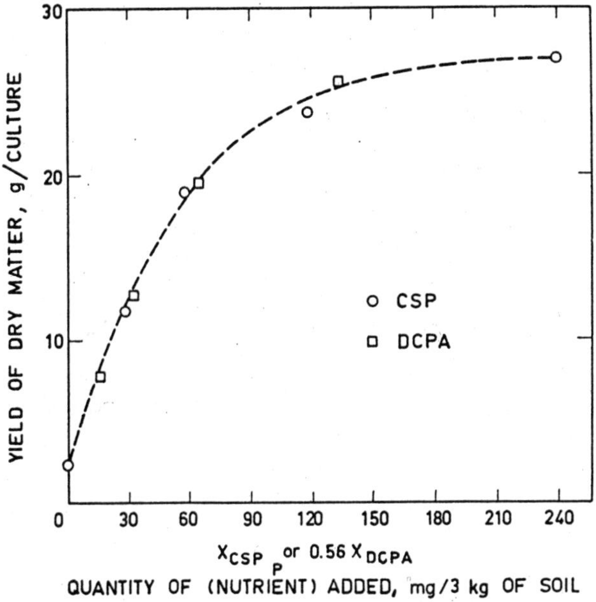

FIG. 6.16. Yields of dry matter by oats plotted against quantity of P added as concentrated superphosphate (X_{CSP}) or 0.56 times that added as anhydrous dicalcium phosphate ($0.56X_{DCPA}$). From Terman *et al.* (1962).

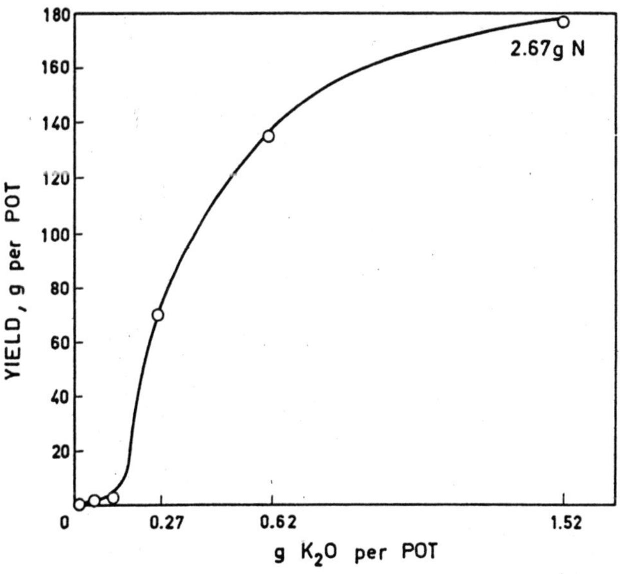

FIG. 6.17. Relationship between applied potash and yield of oats (grain and straw). From Steenbjerg and Jakobsen (1963).

and substituting Eq. (6.14) into Eq. (6.12) results in

$$\log(A - y) = \log A - c(x + b) \tag{6.15}$$

Finally, when $b = 0$

$$\log(A - y) = \log A - cx \tag{6.16}$$

where A is the maximum yield; y the yield when x amount of nutrient is present; and c, a proportionality constant.

If yield and amount of nutrient x are expressed on a relative basis, then A may be given a relative value of 100 and x given a value of 1 when y equals 50. This simply defines a unit of nutrient as that amount of nutrient that will give 50% of a maximum yield. Solving for c when $x = 1$ gives a value of $c = 0.301$ where $A = 100$, $y = 50$, the percentage of the maximum yield, and $x = 1$, the relative amount of nutrient. The percentage of maximum yield can be determined for each increment of x and equals 50, 75, 87.5, 93.75, 96.88%, etc. for 1, 2, 3, 4, 5, . . . increments of x.

The amount of nutrient required to obtain 50% of the maximum yield has been called the "Baule" after the German mathematician who first proposed, in discussing the Mitscherlich approach, that increases in yield should more properly be measured as percentages rather than in absolute units (Baule, 1918).

This quantitative approach and the variation by Spillman and Lang (1924) differ in the interpretation placed upon the ratio of successive increments of yield. Mitscherlich held them to be the same for all crops and soils, whereas Spillman thought they could differ under different conditions. These theories unleashed an enormous amount of thought and experimental work, not only to test the relationship but also to both modify the law for such things as interacting growth factors, toxicities, etc., and to determine the constant c or the value of a Baule for all of the growth factors such as N, P, K, etc.

From the standpoint of determining soil nutrient supply, Eq. (6.15) can be used as a basis for estimating b when no fertilizer is applied, x in Eq. (6.15) equals zero, the yield is y_0 and

$$b = [\log A - \log(A - y_0)]/c \tag{6.17}$$

where b is the amount of available nutrient in the soil expressed in units of the applied fertilizer material. Graphically, b is the measured distance along the x axis from $x = 0$ to where the growth curve such as the one illustrated in Fig. 6.16 cuts the x axis when extrapolated to zero yield of dry matter.

This distance is then expressed in the same units as that of the x axis. In Fig. 6.16 just referred to, this would amount to some 10 lb of P in the form of concentrated superphosphate (—6 + 10 mesh) or 18 lb of P in the form of anhydrous dicalcium phosphate (—6 + 10 mesh).

If Eq. (6.15) truly described the growth response curve and if the points on the curve are determined without error, b would be a reasonable determination of the amount of nutrient in the soil expressed in units of an added standard fertilizer material. Neither of these precepts hold. The yield response curve is often sigmoid in shape and does not necessarily follow the rigid mathematical relationship of constant increment ratios. Yield response curves always have apprecaible inherent error and any extrapolations from the end of these curves magnifies the error so greatly as to make this approach relatively useless.

The factors c quoted by Mitscherlich and his co-workers, based on an average of thousands of field and pot trials, are said to be 0.2 for N, 0.4 for K, and 0.6 for P where yield is expressed in 100 kg units/ha. Most investigators realize, however, that c is not a universal constant but depends on the set of conditions at which each experiment is carried out (Boguslavski, 1958; Rippel, 1931; Pauw, 1952, 1955; Ulrich, 1961).

The Mitscherlich equation might hold if the various yield factors were constant and independent of each other. Moreover, yield factors change during the development of a crop and, furthermore, when fertilizer is applied, it is not possible to increase the rate of application of a single nutrient without increasing the application of accompanying elements, e.g., higher doses of superphosphate mean higher doses of Ca and P at the same time. When the variability of the data and the inability to obtain useful data when yield response is low is added to those fundamental difficulties, the limitations of the method become apparent.

The original Mitscherlich trial required ten pots, but a simplification was introduced by Boguslavski requiring only eight (Boguslavski, 1952). From the yield data, A can be determined from the Mitscherlich equation using the c values originally quoted by Mitscherlich (Eq. 6.16). Since c is not a universal constant, however, more accurate information can be obtained if A is determined from a yield curve by graphical extrapolation (Behrens, 1928). This requires an experimental set-up with several rates of application. When A has been determined, the corresponding c can then be approximated or calculated, using the methods of least squares (Pauw, 1952, 1955; Behrens, 1950). It is doubtful, however, whether these involved calculations are justified in dealing with such variable material, and an approximation method of fitting various x, y values and correcting the A is probably more than adequate. When finally A and c are known, the calculation of the available quantity of soil element b is straightforward from Eq. (6.17).

The actual technique recommended use of specially designed pots. The soil was mixed with sand (2:1). The b values had to be corrected, therefore, to a true soil-weight basis. Further, Mitscherlich assumed that because of the difference between field conditions, where a subsoil is present, and pot cultures, an additional factor of 2 should be applied. The recommended technique already demonstrates the high amount of empiricism that is involved in interpreting the findings of pot experiments in terms of field conditions.

It may be concluded that the Mitscherlich technique is a possible approach for comparing soils in a green-house pot test, but extrapolation to field conditions is fraught with difficulties. The method is only of use when a definite yield response to fertilizer can be obtained. It may well be that the average c values found for German conditions fluctuate little, but as pointed out by Ulrich (1961) in his discussion of the Mitscherlich approach, "neither the farmer nor the soil scientist is dealing with a statistical average of soil types, but with their individual soil types for which there is no law that states that their characteristics should fall within the 'confidence' limits set by statistics." In other words, what is valid for the mean need not be valid for the individual case.

2.2 Yield of Nutrient Relationships

If the primary objective is the determination of the soil nutrient supply of a given element, the measurement of yield response is only a secondary effect. The primary effect is the uptake of nutrient element by the plant. The use of yield of nutrient relationships to determine soil nutrient supply is, therefore, basically more sound than the use of yield of dry matter relationships. The Neubauer technique (Neubauer and Schneider, 1923; Roemer and Scheffer, 1929; Deger, 1934; Thornton, 1935; Kohnke, 1937; Mooers, 1938; McGeorge, 1946; McGeorge and Breazeale, 1956) was one of the first generally accepted techniques that utilized a method dependent upon nutrient uptake. The principle is simply based on the determination of the amount of nutrient extracted from a small amount of soil by a large number of plants. The soil is diluted with sand to offer a suitable physical medium for the plant roots. The amount in the seed must be deducted. From empirical calculations and comparison with pot and field experiment, "Grenzzahlen" (limiting quantities) were established for each crop, giving the quantities of an element in the soil below which the supply of the element was insufficient.

Table 6.6 gives the Neubauer "Grenzzahlen" as established by four investigators and shows that the crops which need much nutrient such as potatoes, beets, and alfalfa generally have higher "Neubauer limiting" values than cereals. The difference in the values obtained by various investigators is not surprising. The conditions under which the experiments are done deviate considerably from those in the field and are greatly affected by the experimental procedure chosen.

Unfortunately, the technique suffers from certain fundamental errors owing to the difference in growing conditions in the normal soil situation and in the Neubauer technique. First, the intensity factor (i.e., the concentration of the ion in the soil solution) is changed, thus changing the rate of nutrient uptake by the plant. Secondly, even though growth

TABLE 6.6

NEUBAUER LIMITING QUANTITIES ("GRENZZAHLEN") FOR P_2O_5 AND K_2O[a]

	Neubauer (1931) high yields		Lemmermann (1932) medium yields		Hazelhoff (1931)		Roemer (1927)	
Crop	P_2O_5	K_2O	P_2O_5	K_2O	P_2O_5	K_2O	P_2O_5	K_2O
Barley	6	24	5	19	6	14	6	30
Oats	6	21	4	16	6	17	4	20
Wheat	5	20	4	15	8	17	6	30
Rye	5	17	4	13	8	17	4	20
Potatoes	6	37	5	28	9	37	6	40
Sugar beets	6	25	5	19	10	33	8	30
Fodder beets	7	39	6	29	12	47	—	—
Rape	9	18	7	15	15	18	—	—
Alfalfa	9	35	7	25	15	35	—	—

[a] Data given in mg/100 g of soil. From Bergman (1958).

is over a shorter time period, the capacity factor is much more severely tested than under field and usual greenhouse conditions. This severe testing of the capacity factor may even result in the rate-limiting step in nutrient uptake becoming the rate at which the capacity factor is renewed from within the soil mass. Third, Neubauer values are differences in amount of nutrient in the plant and the seed, and each amount is determined with considerable error. If this difference is equal in magnitude to the amount in the seed, the magnitude of the error of estimate of the Neubauer value is doubled or, if this difference is half the amount in the seed, the error is quadrupled, etc. Errors can, of course, be reduced by doubling or even quadrupling the number of replications, although this adds considerably to the effort necessary. Fourth, in certain situations, particularly with so-called high phosphate-fixing soils, the amount of nutrient in the plant at the end of the experiment is less than originally in the seed, showing an even possible outgo of nutrient under exceptional conditions. Fifth, the test is of no value for N because the conditions for mineralization of N in the pots is entirely different from those in natural soil (Blank and Scheffer, 1925; Thun, 1930; Neubauer, 1931). In addition, the high N content of the seeds is sufficient to maintain seedling growth and is responsible for the small differences between the amount taken up by the plants and the amount in the seed. Even attempts to use seeds with little endosperm (Wagner, 1930) failed because of the entirely different rates of mineralization of N in pots and in the field.

The Stanford and De Ment short-term nutrient absorption technique (Stanford and De Ment 1957, 1959, 1959a) has much in common with the Neubauer technique, but succeeds in overcoming some of the fundamental objections. This technique involves the initial growth of a mass of seedlings in pure sand to form a mat of roots. These are then placed in contact with the soil under study and the amount of nutrient taken up by the plants in 3 to 7 days is measured by difference. Thus, the intensity and capacity factors for the soil have not been modified. The large mass of roots, however, does set up an artificially high root to soil ratio at the point of contact, that may be a serious factor in certain types of investigations, including the determination of soil nutrient supply.

Both the Neubauer and Stanford and De Ment procedures measure the amount of nutrient supply in terms of quantity of nutrient taken up. It is very difficult to set up environmental conditions that will result in good reproducibility of the amount of nutrient taken up in different experiments. In Chapter 6, Section 2.1 it was pointed out that the Mitscherlich techniques measure nutrient supply in terms of a fertilizer standard. Regardless of actual total uptake of the nutrient, the amount contributed by the soil (other things being equal) bears the same relationship to the standard, thus giving the same value of nutrient supply under different conditions of total nutrient uptake. Only if the soil nutrient supply is actually changed by the new conditions will the value change in relation to the standard.

Yield of nutrient curves can also be used to evaluate soil nutrient supply in terms of a fertilizer standard. This has been suggested by Dean (1954) in his published yield-of-phosphorus curves. The principle utilized was that the normal yield-of-phosphorous curve is an ascending straight line. Steenbjerg (1952) earlier stated this as a fact, while Dean (1954) showed that in more than 100 soils in experiments conducted in Beltsville over a 2-year period approximately 60% of the soils gave this relationship for P. Under these conditions of ascending straight lines and, if the added fertilizer standard does not interact to change the amount of soil phosphorus absorbed, the schematic diagram shown in Fig. 6.18 should hold.

Where BC is the presumed experimentally defined yield of P curve, and AB the extrapolated portion to $y = 0$, AO or a then equals the amount of nutrient present in the soil in terms of fertilizer standard. In this case of the schematic diagram, the amount would equal 50 lb P/acre in superphosphate equivalents.

Figure 6.19 (Dean, 1954) shows examples of the effect of plant species, change in fertilizer standard, and a change in environment (known to

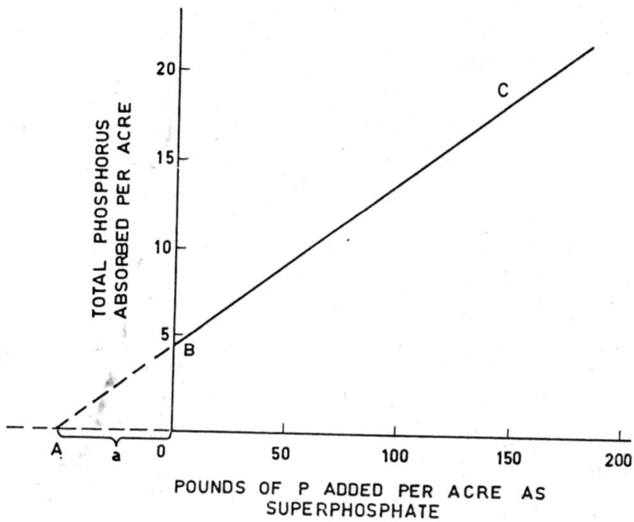

FIG. 6.18. Schematic diagram of the relationship between P supply and uptake by the plant when the relationship is linear.

affect available soil P) on the extrapolated yield of nutrient method for determining soil nutrient supply. The results showed that even though buckwheat and millet took up entirely different amounts of total P, the amount of nutrient available in the soil was the same. When a different fertilizer standard was used, the unit for expressing soil nutrient supply changed and 150 superphosphate equivalents equaled 780 fused tricalcium phosphate equivalents. The results also showed that a change in the soil nutrient supply due to treatment was reflected in the extrapolated value of soil nutrient supply.

In the experiments reported by Dean, radioactive-labeled fertilizers were used and the extrapolated yield of nutrient value could be compared directly with the actual amount of soil and fertilizer P taken up. The results were similar enough to suggest that where isotopes were not available and where ascending straight lines were obtained for yield of nutrient curves, this technique could be utilized to measure plant nutrient supply.

The limitation of the method is, of course, primarily the assumption

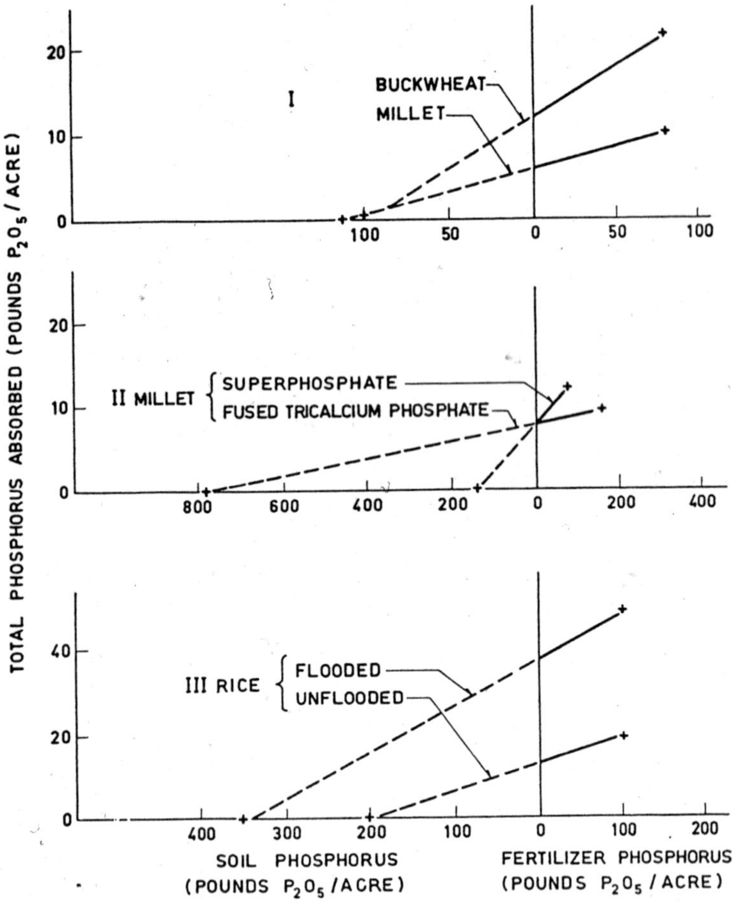

FIG. 6.19. Influence of paired treatments on yield-of-P curves. From Dean (1954).

of an ascending straight line which Dean himself showed did not hold for about 40% of the cases. For nutrients other than P, however, this assumption may indeed have greater validity. Other limitations also involve the large errors associated with extrapolations and the possible influence of yield response on the yield of nutrient and the amount of nutrient taken up from the soil.

2.3. Tracer Method—"A" Value

2.3.1. *General Concept*

Fried and Dean first printed their concept of "A" value in a USDA Research Report (Fried and Dean, 1950) which was later condensed (Fried and Dean, 1952). The authors stated that the amount of available soil nutrient could be measured in terms of a standard if an appropriate definition of available nutrient was chosen. Thus they stated that "it is assumed that the definition of *available nutrient in the soil* implies that when two sources of a given nutrient are present in the soil the plant will absorb from each of these sources in proportion to the respective quantities available." This then was a definition of available nutrient, just as one might assume that parallel lines never meet and examine the consequences of this assumption.

The authors then pointed out some of the consequences of their assumption. If a plant was confronted with two sources of a nutrient, one of which was the soil and the other a known amount of a fertilizer standard, "the amount of available nutrient in the soil can be determined in terms of a standard, provided the proportion of the nutrient in the plants derived from this standard is determined." This is a mathematical consequence of their definition of available nutrient. The units of measurement are units of fertilizer standard, and the amount of available nutrient was referred to as "A." Therefore "A" is an amount of available nutrient measured in units of the fertilizer standard. The authors clearly indicated that "A" did not refer to a single chemical entity, but was a necessary integration of different chemical entities and what might be called their coefficients of availability.

The concept does not include the method of measurement. Thus, it is left to the ingenuity of the investigator as to how to present two sources of a nutrient to a plant in such a way that the plant will take nutrient from each of these sources in direct proportion to the amounts available. If one of these sources is soil, one might think of split-root technique, coating the fertilizer standard to prevent interaction with the soil, intermittent layers of soil and standard, large particle size of the standard, etc. Since the plant must take nutrient from each of the two different sources, ideally there should be no interaction between the two sources. The fertilizer standard need not be soluble or reactive with the soil system. Furthermore, the two sources need not include the soil but may be two or more different fertilizer sources (Fried, 1954) thus allowing a quantitative evaluation of the relationship between fertilizer sources. The use of labeled standard is not a necessary requiremen for determining "A" values since the methodology is not stated in the definition. Thus, if the amount of nutrient taken up by the plant is linearly related to the amount present in the medium, the amount of nutrient taken up by the plant from a given standard can be obtained by deducting the amount obtained in a check treatment in the absence of added standard from the amount obtained in the presence of the added standard. This gives enough information to calculate an "A" value without the use of radioactive isotopes and is essentially the way A in Fig. 6.18 is calculated. The "A" value, then, is the amount of available nutrient in a particular source measured in terms of fertilizer standard and based on the definition that if a plant is confronted by two sources of a nutrient it will take up nutrient from each of these sources in direct proportion to the amounts available. The "A" value does not refer to any one nutrient, but has already been used extensively for P (Lathwell *et al.*, 1958; Fine *et al.*, 1955; Fried, 1955; Hunter *et al.*, 1961, Fitts *et al.*, 1956; Oberländer and Zeller, 1961; Knight and Williams, 1956; Hawkes and Fried, 1957; Fried, 1953; Mistry, 1962; Mellado, *et al.*, 1962; Mistry, 1962; Rennie and Spratt, 1960; Shapiro, 1958; Armiger and Fried, 1958; Oberländer and Zeller, 1962; Beaton *et al.*, 1962; Rennie and Soper, 1958; McConaghy and Stewart, 1963; Datta and Goswami, 1962; Smith and Pesek, 1962; Suzuki *et al.*, 1963; and Eid *et al.*, 1962). It has also been used for Zn, (Shaw *et al.*, 1954) Ca (Fried and Dean, 1952; Smith *et al.*, 1963), N, (Legg and Allison, 1959) and S (Harward *et al.*, 1962; Nearpass *et al.*, 1961).

2.3.2. *Applicability*

Fried and Dean (1950, 1952), on the basis of the assumption that plants absorb nutrient from two sources in direct proportion to the amounts available, developed a mathematical expression for calculating the amount of available nutrient in soil in terms of a standard fertilizer material.

$$A = B(1 - y)/y \tag{6.18}$$

where A is the amount of available nutrient in the soil; B the amount of fertilizer nutrient (standard) applied; y the proportion of nutrient in the plant derived from the standard.

As these authors state, the method simply "involves the introduction of a known quantity of a standard source of nutrient under consideration into a soil, growing the desired crops, determining the proportion of the total nutrient absorbed that was derived from the standard source, and calculating the amount of available nutrient in the soil, A, by use of Eq. (6.18)." Although the use of isotopes is not inherent in this concept, using labeled materials is the only way known at present of directly determining the proportion of the total nutrient that was derived from the standard source. This method has certain advantages over nonisotopic procedures. A in Eq. (6.18) is independent of the amount of nutrient applied. Thus, one rate of fertilizer application is sufficient to obtain the desired result in contrast to the large numbers of rates needed to define a curve. Second, the plants do not have to be grown under abnormal conditions of severe nutrient stress. This means that the indicator plants are healthy normal plants. A third consideration is that the value obstained is not an extrapolated value, and thus the error of estimate is not appreciably increased. A fourth and major consideration is that the amount of growth does not affect the results. Insofar as carrying out a technique for measuring A values, one of the major difficulties is setting up a system in which the fertilizer standard does not interact differently in the different systems being measured.

TABLE 6.7
EFFECT OF RESPONSE TO PHOSPHORUS AND RATE OF SUPERPHOSPHATE APPLICATION ON A VALUE OBTAINED WITH MILLET[a]

Rate of fertilizer application (lb P_2O_5/acre)	Fort Collins loam		Ida silt loam	
	Yield (g/pot)	A Value (lb P_2O_5/acre)	Yield (g/pot)	A value (lb P_2O_5/acre)
25	8.5	59	4.5	22
50	9.8	50	7.1	22
75	10.6	58	7.8	24
100	12.5	61	10.3	23

[a] From Fried (1957).

If an environmental change affects growth but not the amount of available nutrient, the A value will not change. The A value presumably changes only when the environment affects either the actual amount of available nutrient in the soil or the availability of the nutrient standard.

Tables 6.7–6.10 (Fried, 1957) illustrate that A values are relatively independent of both the rate of standard and various environmental factors. Marked changes in either of these factors do not affect the measurement of available P by the A value technique. Similar data have been presented by others (Grunes, *et al.*, 1955; Fried and Dean, 1950, 1952).

Table 6.8 shows the effect on *A* values of a change in yield due to other nutrient deficiencies than the one under study. The results indicate that *A* value for P remains relatively constant when yield varies due to the level of K or N in the soil system.

Table 6.9 indicates the effect of pot size on the *A* value obtained. Even a tenfold variation in pot size does not appreciably change the measured phosphorus *A* value.

TABLE 6.8
EFFECT OF YIELD AND NITROGEN AND POTASSIUM ADDITIONS ON PHOSPHORUS *A* VALUE

Nutrients added[a]	Chester loam		Orangeburg silt loam	
	Yield (g/pot)	*A* Value (lb P_2O_5/acre)	Yield (g/pot)	*A* Value (lb P_2O_5/acre)
NP	15	82	8.7	150
KP	19	94	6.8	160
NKP	20	90	11.9	160

[a] N = 150 lb N/acre; K = 150 lb/acre; P = 100 lb P_2O_5/acre.

TABLE 6.9
EFFECT OF POT SIZE ON *A* VALUE OBTAINED

Weight of soil per pot (g)	*A* Value[a]	
	Davidson clay loam	Norfolk fine sandy loam
290	30	52
760	26	49
2700	28	48

[a] Data given in lb P_2O_5/acre.

Finally, Table 6.10 indicates the effect of plant species on the *A* value. When monocalcium phosphate is the fertilizer standard, similar results are obtained for a variety of plant species. If rock phosphate were present in the soil, however, a wider variation might be expected (Fried, 1953).

Another advantage of *A* values is that they can be added and subtracted. Fried (1954) has pointed out how this can be used to evaluate processed phosphates, natural phosphates, and residual P in soil. Rubins (1953) has shown that *A* values reflect the actual removal of P by successive cropping of a soil. Prince (1953), Olsen *et al.*, (1954) and Grunes *et al.* (1955) have used *A* values to measure the residual value of applied phosphates.

The available nutrient as measured by "A" value, the Mitscherlich *b* value, and the yield of nutrient *a* value is a measure of the amount of nutrient in the soil in units of a fertilizer standard. If all the assumptions of the three methods were valid and the same fertilizer standard were used, the results obtained would be identical. This is not to be confused with the *L* value which is a means of determining the amount

TABLE 6.10
EFFECT OF PLANT SPECIES ON *A* VALUE OBTAINED WITH MONOCALCIUM PHOSPHATE AS FERTILIZER STANDARD

Plant species	*A* Value[a]	
	Norfolk fine sandy loam	Davidson clay loam
Alfalfa	90	200
Crotalaria	100	190
Perennial rye grass	90	190
Oats	100	190
Buckwheat	100	160

[a] Data given in lb P_2O_5/acre.

of surface soil nutrient in equilibrium with the nutrient in the soil solution and represents a more valid method of determining *E* value (see Chapter 6, Section 1.3.2 and Fried, 1964).

3. *Field*

Field measurement of plant nutrient supply are more troublesome and less precise than greenhouse and laboratory measurements because of the inability to control a myriad of variables. They also involve large investments of time, labor, and money. However, they are uniquely capable of reflecting the actual soil–plant relationship in the farmer's field. Greenhouse measurements reflect the amount of plant-available nutrient per unit volume of soil, whereas field measurements reflect the actual volume of the soil from which the plant obtains its nutrient.

Field experiments depend primarily on yield response for obtaining information, although in recent years the use of tissue testing and plant analysis has been increasing. From the standpoint of determining information about the nutrient under study these experiments are subject to the same limitations as greenhouse experiments. In addition, neither the environment nor the soil variation is controlled. A further difficulty with field experiments is the inability to measure a nutrient level in the soil unless marked yield response is obtained. Unfortunately, plant disease, insect pests, and weather often are responsible for reducing or eliminating this response. The oft-repeated observation that a field experiment, because of its lack of precision and control over variables, gives useful information only for a particular season on a particular plot area, which has already been changed by the experimental treatments, is not entirely facetious.

All of the measurements made in the greenhouse involving yield, yield of nutrient, and tracer methods can and are used in the field, although field experiments are never set up to determine the plant nutrient supply only.

3.1. Yield Response

Although this attempt to quantitatively generalize the relationship between amount of growth factor and yield does not stand the test of detailed experimentation, the concept of decreasing response to each additional increment of growth factor is still a most useful generalization, providing it is realized that the growth curve, in general, tends to be sigmoidal and that the interaction of growth factors must be taken into account. For most agricultural soils in areas where fertilizers are used, however, the inflection point in the sigmoid yield-response curve has been passed and, in the absence of other information, the mathematical relationship described by a Mitscherlich type equation is not unreasonable (Russell, 1950). Crowther and Yates put together all the results of fertilizer experiments made on various crops in Great Britain and from these, using the Mitscherlich approach, were able to formulate a suitable national wartime fertilizer policy for their country. In countries where the use of fertilizers is rather sparse and the yield level rather low, however, the possible sigmoid nature of the response curve could result in the development of a completely erroneous fertilizer policy by using the same procedure as Crowther and Yates. In any event, for the individual case, it is still advisable to utilize the best-fitting yield-response curves rather than try to make the data conform to a predetermined relationship.

In actuality, the addition of a nutrient to the soil system has an effect on both the capacity of the soil to supply nutrient to the plant and the intensity of this supply. In the case of nonadsorbed nutrients such as nitrate nitrogen, no other complications arise to any appreciable extent. The supply of nutrient can be said to have increased by the amount added, assuming no leaching losses, chemical fixation, or temporary microbiological immobilization. Indeed, this is probably the basis for the conviction of Willcox (1954, 1954a) that there is such a thing as an inverse N law, although Bray (1963) claims this is just the reason why N follows Liebig's law of the minimum rather than the Mitscherlich law.

Most other nutrients, however, react with the soil and with each soil differently. If the capacity factor is all-important, if the reaction of the nutrient with the soil does not take any of that nutrient out of equilibrium with the soil solution, and if the interaction of the added nutrient with the soil does not affect the amount or intensity of some other nutrient or factor in the system that may affect yield, then it can be said that the added nutrient has only increased the supply by a given increment and simple predictable responses are more likely to be valid. The corollary also holds that if intensity is all-important, simple predictable response curves generalizing the results for different soils are not likely to be particularly useful because the subsequent intensity owing to addition of the nutrient is a result of a difference in interactions in different soils. The addition results in some removal of the nutrient from the soil-solution system or in a change in the amount of intensity of supply of some other factor (or nutrient) that may affect yield.

Interactions not only take place in the soil but also in the plant. These plant interactions will surely be different for each plant species and to some extent for each variety within a species. Furthermore, the interactions will be different for different amounts and proportions of other elements present and for other environmental factors that affect yield.

Predictions concerning yield can be made on the basis of cause and effect when enough knowledge is available concerning the plant factors and their interrelationships, the soil factors and their interrelationships and the effect of those on the various steps in the continuum starting with nutrient on the soil and finishing with nutrient utilized by the plant in its metabolism. Otherwise, it is advisable to assume that each location and crop is a unique situation and the more of its uniqueness that becomes known, the greater the potential for maximum efficient utilzation of supplemental nutrition.

3.2. Plant Analysis

Essentially, all that was said for yield response holds for yield of nutrient, except that yield of nutrient is less dependent on climatic conditions and can be extended past the yield response level. However, the determination of yield of nutrient involves a much greater expenditure of time, labor, and money. Plant analyses, however, are used a great deal in the field in the form of determination of the concentration of the element in certain plant organs such as leaves or stems, on the assumption that this provides information on the nutrition status of the soil prior to sampling time.

Plant analysis can only indicate that at a particular moment during the development, the supply of one or more elements may be inadequate. It may sometimes be found that Mg is deficient, although the cause of the Mg deficiency might be a temporary wet period in the spring.

In those cases where plant nutrient needs can be met by later applications of appropriate fertilizers, plant analysis may be of real practical importance. There is only a limited advantage in using the results of plant analysis as a basis for fertilizer recommendations for the subsequent crop, both because of the difficulty in interpretation and because of the soil intensity factors may have altered considerably in the next growing season and only broad qualitative judgement can be made. Therefore foliar analysis has been applied most frequently to tree cultures, perennials, and annual crops in monoculture.

The basis for the application of plant analysis as a diagnostic tool was established by Lundegårdh (1938, 1943, 1945), Lagatu and Maume (1928, 1930) Thomas (1937, 1945), Beauchamp (1940), and Chapman (1941). For critical review of methods of sampling and interpretation of results of plant analysis, the reader is referred to Goodall and Gregory (1947) and Bergman (1958).

3.2.1. *Tissue Test*

For a measure of the nutritional status at a particular moment usually transport organs such as petioles or stems are extracted with dilute acids or salt solutions. In this way an analysis is made of elements in the plants that have been taken up recently and have not yet become incorporated into metabolic products of the plants. The tests are usually carried out in the field and are known as "tissue tests." One of the main reasons for the immediate analysis in the field is that during the transportation of samples from field to laboratory, the proportion between readily soluble and nonreadily soluble fractions may change (Thornton, *et al.*, 1934; Ulrich, 1943; Emmert, 1942; Morgan, 1937; Carolus, 1938; Nicholas, 1948; Plant *et al.*, 1944; Nicholas, 1955). One of the reasons for the extremely variable results of tissue tests is that they reflect the intensity factor in the soil which itself is markedly affected by moisture conditions.

3.2.2. *Foliar Analysis*

In leaf or foliar analysis, samples are usually transported into the laboratory. The analysis of elements is done after dry ashing or wet digestion. Expression of the results may be a percentage of dry matter, fresh weight, ash, or leaf area (Lindner, 1944; Ginniken, 1943; Chapman and Gray, 1949; Prevot and Ollagnier, 1951).

Two factors tend to interfere with the applicability of foliar analysis: (1) the chemical composition of plant tissues depend on the physiological age or morphological position of the organ; and (2) daily, weekly, or monthly changes in climatic conditions have a pronounced effect on the chemical composition of the plant tissues.

It is therefore essential to standardize the sampling procedure with regard to position of the organ and time of sampling (Broeshart, 1956; Hanway, 1962).

The interpretation of leaf analysis data is usually based on so-called "critical" levels, i.e., it is assumed that below a certain percentage of element in the plant sample its supply from the soil has been insufficient.

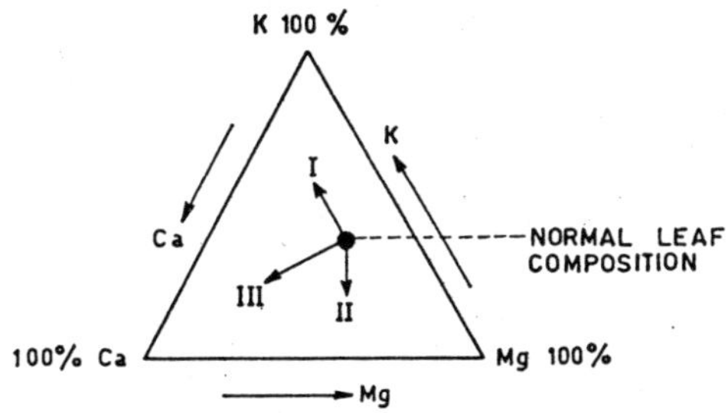

FIG. 6.20. The triangular representation of nutrient interrelationships used as a basis for interpretation of leaf analysis data. I, Deficiency of Mg; II, deficiency of K; III, deficiency of K + Mg.

Most investigators have found, however, that the deficiency of a particular element may induce changes in the contents of other nondeficient elements (Chapman and Gray, 1949; Prevot and Ollagnier, 1951; Lundgårdh, 1945; Broeshart, 1957). Instead of critical levels, the ratio between elements has therefore often been used as an indication for nutritional status, such as N/P, K/Ca, or K/Ca + Mg.

For a particular crop sampled according to a standard method and time, there is a range of "normal" chemical composition. As soon as one or more elements become deficient, characteristic deviations from the "normal" composition that might serve as a basis for diagnosis may be detected. The changes in composition are often represented in triangular diagrams (Thomas, 1945; Prevot and Ollagnier, 1951; Broeshart, 1955; Broeshart and Redlich, 1961) in which the contents of three elements are expressed as the percentage of their sum total (Fig. 6.20).

Great care, however, should be taken when the diagnosis is based on nutrient ratios only, because other elements may cause a change into direction I, II, or III of Fig. 6.20. Phosphorus deficiency may change the cationic composition similar to Mg deficiency, whereas N deficiency has a tendency to displace the cation ratios into the K-deficient or K + Mg-deficient direction (Broeshart, 1955).

In practice, however, the magnitude of the changes in concentration of an element in the leaf are such that a modification of the "critical value" concept can sometimes be applied, provided the "range" of optimal values which may be dependent on climatic conditions is known (Broeshart and Schouwenburg, 1961). This may be illustrated by the recent work on leaf analysis of corn in the United States. The optimal level of total N in the leaves that was found to be associated with maximum corn yields was 2.4 to 2.5% in North Carolina (Baird *et al.*, 1962). In other locations, a figure of 2.9% was suggested (Tyner, 1946) probably owing to a difference in climate.

The climatic influence on "critical" level may be demonstrated by data obtained from corn leaf analyses during subsequent years in North Dakota (see Table 6.11).

TABLE 6.11
MEAN MONTHLY TEMPERATURES AND CRITICAL N PERCENTAGES FOUND IN CORN LEAVES AT TIME OF POLLINATION FOR DIFFERENT YEARS IN NORTH DAKOTA[a]

Year	Critical N level (%)		Mean temperature (°F)	
	Grain	Forage	June	July
1953	3.35	3.08	63.8	68.4
1954	3.61	3.30	60.9	69.4
1955	3.12	2.81	62.5	70.6
1956	2.87	2.99	68.6	66.4
1957	2.75	2.38	61.6	74.2

[a] From Reichman *et al.* (1959).

Leaf analysis has been applied in Sweden for cereals (Lundegårdh, 1938, 1943, 1945), in the United States for tomatoes (Emmert, 1942; Reichman *et al.*, 1959; Malcolm, 1955, 1959; Thomas, 1937), and in Africa for groundnuts (Prevot *et al.*, 1951, 1953). The application of leaf analysis for annual crops remains, however, of limited significance (Scharrer and Lemme, 1952; Steenbjerg, 1951; Bould, 1955; Viets *et al.*, 1954; Clover, 1953) because of the many interfering factors other than nutrient supply that cause variability in leaf composition and because generally little or nothing can be done for the current crop when a low level of nutrient is found.

It has been shown that it was possible to analyze cereals at an early stage of development, e.g., 2 to 3 weeks after germination in order to try to correct nutrient deficiencies with top dressing (Broeshart and Schouwenburg, 1961). In practice, the time between sampling and interpretation of analytical data is usually too long to be of any value for a top dressing, particularly if large areas and large numbers of samples are involved. Furthermore, the principle may not be applicable under widely fluctuating climatic conditions during the early development of the crop.

In contrast to the concept of cation balance used as a basis for leaf analyses, de Wit *et al.*, (1963) have suggested the necessity for considering the cation-anion balance. They have indicated that in practice the cation content of the plant minus the inorganic anion content equals the organic anion content, e.g., in normal pasture plants this is about 1000 meq/1000 g of dry matter. Corrections of abnormality in the cation-anion balance must take into account both the level of the nutrient in the plant and the relative rate of uptake of different cations and anions.

As opposed to annual crops, the use of foliar diagnosis has proved very useful for perennial and tree crops. Once a standard sampling procedure has been worked out, it is not only possible to correct nutritional disorders of the same trees, but the effect of such treatments can again be followed up by means of leaf analysis. This is of particular importance for tropical and subtropical plantation crops such as citrus, coffee, rubber, and oil palm (Broeshart, 1955; Chapman, 1941; Culot *et al.*, 1958).

It may be concluded that plant analysis is not particularly useful in determining supply factors of nutrient, i.e., capacity factors. It is, however, useful for testing whether at a particular moment the supply of elements is sufficient or not. In other words, the methods give information about the nutritional status of plants but no quantitative information on the nutrient status of soils.

3.3. Tracer Methods

Field experiments with tracers are never set up primarily to determine plant nutrient supply. Nevertheless, the plant nutrient supply can be calculated from tracer experiments performed for other purposes. Again, as in the greenhouse, if the measurement is made in terms of some fertilizer standard, values are obtained which are quantitative and can be directly compared. Under the usual conditions, however, where the fertilizer standard cannot be mixed throughout the soil, the uptake from the standard will be affected by the nature of the root system as well as by the plant nutrient supply. Thus, to be strictly comparable, the nature of the root system with respect to the standard must be relatively constant when measuring levels of soil nutrient. In a given soil with the same nutrient supply, however, this apparent complication can be used to quantitatively estimate the effect of different root systems.

4. Evaluation of Soil-Testing Methods

Only the plant can determine the amount of plant-available nutrient, and chemical soil tests can only be evaluated when accompanied by appropriate observations on the plant. The validity of the evaluation depends entirely on the choice of the plant test. There is no universal agreement as to the most valid plant test. Although many evaluations are made with only one of the plant tests, others utilize two or three and assume that if the same extractant gives the highest correlation with all the plant tests, then it is the extractant of choice. This appears to be the best approach at present.

The plant tests referred to above are such tests as per cent maximum yield, Mitscherlich *b*, total uptake of the nutrient studied without addition of the nutrient, extrapolated *a* value, and the Fried and Dean *A* value. Of course, per cent maximum yield and Mitscherlich *b* are not independent variables nor are uptake of nutrient and a. Therefore, one of each of these two groups plus the *A* value when feasible, represents a reasonable basis for comparison. A fourth method, the Neubauer, is not considered to be as reliable as the previously mentioned three, but can be added as a fourth plant test. Indeed, this was the procedure utilized by the Soil Test Work Group in their attempt to find the most reliable quick test for P and K (Fitts *et al.*, 1956). The assumption has generally been made that the ultimate criterion for the most valid soil test is the extent of yield response in the field. However, this assumption is not necessarily valid. Yield is a function of a myriad of factors of which the uptake of the nutrient under investigation is only one. Unfortunately, these other factors cannot be controlled, and if a field measure were considered the ultimate criterion, it would have to reflect the amount of the nutrient available in the soil. This could involve direct measurement or measurement in relation to some standard. Although greenhouse tests do not measure the root-soil volume relationship as they occur in the field, neither can any soil test.

Greenhouse experiments are much cheaper and much more precise, and the soil test which effectively measures the plant nutrient supply under greenhouse conditions is likely to be the same test that effectively measures the supply of nutrients to the plant in the field. It is often more important to compare a greater variety of soils under the same environmental conditions than to make the tests out in the field. Greenhouse tests may be relatively useless for determining the amount of nutrient to apply, but they should be fairly effective in determining which laboratory soil test most effectively describes the plant nutrient supply. A few observations which respect to actual soil tests are recorded below.

4.1. Nitrogen and Sulfur

Incubation experiments for determining soil N supply are often significantly correlated with plant measurements of the availability of N to crops. The response of winter wheat to N was shown to be predictable on the basis of N release values from incubation experiments (Eagle, 1963). The correlation between the release values and the yield response depend on the plow depth. Figure 6.21 shows the relationship between

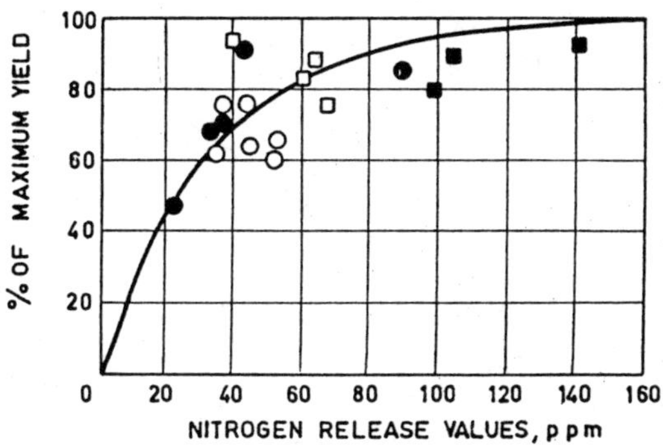

FIG. 6.21. Relation between yield (y, % of maximum) and nitrogen release values, ●, 1959; ○, 1960; ■, 1961; □, 1962; $\log(100\text{-}y) = \log(100\text{-}0.0133b)$; $r = 0.66$. From Eagle (1963).

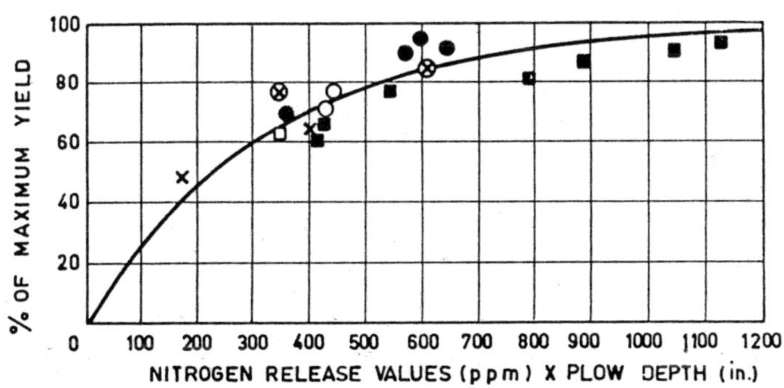

FIG. 6.22. Relation between yield response (y), soil texture, and products (b) of nitrogen release value and plow depth. ●, Loamy sand; ○, sandy loam; □, fine sandy loam; ■, silt loam; ×, clay loam; ⊗, clay; $\log(100 - y) = \log(100 - 0.00130b)$; $r = 0.76$. From Eagle (1963).

release values and yield response to N applications during 3 consecutive years. Figure 6.22 shows the same yield plotted against the N release values corrected for plow depth. In this case, the N status accounted for 60% of the yield variance and was not affected by different soil types. A poor correlation was improved by taking in a variable not normally considered, however, even with this improvement the correlation is relatively poor.

Much less variable results with incubation experiments are obtained if the chemical methods are compared with the results of greenhouse experiments. Grunes and co-workers (1963) compared different N laboratory tests with yield of dry matter and yield of N by plants growing

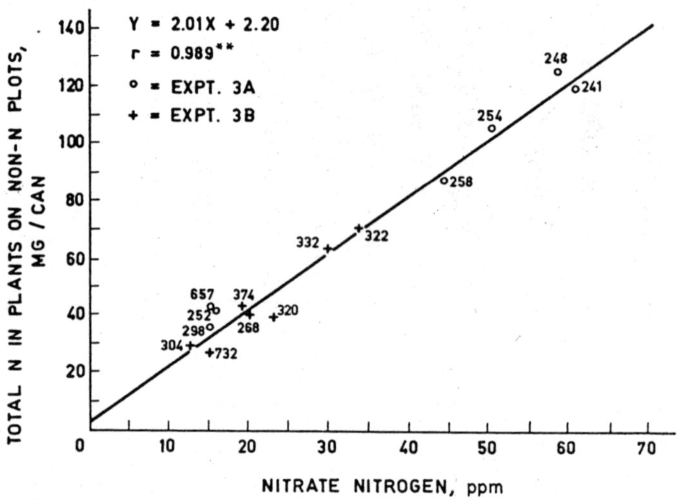

FIG. 6.23. Relation between total N in barley, grown in soil not fertilized with N, and nitrate-N in the soil following 3-weeks incubation with no $CaCO_3$ added. From Grunes *et al.* (1963).

in growth chambers where external conditions should be controlled. From a large number of different soils of 13 western states, total N, initial nitrate, and nitrate N after incubation were determined. The best correlation between growth response or N yield was with the sum of nitrate present before incubation and the nitrate release during incubation. The best results were obtained for a 3-week incubation time without addition of $CaCO_3$. The close correlation they obtained between the incubation method and yield of N in barley is shown in Fig. 6.23.

Similar to the supply of N, the soil supply of S is closely related to the release of S after incubation. Harward and co-workers (1962) investigated the relation between *A* values, S percentage in plants, and extractable S plus sulfate release on incubation with the yield of alfalfa in the greenhouse. The highest correlation was between yield and A values ($r = 0.925$). The S percentage in the plant gave a significant indication of the S status of the soil (correlation coefficient $r = 0.787$). A correlation coefficient of $r = 0.749$ was found for the extractable and released S during incubation vs. yield, whereas vs. A value r was 0.818.

4.2 Potassium

For K, the relationship between the results of chemical extractants and availability determinations in greenhouse and field depends on the soil type and many other factors such as pH, percentage of K saturated, magnitude of exchangeable K and Ca, etc.

In the Netherlands the extraction of soils for K is done with 0.1 N HCl. For sandy soils the K values are corrected for the humus content and the K number is represented by:

$$\text{K number} = (20 \times K_{HCL})/(\%\ \text{humus} + 10)$$

Similarly, the K content as determined by 0.1 N HCl is corrected for pH and clay content in the case of high clay content soil, (Pauw and Ris, 1960, 1962). Although the deviations from the yield-chemical extraction relationships are frequent, a usable criterion has been established (see Fig. 6.24).

Semb and Uhlen (1955) have found extraction with boiling 1 N nitric acid the best method of K extraction as judged by crop yields on Norwegian soils of different texture. This is a good example of a high correlation, even though the entity measured is not the source of the nutrient. Boiling 1 N nitric acid extracts much more K than is in equilibrium with the soil solution. Therefore most of what is extracted does not represent plant-available nutrient.

Others have found that water-soluble K gives higher correlations with yield than exchangeable K (Halstead and Heeney, 1959; MacKay and De Long, 1955; Hood *et al.*, 1956), and still other extracting solutions such as sodium acetate, water-soluble K, nitric acid-soluble K, and NH_4 acetate are claimed to give suitable results (MacLean *et al.*, 1957). The most common extractant for exchangeable K is probably 1 N NH_4 acetate at pH 7, and good correlations with plant data have been obtained by many investigators (Schmitz and Pratt, 1953; Pratt, 1951; Nelson, 1959; Sutton and Seay, 1958; Pearson, 1952; Bishop *et al.*, 1954), although other methods may give equally good results (Garman, 1957; Barber and Humbert, 1963). No single value obtained by a particular extraction method, however, is adequate to describe the supply of so-

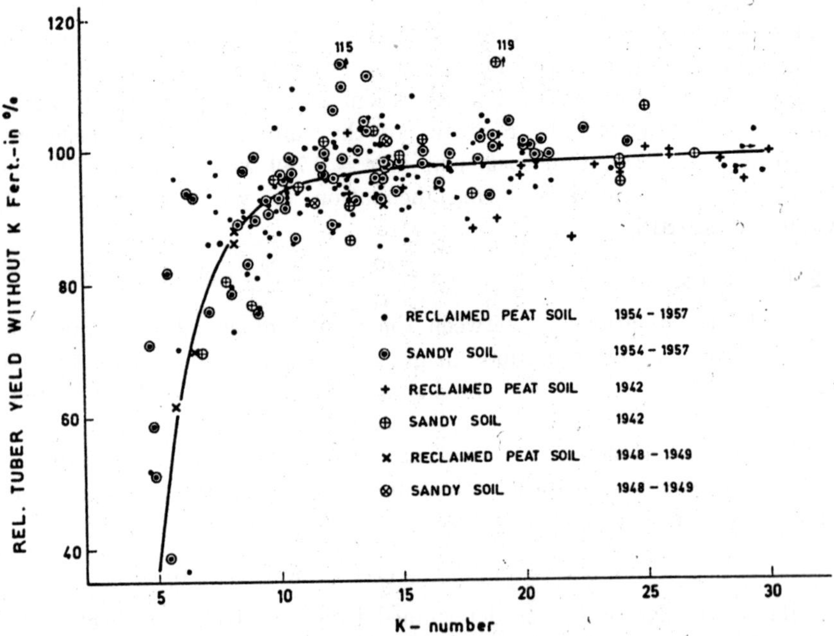

FIG. 6.24. Relation between K number and relative tuber yield in the absence of K applications. From van der Pauw and Ris (1960).

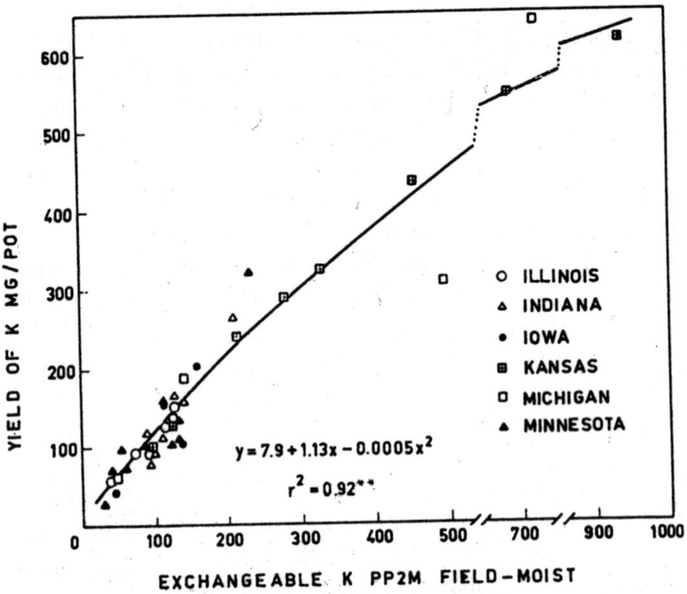

FIG. 6.25. Exchangeable K pp2m field-moist.

K. Generally, it is possible to determine values for each soil type below which the K supply is insufficient.

Much closer relationships are found when the chemical methods are tested against the yield of K in greenhouse experiments in contrast to field data. This is illustrated by results obtained by Barber and coworkers (1961) who used millet as a test crop to study the relation between exchangeable K and yield of K of a number of midwestern soils. Figure 6.25 illustrates their results, and a much better criterion for testing the method of exchangeable K determinations is the yield of K in a pot, rather than crop response in the field, insofar as precision is concerned.

4.3. Phosphorus

A large number of methods are used for the determination of level of soil P. The various plant test criteria are correlated with each other (Dean, 1954; Mattingly and Talibudeen, 1960).

Table 6.12 shows the high correlation between uptake of P and *A* value.

TABLE 6.12

Linear Regression Equations and Correlation Coefficients between Uptake of P by Rye Grass in the Greenhouse (y) and *A* Values of Soil (x)[a]

Soil	No. of soils	Regression equation	Correlation coefficient
Rothamstead, calcareous	9	$y = 0.546x - 0.52$	0.998
Forest Hursenis, acid	12	$y = 0.668x - 0.37$	0.986
Rothamstead, calcareous	7	$y = 0.217x - 0.42$	0.977
Saxmundhan, calcareous	9	$y = 0.290x - 0.41$	0.991
Acid and arable (calcareous) soils from C and E England	15	$y = 0.418x - 1.31$	0.870
Acid and very acid soils from England, Wales, N. Ireland, and Japan	24	$y = 0.335x - 1.15$	0.879

[a] Data given in mg P/100 g of soil. From Mattingly and Talibudeen, 1960.

From the many comparisons that have been made between chemical extraction techniques for P and the yield of P or *A* values, the plant nutrient supply of P in the soils seems to correspond rather well with the quantity of readily exchangeable phosphorus, as illustrated in Table 6.13. Surface P in this case was determined by adding P^{32} to an equilibrium soil suspension in water and determining the amount of rapidly exchanging P by means of isotopic dilution. The $NaHCO_3$ and surface

P gave the best agreement with plant-available P, and the Bray was almost as good. Resin extractions which measure capacity are also effective, as indicated by Lathwell *et al.* (1958) who compared resins and the extraction methods of Peech and Bray with *A* values. The highest correlation was obtained for the resin method ($r = 0.8$). Moser *et al.* (1959) found the order of precision in predicting the yields of P the highest for the anion resin method, followed by the 0.5 *M* $NaHCO_3$ method, and van Diest *et al.* 1960 (490) found the anion-exchange resin method to give a more precise estimate of the yield than the extraction method of Bray and Kurtz. The latter method of van Diest is a good example of an approach in which the findings from kinetic phosphate-uptake studies are used to improve the scope of a particular method of extraction.

TABLE 6.13
CORRELATIONS OF *A* VALUES WITH SOLUBLE P BY VARIOUS METHODS AND WITH SURFACE P[a]

Method	r
CO_2	0.128
H_2O	0.411
Bray	0.786
$NaHCO_3$	0.860
Surface P	0.952

[a] From Olsen (1952).

The amount of literature on the comparison between chemical extractants and availability tests in greenhouse or field is continuously increasing. With regard to the determination of the P supply, the amount of work that is still being done seems to suggest that general agreement on "the best extraction technique" does not yet exist. So much research has been done on many different phosphate-extracting methods that it has become possible to find data to prove that either water (Hagin *et al.*, 1963; Thompson *et al.*, 1960), acid extraction (Susuki *et al.*, 1963; Okrussko *et al.*, 1962), "Bray" (MacLean *et al.*, 1955; Smith and Pesek, 1962; Susuki *et al.*, 1963), "Olson" (Beaton *et al.*, 1962; Sen Gupta and Cornfield, 1963; Grunes *et al.*, 1963), or resin extraction methods (Moser *et al.*, 1959; Hagen and Hopkins, 1955) would give the best estimate of the soil P supply.

It will always be possible to extract a particular soil in such a way that high correlation coefficients are found between extracted values and field or pot tests. It does not seem likely that correlations between empirical soil extractants and availability tests will ever result in the adoption of a suitable extraction technique for P with a general applicability for all soils. Yet some extractants have wider application than others.

In general, it may be concluded that the plant nutrient supply of P seems to be related to surface, readily exchangeable P as determined by Olsen, and resin or isotope exchange methods. The extraction with water sometimes gives very good correlations with availability of P, but the results depend very much on the extraction procedure that is chosen, i.e., successive extraction or excess extractant, etc.

For particular soil types or conditions, other extraction techniques such as Truog, Morgan, Eger, or Bray may give excellent results, particularly when the dominating factors that are associated with P supply are known (such as nature of the cation associated with P, pH, etc.). Then it is possible to assess criteria for each soil type for a particular extractant and even, in some cases, to introduce a correction factor and use the method for all soil types.

4.4. In the Future

Most of the present chemical methods are not independent of soil types because they do not measure the capacity, the intensity, or both of plant nutrient supply in the soil. On the other hand, unless the future soil-testing methods are directed toward the development of techniques for the determination of the capacity and intensity factor, there is less chance that a soil-testing method will be developed in the future that is independent of soil type. Methods that can determine the capacity or intensity factor or both in the soil (e.g., simple isotope dilution techniques or resin extraction) are probably the most valid, but not necessarily ideally suited for routine testing. The decision of whether the increased accuracy is worth any extra effort is a local one.

CHAPTER 7

Principles of Fertilizer Application

1. Introduction

1.1. Trends

Once the nutrient supply of the soil has been determined, a decision will have to be made on fertilizer application. The object of fertilization is to obtain the highest possible yield with a minimum of labor and expense, including fertilizer. In order to fertilize in the most efficient way, a wide selection of fertilizers having different chemical and physical characteristics is available and these fertilizers may be applied at different times and by various methods. Each choice has its accompanying economic implications such as cost of fertilizer and power requirements. The actual fertilizer recommendation for a particular site will be governed by nutrient status of the soil and by specific soil, plant, and climatic factors which in the last analysis can only be determined by field experimentation (van der Pauw, 1962; Vanstallen, 1959; Riehm and Hofmann, 1955). Therefore, there is a limit to the applicability of the results of field tests for the prediction of the rate of fertilizer application. Irrespective of this limitation, findings with regard to time, placement, and nature of the fertilizer and nutrient interactions which can often be determined with much greater accuracy remain valid economic considerations.

With regard to current trends in fertilization practices, as stated by Vichar *et al.* (1963) there is a tendency to reduce the cost of labor involved in fertilizer application. This is particularly pertinent in those countries where fertilization is a normal procedure in agricultural production. In addition, the number of small farms is decreasing and farm size is increasing. At the same time, in some countries the total quantity of arable land is decreasing at the expense of industrial development and expansion of cities. As a consequence of these generalizations, the costs of application of fertilizers and fertilizer transport are being reduced to a minimum, and more fertilizers of high nutrient content are going to be applied. In the United States alone approximately 84% of the K, 80% of the P, and 32% of N are bought directly as mixtures, as shown in Table 7.1.

The remaining percentage represents situations in which only one nutrient is added owing to particular agronomic requirements such as top dressing of pastures with superphosphate or side dressing of maize with N, or large soil and climatic areas where response to only one nutrient (usually N) is clearly established. From the standpoint of plant nutrition, the application of fertilizers of higher nutrient content is also already tending to become a general practice throughout the world. In countries where labor is not a major cost factor but fertilizer is, efficient utilization of each nutrient applied is crucial. But even in these countries the plans for the future take into account this general trend.

TABLE 7.1
K_2O, AVAILABLE P_2O_5, AND NITROGEN CONSUMED IN THE UNITED STATES FOR ANNUAL PERIODS[a]

Nutrient	1960–1961		1961–1962		1962–1963	
	Mixtures	Total	Mixtures	Total	Mixtures	Total
K	1,883,111	2,168,533	1,973,149	2,270,537	2,170,112	2,536,366
P	2,069,425	2,645,085	2,219,444	2,807,039	2,494,843	3,092,070
N	1,071,224	3,030,788	1,147,266	3,369,980	1,261,375	3,903,629

[a] Data given in tons; year ending June 30, 1963. From Scholl *et al.*, (1964).

Intensive agricultural systems are the natural consequence of a continuous removal of all plant nutrients. Sooner or later, amendments for all plant nutrients have to be made to each soil that is intensively cropped. In general, it may be stated that fertilization for a high standard of agriculture eventually means regular (i.e., annual) applications of most major plant nutrients, N, P, K, Ca, Mg, and possibly S, and in some areas microelements.

1.2. FERTILIZER AND SOIL FACTORS

If no interaction between soil and fertilizers took place, the most efficient way to apply fertilizers would be in the form of liquids or highly soluble mixed or compound materials applied at the appropriate time and by appropriate methods. In this way the soil solution in close proximity to the plant roots could be maintained in an adequate state to obtain maximum growth or yield. In practice, this situation rarely occurs because the water solubility of highly soluble or liquid fertilizers is generally changed after their application to the soil. Furthermore, even the nonreacting nutrients may move away from the root zone, in addition to other economic considerations. The supply of plant nutrients from fertilizers is therefore affected by the extent to which soil and climatic factors interact with the added fertilizer. Consequently, the efficiency of fertilizers cannot always be forecast from their physical and chemical characteristics, but has to be experimentally determined either in greenhouse experiments or, when possible, in the field.

The major interfering soil and climatic factors that may modify the capability of fertilizers to supply nutrients to plants are: (1) losses due to leaching, (2) losses in gaseous form; (3) immobilization by chemical, physicochemical, and microbiological processes; (4) chemical reactions between various components in fertilizer mixtures; (5) changes in the capacity to supply nutrients due to fluctuations in rainfall and temperature; and (6) unfavorable effects associated with fertilizer applications to the extent that certain soil characteristics are modified in an unfavorable direction (pH, salt effects, toxicity, etc.).

The means that practical agriculture can use to reduce the extent of interfering soil factors are: (1) selection of the right kind of fertilizers of appropriate solubility and particle size; (2) application of fertilizers at the time that it is most required by the crops, in order to reduce the time of soil-fertilizer interaction to a minimum; and (3) placement of the fertilizer where it can be most efficiently taken up by the root system, in order to reduce the competition between soil and crop for the fertilizer nutrient.

1.3. Plant Factors

Special crop requirements may call for particular fertilizers, times, or methods of application. In the case of a rapidly growing annual crop, the rate at which nutrients are taken up by the plant reaches its maximum before vegetative development is at its maximum. This may result in a rapid removal of nutrients during a particular time, which becomes ever more extreme if the root system of the crop is not very extensive.

1.4. Amount to Apply

It is evident that there is no simple answer to the question: "How much fertilizer should be applied?" The actual quantity to be applied is a function of many factors which might be indicated in the form of an equation.

Quantity of fertilizer to apply $= F(s, p, n, i, t, m)_c$ where s is the soil nutrient supply; p, plant nutrient requirement; n, nature of the fertilizer (e.g., water solubility, particle size, acidity, etc.); i, interaction of fertilizer and soil, t, time of application; m, method of application; and c, cost of the application.

The constant, c, is to indicate that all the other factors are efficiency factors even if the cost of application remains constant. Several factors in this equation are related to each other, which explains why extensive experimentation is required to arrive at an optimal fertilizer recommendation for a particular location. Also p takes into account the nature of the plant and climatic factors determining the plant nutrient requirement.

One of the main problems in field experimentation to find the most efficient fertilization procedure is that often the criteria for comparing different treatments are not sensitive enough to detect even large differences. Total dry matter produced or yield of product is generally a rather unsatisfactory criterion for efficient fertilizer testing. If two treatments, which may be two chemical compounds of different solubility or two methods of placement, are compared, using yield of product as the criterion, the ability to detect differences between treatments or materials is so difficult that this approach is the least desirable, even though yield of product is the ultimate goal.

For the determination of the efficiency of a fertilizer treatment, a yield of nutrient approach is often used. This means that, for any treatment, the amount of nutrient in the plant derived from the fertilizer is the difference between that amount of nutrient in the plants receiving the treatment and those untreated. Apart from the high experimental error usually associated with differences between treatments, the vegetative development of check and treatment plants may differ greatly. It thus becomes difficult to accurately assess the uptake of the element from the fertilizers as compared to the uptake from the soil as additional errors are introduced.

To evaluate the efficiency of a particular fertilizer treatment, the direct determination of the quantity of element in the plant derived from the fertilizer would be the most obvious approach. In principle, this is only possible when radioactive or stable isotopes are available to label the fertilizers. In practice, this is the case with P for which the isotope P^{32} is of suitable half-life. In principle, the direct method of determining fertilizer efficiency can also be applied to N, although at present little field experimentation has been carried out with the stable isotope N^{15} due to its supposedly high cost. Suitable isotopes are available for Ca and S, but no suitable K isotope exists. The half-life of radioactive K^{42} is so short and enrichment of K with the natural radioactive isotope K^{40} is so expensive that even greenhouse and pot experimentation becomes impossible. In some cases Rb may be used as a label for K, but it is evident that a number of assumptions, not necessarily valid, have to be made with respect to the similarity in behavior of Rb and K in soil and plant systems.

2. Fertilizer Evaluation Methods

Once the nutrient supply of a soil has been evaluated, the most efficient fertilizer treatment has to be determined. Efficiency is defined here in the sense that plant nutrients are applied in such chemical and physical forms and at such times and placements that the maximum yield is obtained with the minimum possible amount of fertilizer nutrient application.

The final fertilizer recommendation in terms of cost and profit is not necessarily the most efficient, as defined above, as other cost factors such as material and labor must be considered in the final recommendation. Nevertheless, if a quantitative evaluation of the efficiency of each plant nutrient in the fertilizers is known, a basis for the choice of the most economic fertilizer treatment would be present even though the fertilizer treatment may finally be determined by overwhelming factors other than a soil–fertilizer–plant relationship.

The amount of fertilizer to be applied must be determined by field experiments, but only after the efficiency of the fertilizer in terms of nature, time, and method of application has been evaluated. The latter evaluation also normally requires field experimentation.

The object of a fertilizer evaluation is therefore to obtain a quantitative measure of the ability of the fertilizer to supply nutrients to the plant. Such an evaluation can be direct, in which a determination of the quantity of an element in the plant derived from the fertilizer is possible (as by means of isotope techniques), or indirect, in which the quantity of element in the plant derived from the fertilizer is determined from the difference in plant or nutrient yield between treated and untreated, or two different treatments.

2.1. Direct Methods

By means of isotope techniques it is possible to discriminate directly between that quantity of element in the plant derived from the fertilizer and that quantity of element derived from the soil. There are essentially two ways of utilizing this information to rate quantitatively the ability of two materials to supply nutrient to the plant, namely, calculation of per cent utilization of applied nutrient and calculation of the per cent of the total element in the plant derived from the fertilizer.

2.1.1. *Per Cent Utilization of Applied Nutrient*

The per cent utilization of applied nutrient is a direct comparison between amount of nutrient taken up from the fertilizer and amount of nutrient applied. Thus, if the amount of nutrient applied in the form of two different fertilizers is B and D and the amount of nutrient taken up from each source is M_b and M_d, then the per cent utilization of B, U_b, and D, U_d are shown in Eq. (7.1)

$$U_b = M_b/B \times 100, \qquad U_d = M_d/D \times 100 \tag{7.1}$$

The relative rating of the two materials is then:

$$U_b/U_d = M_b/B \div M_d/D = M_b/M_d \times D/B \quad (7.2)$$

provided the relationship between nutrient supply and per cent utilization is linear. This relationship, however, is not linear but parabolic and only approaches linearity at a low per cent utilization. A low per cent utilization is normally obtained when the soil is high in the nutrient in question. In practice, the rate of application of the two nutrients, B and D, is chosen at the same level and the per cent utilization is then compared.

Another difficulty in using per cent utilization is that the addition of the fertilizer itself may result in a yield response and an over-all increase in the uptake of the nutrient not only from the soil but also from the fertlizer. Thus, the increase in the per cent utilization of a given material may be confounded by an increase in growth resulting in further uptake from the added fertilizer material. Only when there is no physiological response to the applied fertilizer can this difficulty be overcome. Nevertheless, per cent utilization of applied fertilizer is a direct measure of fertilizer efficiency under a given set of conditions. This direct measure of amount of, or per cent of, the fertilizer nutrient utilized cannot effectively be performed in any other way.

2.1.2. *Per Cent of the Total Element in the Plant Derived from the Fertilizer*

When a nutrient element, M, in the fertilizer is labeled with an isotope of the element, M^1, which may be either a stable or radioisotope, the fraction of M in the plant derived from the fertilizer, F_m, is found as the ratio of the specific activities of the element in the plant and in the fertilizer, i.e.,

$$F_m = M^1(\text{plant})/M(\text{plant}) \div M^1(\text{fertilizer})/M(\text{fertilizer}) \quad (7.3)$$

where M is the amount of a given unlabeled nutrient ion; M^1, amount of a given labeled nutrient ion; and F_m, the fraction of the nutrient ion in the plant derived from the fertilizer. F_m can be expressed as the per cent of the nutrient ion in the plant derived from the fertilizer by simply multiplying by 100.

Equation (7.3) may seem an indirect way of expressing the data, but it has, in fact, certain advantages.

First, it indicates at a glance the relative contribution of the soil and fertilizer to the nutrition of the plant and, second, it permits an evaluation of the fertilizer source in relation to the soil source of the nutrient, thus giving a base line to the evaluation. If two materials are compared, they are then compared to the same base line (i.e., the soil) and thus given materials B and D, Fm_B and Fm_D can be compared to give a relative rating of the two materials. However, again the rating is only valid where the relationship between the per cent of the nutrient in the plant derived from the fertilizer and the supply of nutrient in the fertilizer are linear for both materials. Unfortunately, this relationship is not linear but parabolic, as shown in Figure 7.1.

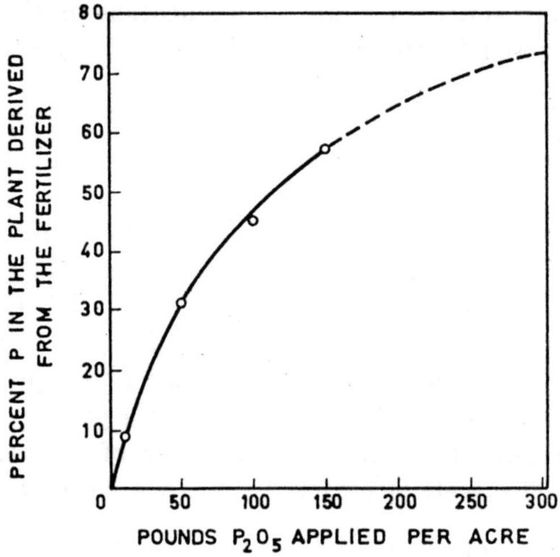

FIG. 7.1. Evaluation of phosphate fertilizer. From Fried (1954). Reprinted with permission of copyright owner, the American Chemical Society.

This parabolic relationship can be transformed to a linear relationship (Fried, 1954) by expressing the relative ability of the two fertilizers to supply nutrient in terms of the soil supply of nutrient which has remained a constant. This transformation is the A value transformation as discussed in Chapter 6, Section 2. Knowing that the relationship is parabolic, however, permits one to make such a transformation irrespective of the assumed definition of the A value. The calculation is shown below:

$$A_B = B(1 - F_{M_B})/F_{M_B}, \qquad A_D = D(1 - F_{M_D})/F_{M_D} \tag{7.4}$$

where A is the amount of nutrient available in the soil expressed either in terms of units of B, A_B or units of D, A_D; B, amount of fertilizer B applied; D, amount of fertilizer D applied; and F_M, proportion of the nutrient in the plant derived from either B, F_{M_B} or D, F_{M_D}.

Since the amount of nutrient available in the soil is a constant, a solution of the ratio of A_B to A_D at any point gives the relative size of the units, i.e., the relative ability to supply nutrients to the plant.

Dividing A_B by A_D

$$A_B/A_D = B/D \times (1 - F_{M_B})/(1 - F_{M_D}) \times F_{M_B}/F_{M_D} \quad (7.5)$$

when $F_{M_B} = F_{M_D}$,

$$A_B/A_D = B/D \quad (7.6)$$

The ratio A_B/A_D is the ratio of the amount of fertilizer B and D necessary to supply the same amount of nutrient to the plant relative to the soil supply, i.e., the same F_M. Since the larger the amount of fertilizer necessary to supply a given amount of nutrient the less efficient the fertilizer, the rating of the fertilizer is inversely proportional to the A values. Furthermore, since A values are relatively independent of rate of application, neither the same rate of application of nutrient nor of fertilizer need be used to obtain a quantitative evaluation of the two materials.

Although this derivation has been made in terms of the A value, the same derivation would be applicable if a linear transformation approximating the conversion of the parabolic curve of Fig. 7.1 to a linear relationship, or even a semilogarithmic relationship, were utilized.

The calculation of per cent of the nutrient derived from the fertilizer in comparing fertilizer materials and its conversion to A values or, mathematical transformation, has the added advantage of canceling out an effect that the fertilizer material has upon response to the plant. Since this response will result in an increase in uptake of both soil and fertilizer nutrient, there is no compounding of the higher effectiveness of one material compared to another as is the case with per cent utilization. In other words, since the effectiveness is measured in terms of soil nutrient and this is a constant, the amount of growth has no appreciable effect on the relative ability to supply nutrient from the soil or the fertilizer.

2.1.3. *Additional Methods*

a. *Competitive Ions as Tracers.* Unfortunately, suitable isotopes of each of the nutrient elements do not exist. This has suggested the use of competitive ions as tracers. Of the known competitive ions of nutrients without suitable tracers (Chapter 3, Section 2.3) the competition of Rb and K is the most important. There is no suitable isotope of K for use in labeling fertilizers. Potassium 42 ($t_{1/2}$ = 12.5 hours) and K^{43} ($t_{1/2}$ = 22 hours) have too short a half-life to be of practical value in pot or field experiments that usually last from several weeks to months. Enrichment of the natural radioisotope of K, K^{40}, or enrichment of the stable isotope K^{41} are so expensive at present as to be prohibitive in cost.

Since Rb and K are indistinguishable insofar as ion uptake is concerned (Chapter 3, Section 2.3), Rb has been tested as a tracer of K in soil studies. The results of the classic experiment which compared the use of Rb as a tracer to a true tracer of K, namely K^{41}, are shown in Tables 7.2 and 7.3.

The results in Table 7.3 illustrate that when Rb data are used to calculate the per cent K in the plant derived from added K, completely erroneous results are obtained. Not only is the magnitude of the results in error, but the soils are not even rated in the same order. Thus, the plant may not be able to distinguish between K and Rb insofar as the uptake process is concerned (see Chapter 3, Section 2.3), but the distinction is made in the soil. As far as distribution in the plant is

TABLE 7.2
EFFECT OF SOIL TYPE AND ADDED POTASSIUM ON THE DISTRIBUTION FACTOR (D.F.) FOR RUBIDIUM UPTAKE[a]

Soil type	Level of added potassium ($lb\ K_2O$/acre) 10	50
Decatur	0.24	0.26
Davidson	0.25	0.25
Grenada	0.30	0.31
Wooster	0.35	0.34
Herrick	0.43	0.38
Hagerstown	0.46	0.45

[a] D.F. = Rb/K (plant) ÷ Rb/K (soil). From Fried and Hawkes (1959).

concerned, K:Rb ratio can vary (Mackie and Fried, 1955). The distribution factor, i.e., the ratio of Rb/K in the plant to Rb/K in the soil, is constant for a given soil-plant system, but differs from soil to soil (Table 7.2). This means that Rb can be utilized as a tracer for K in a given soil with respect to kind, time of application, or placement with valid results. It is not valid, however, to make a comparison among soils or to attach particular significance to the absolute magnitude of Rb uptake.

b. *Commercial Materials.* It is essential that any comparison with labeled fertilizers or, for that matter, with any fertilizers, be done with materials that reasonably well simulate the commercial product. This is no particular problem with nonlabeled fertilizers as the commercial product can be used directly. Laboratory preparations of labeled fertilizers, however, will seldom be the same as the commercial product unless a special effort is made. This was illustrated by a comparison of a series of nitric phosphate fertilizers containing different amounts of water-soluble P. The entire series of materials was made in two ways, resulting in two sets of materials with identical water and citrate solubilities. A plant test in the greenhouse, however, indicated that the effect of method of preparation was very great and, in fact, much greater than the effect of water solubility (Hawkes and Fried, 1957).

The labeling of fertilizer materials must be done during their manufacture so that all fractions of the fertilizer have identical specific activities. The mixing of fertilizers with solid or liquid labeled material not only results in the labeled material being entirely different chemically than that in common commercial fertilizers but also the labeling—in particular with elements of low mobility such as P^{32}—will be very heterogeneous.

TABLE 7.3

EFFECT OF TRACER USED ON THE CALCULATION OF PERCENTAGE OF ADDED POTASSIUM IN THE PLANT DERIVED FROM THE FERTILIZER[a]

Soil type	Isotope used: K^{41} (%)	Isotope used: Rb^{86} (%)
Decatur	19	4.8
Wooster	18	6.1
Grenada	15	4.5
Hagerstown	13	5.9
Herrick	13	5.0
Davidson	12	3.1

[a] From Fried and Heald (1960).

c. *Natural Products.* Generally, processed materials can be labeled, provided suitable isotopes are available. Natural products cannot be labeled because any treatment of the fertilizer will drastically change the chemical characteristics of the fertilizer and its interaction with soil (Fried, 1954). To overcome this, Fried and MacKenzie (1949) irradiated rock phosphate directly in a neutron pile and attempted an evaluation in soil. This was later shown to be invalid, owing to the recoil effect on P atoms that absorbed the neutrons, resulting in the presence of labeled nonorthophosphate P (Fried and MacKenzie, 1950; MacKenzie and Borland, 1952).

Natural products, however, can be compared with each other or other sources by comparing each against a standard source. The residual supply of P from past fertilizer treatments can be measured in the same way. The evaluation is based on the thesis that A values are quantitative measures that can be added and subtracted since they are measures of the amount of available nutrient in terms of some standard unit of measure. The technique involves determining the A value of the soil alone while, at the same time, determining the A value of the soil plus the residual material or the natural fertilizer. As long as the same standard is used, the two A values may be subtracted from each other. The calculations are shown below:

$$A_1 = B_1(1 - F_{M1})/F_{M1} \tag{7.7}$$

$$A_1 + A_2 = B_1(1 - F_{M2})/F_{M2} \tag{7.8}$$

where A_1 is the amount of available nutrient in the soil; A_2, amount of available nutrient in the natural fertilizer material (or residual material); B_1, amount of nutrient applied as standard; F_{M1} and F_{M2}, proportion of the nutrient in the plant derived from the standard. Subtracting Eq. (7.7) from Eq. (7.8)

$$(A_1 + A_2) - A_1 = A_2 = B_1/F_{M2} - B_1/F_{M1} \tag{7.9}$$

gives the amount of available nutrient in the applied natural fertilizer (or residual material) in terms of the standard.

TABLE 7.4
COMPARISON OF VARIOUS PHOSPHATE SOURCES WITH SUPERPHOSPHATE AS A STANDARD[a]

	Soil			
Fertilizer	Thailand (pH 5.1)	Brazil (pH 6.2)	Hungary (pH 6.6)	Pakistan (pH 8.2)
Olinda	640	98	66	2
Araxa	466	110	126	12
Araxa thermo	1000	1758	1280	1230
Tunis rock	927	649	88	82
Florida rock	866	420	27	0
Basic slag	848	1345	690	719
Bone meal	852	735	244	173

[a] The data give the grams of superphosphate equivalent to 1000 g of phosphate fertilizer (average of three replications).

If B_2 is the total amount of nutrient in the applied natural fertilizer, than A_2/B_2 is the relative amount of each material (standard and natural product) needed to supply the same amount of plant available nutrient. The relative efficiency of the two materials is the inverse of this ratio.

The results are essentially identical to those obtained under Chapter 7, Section 2.1.3 since the property of the ability to add and subtract A values is utilized to obtain a relationship that can be obtained directly when materials can be directly labeled. Thus, any number of materials may be compared or residual values of applied materials determined. This is illustrated in Table 7.4 (IAEA, 1964) which compares various natural phosphate sources in terms of superphosphate standard.

2.2 Indirect Methods

The rating of fertilizers using the indirect method involves comparisons of particular fertilizer treatments with a control in which no fertilizer has been applied. A more sophisticated method involves as a control enough different rates of application of a standard material that the whole response curve is determined. The yield of dry matter or nutrient obtained with the fertilizer material tested is then located on this standard curve and the equivalent amount of standard read off the abscissa. Where isotopes are not available, this is the method of choice and has been used by various investigators (Terman *et al.*, 1962; Armiger and Fried, 1957, 1958; Black and Scott, 1956; White and Kempthorne, 1956).

There are certain inherent difficulties in the use of yield response or yield of nutrient to determine the efficiency of applied fertilizers. If the plant is provided with a nutrient element from two materials, A and B, in which B when applied at the same rate as A will supply twice as much nutrient to the plant, there are three possibilities in yield response comparisons of the two materials: (1) no yield response, because the nutrient level in the soil was already so high or because other factors of nutrition, temperature, moisture, disease, or insect pests limited yield; (2) slight yield response to both materials because the yield level is close to the maximum for the conditions of the experiment; and (3) marked yield response to both materials with a slight difference between them.

Any difference between material is indistinguishable in cases (1) and (2). Case (3) is illustrated in Fig. 7.2, where Y_C, the yield of the check, is only 50% of the maximum yield. Nevertheless, the difference between the yield of A, Y_A and the yield of B, Y_B may only be a difference of some 10 to 15%, a difference which is not even detectable if a 5% level of significance is sought. Yet B is very much more efficient that A in supplying nutrient. Even if a significant difference were found, it would be next to impossible to give a quantitative estimate of the relationship between A and B. Field experiments using yield response as a criterion quantitatively evaluate materials, placements, or time of application with difficulty. In theory, quantitative evaluation is valid only when the check yield and both treatment yields are on the linear response part of the curve that is between zero yield and below Y_C. In practice, this requires the use of a soil almost devoid of the nutrient and a comparison at yield levels that are not usually found in agricultural practice.

The low sensitivity of the yield approach for testing fertilizer efficiency may be illustrated by results from a large number of experiments on comparisons of various water-soluble and citrate-soluble P fertilizers (Terman, 1960, 1961). From 298 experiments in which yield responses to fertilizer were significant at the 5% level, less than 50% showed a statistically significant difference between the standard source and the sources under comparison. "In only a few of the tests (author's

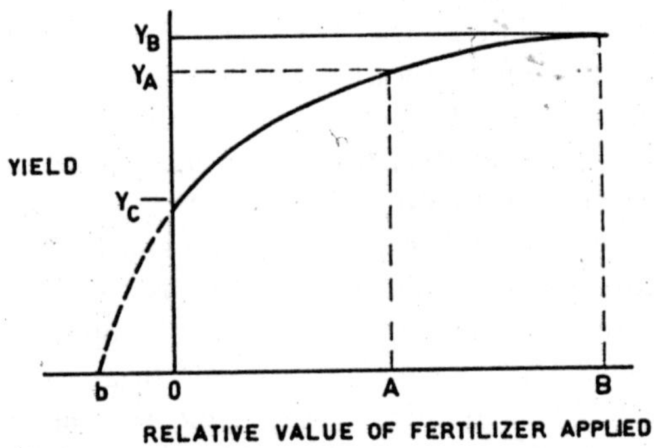

FIG. 7.2. The yield response to two fertilizer materials or two placements which differ markedly in ability to supply nutrient (schematic).

note: less than 10%) with various crops showing yield response to P was it possible at the 5% probability level to measure differences among citrate- and water-soluble sources which were from one-half to twice as effective as CSP."

Terman *et al.* (1962) list a number of practices that may reduce the experimental errors in fertilizer source-evaluation experiments in the field. These practices include proper selection of experimental sites, adequate number of replication, uniform preparation and application of fertilizers, control of limiting yield factors, and sampling and harvesting techniques. There is still a fundamental difficulty, however, in quantitatively evaluating fertilizer materials. Replications in the field cannot be increased *ad infinitum,* nor is it valid to make tests only on the homogeneous soil situations which tend to occur only on relatively level areas of similar microclimate. No amount of statistics can substitute for the basic difficulties in the method.

3. Nature of the Fertilizer

A large number of different materials are available on the fertilizer market. Apart from the chemical composition of single, compound, or mixed fertilizers, the materials may have widely different solubility in water and may interact with soils in different ways. They may differ not only chemically but physically. Table 7.5 lists the most common chemical fertilizers and their composition. In addition to chemical fertilizers, a wide variety of natural organic materials is available from animal by-products, excreta, and plant residues (Uexküll, 1963). A de-

TABLE 7.5
CHEMICAL FERTILIZERS: NITROGEN, PHOSPHATE, POTASH

Material	Formula	Total nitrogen (N) %	Units of material needed for one unit pure nutrient
Ammonia, anhydrous	NH_3	82	1.2
Ammonia solution, aqua ammonia	$NH_4OH + H_2O$	24	4.2
Ammonium chloride	NH_4Cl	24	4.2
Ammonium chloride, calcium carbonate	$NH_4Cl + CaCO_3$	15	6.7
Ammonium nitrate	NH_4NO_3	33	3.0
Ammonium nitrate, ammonia solution	$NH_4NO_3 + NH_4OH + H_2O$	38	2.6
Ammonium phosphate[a] (monoammonium)	$NH_4H_2PO_4$	11	9.1
Ammonium phosphate[a] (diammonium)	$(NH_4)_2HPO_4$	21	4.8
Ammonium sulfate	$(NH_4)_2SO_4$	20.5	4.9
Ammoniated superphosphate[a]	$(NH_4)_2HPO_4 + CaHPO_4 2H_2O$	5	20.0
Ammonium phosphate sulfate[a]	40% $NH_4H_2PO_4$, 60% $(NH_4)_2SO_4$	16	6.3
Calcium cyanamide	$CaCN_2$	21	4.8
Calcium ammonium nitrate	$NH_4NO_3 + CaCO_3$	20.5	4.9
Calcium nitrate	$Ca(NO_3)_2$	15.5	6.5
Potassium nitrate[a]	KNO_3	13	7.5
Sodium nitrate	$NaNO_3$	16	6.2
Urea	$CO(NH_2)_2$	45	2.2
Urea-ammonia liquor	—	45.5	2.2
Urea-formaldehyde (urea-form)	—	38	2.6
Urea-acetaldehyde (urea-Z)	—	33	3.0

TABLE 7.5 (*Continued*)

Material	Formula	Potash content (K_2O) %	Units by weight of material needed for one unit of potash (K_2O)
Manure salts	60% KCl	40	2.5
Muriate of potash	80% KCl	50	2
Muriate of potash	96% KCl	60	1.7
Nitrate of potash[a]	KNO_3	44	2.3
Potassium magnesium sulfate[a]	50% K_2SO_4, 30% $MgSO_4$	26–30	3.4
Potassium sulfate	90–96% K_2SO_4	48–52	2
Potassium bicarbonate	96% $KHCO_3$	45–47	2.2

Material	Formula	Total phosphoric acid (P_2O_5) %	Available phosphoric acid (P_2O_5) %	Units of material needed for one unit of pure nutrient
Ammonium phosphate sulfate[a]	40% $NH_4H_2PO_4$, 60% $(NH_4)_2SO_4$	20	20	5.0
Ammoniated superphosphate[a]	—	14–20	13.5–12.5	7.4–5.1
Ammoniated triple superphosphate[a]	—	43–49	42.5–48.5	2.3–2.1
Basic slag	—	18	16	6.3
Calcium metaphosphate	$Ca(PO_3)_2$	64	53	1.9
Diammonium phosphate[a]	$(NH_4)_2HPO_4$	53	53	1.9
Dicalcium phosphate	$CaHPO_4$	40	39	2.6
Monoammonium phosphate[a]	$NH_4H_2PO_4$	48	48	2.1
Rhenania phosphate	—	27	25	4.0
Roechling phosphate	—	20	18	5.6
Phosphoric acid (liquid)	H_3PO_4	54	(54)	(1.9)
Rock phosphate[b]	—	—	—	—
Superphosphate, single	50% $CaSO_4$, 30% $CaH_4(PO_4)_2\ 2H_2O$	16–20	16–20	6.3–5
Superphosphate, double or triple	$CaH_4(PO_4)_2 2H_2O$	45–50	43–49	2.3–2.0

[a] These materials contain other plant nutrients in addition to potash. From Jacob and Uexküll (1963).

[b] No figures for composition of rock phosphate are given, because the variations in the phosphote content and the response of crops to dressings of the ground phosphate rock are extremely variable, depending on soil, climate, kind of crop, and type of rock. Analysis figures would therefore be entirely misleading.

tailed description of the production and chemical nature of fertilizers is given by Collins (1955), Vickar *et al.* (1963), and Jacob (1953).

To a certain extent, field and greenhouse experimentation is required to make a choice of fertilizers for a particular location and crop. There are a number of soil and plant factors, however, which, when known in advance, can give an indication on the likely nature of the fertilizers that should be used.

3.1. Fate of Fertilizer in Soil

3.1.1. *Leaching Losses*

a. *Nitrogen.* Nitrogen in the form of NO_3^- is particularly subject to leaching. The extent of leaching is greatly reduced when the soil is cropped, but losses may become appreciable on bare soils (Chapter 1, Section 2). Losses due to leaching may also be appreciable on sandy soils because the nitrate movement is a function of the amount of water that moves down the profile. In clay soils with a high water-holding capacity, the movement of nitrate will be much less than in sandy soils (Bates and Tisdale, 1957). Nitrate losses can be reduced by a suitable choice of time and method of application (Dreibelbis, 1957; Harrold and Dreibelbis, 1951; MacIntire *et al.*, 1962; Kilmer *et al.* 1944). Losses of N are also reduced when the fertilizer is applied in the form of ammonium fertilizers, as calcium cyanamide or urea. The NH_4 ion which is introduced directly or after hydrolysis of the fertilizer becomes adsorbed to the exchange complex of the soil. In acid sandy soils with excessive leaching, even losses of NH_4 may become appreciable because these ions will not be able to replace the strongly adsorbed Al ions. Under such conditions, a much greater retention of N is obtained when the applications are made in the form of anhydrous ammonia. The NH_3 gas will react directly with the hydronium ions in the soil and with adsorbed Al ions (Baker, 1959).

Ammonium fertilizers, urea, and calcium cyanamide are also important when the N applications have to be made in the fall or early winter, when not only the labor and transport situation but also the physical conditions of soil tillage is more favorable. As long as the temperature after the fertilizer application is below 50°F, microbiological oxidation of NH_4 to NO_3 will then be limited (Dumenil, 1957; Kurz *et al.*, 1961; Shaw, 1962; Moore, 1960). There are still many conflicting observations in this respect however, and considerable losses of N from fall applications have often been reported (Gasser, 1961; Doll, 1962; Gasser, 1962; Olesen and Bent, 1959; Owens, 1959). Calcium cyanamide appears to have an advantage over NH_4 fertilizers for fall applications apparently because both ammonification and subsequent nitrification are temperature controlled and will proceed at a slow rate as long as the temperature remains low (Gliemeroth, 1959).

Nitrates that have leached to deeper horizons need not be lost altogether because deep-rooted crops such as corn (but also oats and soybean) may still profit from nitrate that has leached into the subsoil (Cook and Scarseth, 1941; Scarseth *et al.*, 1944; Dumenil and Nicholson, 1952; White and Pesek, 1959; Kurtz *et al.*, 1961; Boswell and Anderson, 1964). The nitrogen fertilizers in current use are generally highly soluble, and dissolution of nitrate into solution occurs rapidly either directly or after a microbiological oxidation of NH_4 or amides to nitrate. Relatively insoluble nitrogen fertilizers such as magnesium ammonium phosphate ($MgNH_4PO_4$), oxamide ($NH_2COCONH_2$), dicyan-diamide [$NH_2C({=}NH)NHCN$], or thiourea [$CS(NH_2)_2$] have as yet been little used because the cost per unit of N is twice as high as that of ammonium nitrate (Sharp and Powell, 1963). Recently it was found that nitrification inhibitors such as 2-chloro-6-(trichloromethyl)pyridine are able to reduce downward movement of nitrogen from NH_4 fertilizers (Carter *et al.*, 1967). Apparently the fertilizer NH_4 is not oxidized to NO_3 and consequently does not move as rapidly as NO_3.

b. *Other Nutrients.* The leaching of fertilizer K is not a serious problem in most soils. Potassium is adsorbed to the adsorption complex and, as shown in Chapter 5, Section 2, leaching losses during crop growth are relatively small on cropped soils. The leaching losses in acid soils of the humid tropics, can, however, be considerable if the soil is bare (Humbert, 1958). Potassium in most fertilizers has been found to dissolve readily in the soil solution and as a plant nutrient even less soluble materials such as potassium frit (Lunt and Kwate, 1956), potassium metaphosphate (MacIntire *et al.*, 1953), and fused potassium phosphates (De Ment and Stanford, 1959) have been reported equivalent to highly soluble fertilizer salts such as KCl and K_2SO_4.

Virtually no leaching losses of P occur from fertilizers. Most soluble P fertilizers are rapidly removed from the soil solution by reaction with the soil and migrate very little from their point of adsorption.

Sulfate resembles nitrate more closely with respect to reactions in soils. Leaching losses from added fertilizers may be appreciable. When S is applied as a component with N, P, or K fertilizers, leaching losses generally are not of such a magnitude that additional S compounds have to be added to the soil. Relatively insoluble materials such as gypsum still rapidly dissolve in the soil solution and appreciable losses of sulfate owing to leaching are found, depending on the amount of rainfall (McKell and Williams, 1961).

The cations Ca^{2+} and Mg^{2+} are even more strongly adsorbed to the soil exchange complex than K^+. Leaching losses of these elements when applied with fertilizers are of little immediate agronomic significance. Mg^{2+} is slightly more mobile than Ca^{2+} as demonstrated by the relative accumulation of Mg^{2+} in the B horizon.

3.1.2 *Gaseous Losses*

The loss of fertilizer elements owing to formation of gases refers to N fertilizers mainly. The processes that are involved are: (1) direct loss of NH_3 at high soil pH; (2) losses of N_2, N_2O, or NO owing to de-nitrification; and (3) losses of N_2, N_2O, or NO during nitrification of NH_4 fertilizers and urea. They are markedly affected by the moisture content and temperature of the soil. Direct losses of NH_3 from ammonium fertilizers and urea may become pronounced in alkaline or calcareous soils with a low cation-exchange capacity. High temperature as well as the drying of the surface soil favor losses (Donald *et al.*, 1963). When urea is applied, NH_4 is formed by hydrolysis and the pH may rise to high values in local areas. In that case, losses of NH_3 are even possible under continuous moist conditions, particularly when the soil pH is already high (Kresge and Satchell, 1960; Overein and Moe, 1967) or the exchange capacity low (Volk, 1959).

The use of nitrate under conditions in which ammonia losses are likely is often not practicable. Therefore, the reduction of NH_3 losses generally has to be solved by proper placement methods of ammonium fertilizer or urea (Martin and Chapman, 1951).

Losses of N due to denitrification of nitrates generally occur under conditions of reduced soil aeration and in the presence of fresh organic material. The optimal pH for denitrification seems to be between 7 and 8 (Nommik, 1956) and above 25°C (Bremner and Shaw, 1955; Nommik, 1956). Under such conditions the application of N in the form of NH_4 fertilizers or urea is the obvious approach, although the reduction of N losses may be offset by the fact that NH_4 may be less readily taken up by crops than NO_3 (Shalhevet and Zwerman, 1962). In some cases, however, denitrification losses from NH_4 fertilizers can occur. This is the case for soils under submerged conditions in which the top millimeters of the soil are still under aerobic conditions owing to O_2 diffusion through the water and the plant, but where reduced conditions are prevalent at a greater depth. The application of NH_4 salts under such conditions may be followed by an oxidation of NH_4 to NO_3 in the oxidized few millimeters of the soil. The NO_3 may leach to deeper layers where it will be reduced to N_2O, NO, or N_2, depending on the pH of the soil (Donald *et al.*, 1963; Nommik, 1956; Bremmer and Shaw, 1955; Pearsall, 1950; Shiori and Tanaka, 1959; Abichandani and Patnaik, 1958; Greene, 1960). The proper choice of method and time of placement of NH_4 fertilizers and urea are the alternative for conditions where denitrification losses are likely to occur (see Table 7.5).

When appreciable N is lost during nitrification of NH_4 fertilizers and urea applied to acid soils with a low exchange capacity (Gerretsen, 1950; Gerretsen and de Hoop, 1957; Clark *et al.*, 1960; Carter and Allison, 1961), the use of nitrate fertilizer would seem to be preferable. On the other hand, on sandy soils the conditions for nitrate losses due to leaching may also be appreciable.

Another means of reducing losses of N from NH_4 fertilizers is to apply chemicals with the fertilizer which will stop microbiological oxidation of NH_4 to NO_3. One of the chemicals on which favorable preliminary results have been obtained is N serve [2-chloro-6-(trichloromethyl) pyridine (Goring, 1962)]. Application of N serve at a rate of 1% of the N level has been reported to increase the efficiency of NH_4 fertilizers for cotton, corn, and sugar beets (Swerzey and Turner, 1962). However, its use is beset by many practical difficulties, particularly the insolubility in water and high volatilization loss of the product. In rice cultivation N-serve failed to give positive results (IAEA, 1965). More recently, a similar product, Toyo Koatsu "AM," 2-amino-4-chloro-6-methylpyrimidine, was developed in Japan with more favorable physical characteristics (IAEA, 1965).

3.1.3. *Nutrient Immobilization from Fertilizers*

The immobilization of nutrients in soils may be of importance for almost all major and minor elements. Entirely different mechanisms may be responsible for immobilization. A distinction can be made between chemical and physicochemical processes and microbiological processes. Immobilization or "fixation" of nutrients need not be an unfavorable soil characteristic because it will prevent losses by leaching or volatilization. Fixation will only be unfavorable if the applied fertilizers react with the soil to an extent that the growing crop will not profit from it (Barber and Humbert, 1963).

a. *Chemical and Physical Chemical Processes.*

i. *Ammonium and Potassium.* The fixation of NH_4 and K by soils with 2:1-type clay minerals is the classic example of an immobilization process that may have a temporary unfavorable effect on crop development. The fixation phenomenon is important in soils with a high pH containing clay minerals of high charge density such as illites (Ferrière *et al.*, 1960; Marel, 1954; Page *et al.*, 1963), particularly if the top soils are subjected to alternate wetting and drying (McLean and Simon, 1958; Grissinger and Jeffries, 1957). On acid soils the fixation of NH_4 and K is of much less importance because the presence of Al ions in the hexagonal holes results in a greater stability (Rich and Obershain, 1955; Ramamoorthy, *et al.*, 1952) of the lattice. Under such conditions a dehydration of K or NH_4 ions and collapse of the interlayer spaces will not take place (Kunze and Jeffries, 1953; Ramamoorthy *et al.*, 1952). As soon as acid soils containing illite are limed, however, fixation will increase (McLean and Simon, 1958; Grissinger and Jeffries, 1957). The means of coping with fixation phenomena in practical agriculture are proper timing and placement of the fertilizer, application of nitrate instead of NH_4 fertilizer, and the possible application of relatively insoluble K fertilizers.

The fixation of NH_4 by 2:1-type clay minerals does not seem to affect the supply of NH_4 to the plant to such an extent that N deficiencies still occur after application of NH_4 fertilizers (Donald *et al.*, 1963; Legg and Allison, 1959). Experience suggests that after initial high doses of K fertilizers, fixation will be reduced (Page *et al.*, 1963). As an alternative, K silicates have been used to reduce the fixation of K by the soils. Generally, K from a wide variety of silicates becomes gradually available to plants, as has been shown for synthetic K silicates (Barbier and Gouere, 1946; Lunt and Kwate, 1956); granite powder meal (Swanback, 1950; Graham and Albrecht, 1952; Jackson and Burton, 1958; Walsh and Quinn, 1957); diabase powder (Nemec, 1952); pophyllite, biotite, phlogopite, and vermiculite (Graham and Albrecht, 1952); lencite (Nicolić *et al.*, 1959); biotite (Stahlberg, 1963). The disadvantage of K silicates is their relatively low availability of K as compared with soluble K fertilizers such as KCl or K_2SO_4. Therefore, applications of K silicates have to be made at much higher rates to produce the same effect as a soluble fertilizer would produce in the absence of fixation (Jackson and Burton, 1958; Swanback, 1950). The supply of K to plants from silicates is a function of their degree of fineness. The supply increases with a decrease in particle size. Unfortunately, the cost of fertilizers increases sharply when very fine powders are required (Walsh and Quinn, 1957; Leaf, 1959; Lunt and Kwate, 1956; Wilde and Rosendahl, 1945).

ii. *Phosphorus*. Soluble phosphates that are applied to the soil are generally immobilized. Detailed discussion of phosphate-soil interaction may be found in reviews by Seatz and Stanberry (1963) and Huffman (1962). Only some fundamental processes will be referred to in relation to the choice of phosphate fertilizers.

In acid soils, highly soluble phosphate fertilizers such as super-phosphate may react to form much less soluble complexes with Al and Fe. When a particle of monocalcium phosphate is in contact with the soil, an area saturated with monocalcium phosphate and dicalcium phosphate dihydrate will be found around it with a pH of approximately 1.5. This concentrated acid phosphate solution will dissolve cations from the surrounding soil particles and crystalline and amorphous precipitates will form (Seatz and Stanberry, 1963; Huffman, 1962; Brown and Lehr, 1959; Lehr *et al.*, 1959; Lehr and Stephenson, 1959). The chemical composition of the precipitates depends upon the final pH of the surrounding solution (Lindsay and Moreno, 1960). Under acid conditions, a great number of compounds similar to variscite $[Al(OH)_2 \cdot H_2PO_4]$, strengite $[Fe(OH)_2H_2 \cdot NO_4]$, and taranakite $[H_6K_3Al_5(PO_4)_8] \cdot H_2O$ may be formed (Clark and Peech, 1955; Cole and Jackson, 1951; Kittrick and Jackson, 1956; Hemwall, 1957; Fried and Dean, 1955; Hsu and Jackson, 1960; Haseman *et al.*, 1951).

More than 30 different reaction products from phosphate fertilizers in soils have been identified by Lindsay and co-workers (1962). The cause of the reduction of plant availability of fertilizer P in acid soils, however, has not yet been conclusively shown to be caused by the precipitation of slightly soluble phosphate compounds (Culot and Laudelout, 1957; Martin and Laudelout, 1957, 1961). Truog (1916) showed many years ago that freshly precipitated Fe and Al phosphates were good sources of P to plants. Probably only those processes involved in the aging of crystalline phosphate precipitates (i.e., a process that is of a long-term nature) might be responsible for a reduction in the availability of fertilizer phosphate (Martin and Laudelout, 1961).

In fact, there is a good indication that the immobilized phosphate is not removed from the system but continues to function as a part of the capacity factor M(soil) as indicated by the results of Rubins (1953). Rubins cropped six soils with some seven crops of lettuce and beets and measured the amount of P removed by the harvested crops. At the end of this period he added radioactive superphosphate to the same soils both cropped and uncropped and measured the *A* values essentially before and after cropping. He found a concurrence between the difference in *A* value and the amount of P removed as shown in Table 7.6. Since the *A* value was measured in terms of superphosphate equivalents, this essentially means that the superphosphate added to the soil has the same equivalent availability as the P from which the crops removed P, namely, M(soil) which is in continuous exchange with M(solution).

In calcareous soil dicalcium phosphate appears to form rapidly around fertilizer particles (Seatz and Stanberry, 1963; Huffman, 1962). Although under alkaline conditions a rapid immobilization of phosphate takes place, this does not mean that the fertilizer phosphate has been removed from the P supply to plants. Even in calcareous soils, dicalcium phosphate may be readily available to crops (Haddock *et al.*, 1957; Olsen, 1953; Stanberry *et al.*, 1960; Cole *et al.*, 1953), as long as the plant roots are in close proximity to the fertilizer. In practice, this is accomplished by applying fertilizer of appropriate particle size in close proximity to the roots (Cole *et al.*, 1953; Moreno, 1959; de Wit, 1953; Fried and Shapiro, 1960; Lindsay and Stephenson, 1959; Rennie and McKercher, 1959).

TABLE 7.6
COMPARISON OF *A* VALUES BEFORE AND AFTER CROPPING WITH BEETS AND LETTUCE INDICATOR CROP, LETTUCE[a]

Soil number	*A* Values (superphosphate equivalents)			Phosphorus removed by cropping
	Cropped	Uncropped	Difference	
85	215	291	76	98
86	484	612	128	208
87	585	804	219	165
88	612	842	230	218
89	283	373	90	92
90	589	707	122	141

[a] Data given in mg P/pot. From Rubins (1953).

In general, it may be concluded that phosphate immobilization in relation to supplying P to the plant by fertilizer application may be a problem, particularly on acid soils owing to the marked decrease in M(solution) concentration. In such cases there will be an advantage in proper timing and placement if highly soluble fertilizers are used. The rapidity of this immobilization can be greatly reduced if the fertilizers are in granulated form. In calcareous soils, immobilization in terms of the marked solution-concentration decrease is probably of little practical significance since the concentration appears to remain adequately high in relation to the required rate of uptake by the plant to produce relatively high yields, when highly soluble sources are used.

b. *Microbiological Processes.* Biological fixation of fertilizer nitrogen may become a problem when undecomposed organic material of high C/N ratio is present in the soil. For the mechanism of fixation, see Hegarty (1962), Nommik (1961), Nommik and Nilsson (1963), Stewart *et al.* (1963), Cheng and Kurtz (1963), Skyring and Callow (1962), Martin (1962), and Norris (1962). Small N dressings, particularly in ammonium forms, may be temporarily fixed by the microbiological decomposition of organic material. Such an immobilization is usually of a short duration. In practice, these temporary deficiencies in soil N supply must be overcome either by larger additions of N fertilizer in the initial stages of plant growth or by waiting a suitable length of time for decomposition to take place.

3.2. Interaction between Fertilizers

Unfavorable reactions may occur upon mixing different fertilizers shortly before application, such as NH_3 loss in alkaline mixture, the production of a hygroscopic or a hard solid mixture, etc. With the aid of a fertilizer-mixing diagram, farmers can see immediately which kinds of fertilizers can or cannot be mixed. In practice, unfavorable reaction between fertilizer components are of less concern to modern farmers as more and more fertilizers are purchased in the mixed or compound form.

3.3. Unfavorable Effects Associated with Fertilization

The application of an unbalanced fertilizer mixture or a single fertilizer may induce cation displacements in the soil solution which may result in induced deficiencies. This is particularly the case in soils with very low nutrient reserves such as acid latosols in the tropics (Broeshart, 1955; Laudelout, 1959). Such effects, however, can hardly be considered as unfavorable effects of the fertilizers but rather the result of an unsatisfactory evaluation of some nutrient supply factor. Three major problems are of importance with regard to unfavorable actions that fertilizers have on the soil. These are the effect of salt concentration on the soil solution, the effect on the soil pH, and the production of toxic products such as NH_3.

3.3.1. *Salt Concentration*

When the osmotic pressure of the soil solutions in the vicinity of the roots becomes higher than 0.4 atm, many crops show serious growth reductions (Magistad, 1942; Rader *et al.*, 1943). Highly soluble compound fertilizers may be especially injurious when applied in the seed row (Hood and Ensminger, 1959; Lutz *et al.*, 1961; Purdue University, 1961). Proper placement practice is one solution to this problem, whereas the addition of lime or gypsum to precipitate soluble compounds is another (Nelson and Terman, 1963).

3.3.2. pH

One of the consequences of the regular application of ammonium nitrogen such as $(NH_4)_2SO_4$, particularly in the absence of cations such as Ca and Mg, is a drop in pH of the soil. In acid soils this may be accompanied by a reduction in soil P and an Al and/or Mn toxicity (Nelson and Terman, 1963; Smith, 1952). In the fertilized parts of the soil, this induced acidity may also be accompanied by an increase in salt concentration of nitrates and K (Dunton *et al.*, 1954). Due to leaching, these cations may be removed to deeper horizons and as a result the K, Mg, and Ca supply will be reduced for subsequent crops (Donald *et al.*, 1963).

The trend toward using large amounts of high analysis N material in which metallic cations are absent requires increasing attention owing to the rapid changes in acidity that are induced by such materials. Adequate liming is becoming an even more important practice.

3.3.3. *Toxic and Other Effects*

For certain crops the induction of toxicities or deficiencies of trace elements may be of importance. Phosphorus fertilizing may sometimes induce Zn, Cu, and Fe deficiency (Burleson *et al.*, 1961; Ward *et al.*, 1963; Ellis *et al.*, 1964; Brown, 1961).

Toxic effects may be associated with the application of $CaCN_2$ (Rotini and Guerucci, 1961), unless it is applied a sufficiently long time before the crop is sown. As soon as the fertilizer has hydrolyzed into lime and urea, the toxic effect disappears. The toxic effect of cyanamide can even be used to kill weeds before the crop is planted.

Other toxic effects are those associated with the application of urea due to ammonia nitrogen (Foy and Brown, 1964; Stephens and Ward, 1963; Cooke *et al.*, 1962; Low and Piper, 1961) or biuret (Gadet *et al.*, 1959; Skinner, 1962; Strang and Weir, 1961; Wilkinson and Ohlrogge, 1960; Ogata and Aibara, 1961; Impey and Jones, 1960; Shen, 1959). In order to reduce such toxic effects, urea should preferably not be drilled with the seed (Stephens and Ward, 1963); with regard to biuret toxicity, a high quality product with a low biuret content should be used.

Toxic effects from fertilizers may arise when, e.g., NH_4 fertilizers and anhydrous ammonia placed close to germinating plants produce NH_3 gas owing to a high pH in close proximity to the fertilizer particle. A partial vapor pressure of ammonia as low as 0.125 mm Hg was shown to be toxic for germinating corn (Allred and Ohlrogge, 1964).

3.3.4. *Crop–Soil Fertilizer Effects*

Specific crop factors that may affect the choice of fertilizer are the sensitivity to toxic elements or toxic compounds or specific requirements. From sand and water culture work it is known that crops may develop in a wide range of solution, varying in composition and pH (Arnon, 1942). The fact that crops are sometimes classified as having a preference for acid, neutral, or alkaline soils has little to do with a direct effect of pH on plant development, but it is usually associated with other soil characteristics, in particular, tolerance to toxic concentrations of certain ions (Russell, 1961). In acid soil, some secondary effects may be a shortage of Ca, P, Mo, or excess of Al and Mn. Sugar beets and barley, which are known to be susceptible to high Al concentrations, therefore prefer a neutral or alkaline environment where the soluble Al content is low. Potatoes, sweet clover, and brassica crops are not only high Ca consumers, but are susceptible to Mn toxicity and, for that reason, do better in neutral or alkaline soils (Russel, 1961; Hewit, 1947). Wheat, red clover, peas, beans, and soybeans are not very sensitive to Mn and Al toxicity. Foy and Brown (1964) associated Al toxicity of various species with the inability of plants to absorb phosphate in the presence of excess Al. When it is known that Al or Mn toxicity is likely to occur, either nonsensitive crops should be chosen or lime has to be applied to increase the soil pH.

Similarly, on alkaline soils, deficiencies of Fe, Mn, Zn, and B can occur. Fruit trees particularly have a relatively high requirement for these microelements. On alkaline soils a Mn deficiency frequently occurs that can often be cured by adding $MnSO_4$, sometimes with the extra addition of such acid materials as powdered S to ensure a high acidity and solubilization of Mn near the root system (Leonard and Stewart, 1959). Zinc deficiency causes problems because most Zn compounds are only soluble in very acid or very alkaline soils. Mixing of $ZnSO_4$ either with $CaCl_2$ or Zn(EDTA) has been reported to increase the availability of fertilizer Zn (Leonard and Stewart, 1958; Stewart and Leonard, 1962). Generally, however, Zn is found to be soluble under acid, and less soluble under alkaline, conditions. The solubility is greatly affected by the nature of the N fertilizer that is used; ammonia salts tend to bring Zn into solution, whereas nitrate reduces Zn availability (Boawn *et al.,* 1960). The use of chelating agents to increase the solubility of trace elements under acid and alkaline conditions (Abdulla and Smith, 1963) offers some possibilities for the future. In spite of high cost, some suggestions are already being made such as in the form of a row treatment. Difficulties, particularly with the supply of microelements in citrus culture, have also been overcome by means of fertilizer spraying.

3.4. Fertilizer Efficiency as Related to Fertilizer Characteristics

The efficiency of a fertilizer (i.e., its ability to supply a particular nutrient to the plant) is governed by the chemistry of the fertilizer and its interaction with the soil. The interaction with the soil can be modified by changes in the chemical characteristics of the fertilizer and even its physical state. Despite the large amount of information on the effects of chemical or physical condition of the fertilizer on plant growth in the soil, it is difficult to use these data quantitatively. This is primarily due to a lack of good characterization of the fertilizer and the soil and a lack of the sensitivity necessary for predictive purposes. Thus, the information is not transferable to a new environmental condition. In our present state of knowledge, however, certain generalizations might be made.

3.4.1. *Nitrogen*

The efficiency of N supply from common solid sources such as ammonium nitrate, ammonium sulfate, sodium nitrate, or urea have been little investigated, using N^{15}-labeled materials for direct determination of efficiency. Nevertheless, knowing the chemistry involved and on the basis of present data, certain conclusions may be drawn.

Legg and Allison (1959) found that for two different soils $(NH_4)_2SO_4$ and $NaNO_3$ were equivalent if the fertilizers were placed near the surface as indicated by *A* values determined with N^{15}-labeled salt. This seems to suggest that if NH_4 is rapidly oxidized to nitrate, both ammonium sulfate and sodium nitrate applied to these soils supply equivalent amounts of N to plants. Under conditions where NH_4 is slowly oxidized to nitrate, or where there is a strong preference for NO_3 by the crop, the efficiency of NO_3 fertilizers may be higher. It is appreciably lower, however, whenever denitrification occurs. When rice is grown under submerged conditions the utilization of N from $(NH_4)_2SO_4$ is much higher than from $NaNO_3$ (Tanaka *et al.*, 1959b). The similarity in availability of nitrate and ammonium fertilizers in the case of upland conditions may also be inferred from the observations that particle size of the fertilizer has no effect on the effectiveness of the N source (Nelson and Terman, 1963). This confirms that these fertilizers rapidly dissolve and, after nitrification of the ammonia, are essentially all in the soil solution. Actually, the efficiency of utilization of soluble N fertilizers is so high that differences among them (except when due to secondary factors such as leaching or gaseous losses) would normally be insignificant (Legg and Allison, 1959). Under submerged conditions, however, the efficiency of utilization from various sources may differ markedly as illustrated by an N^{15} field experiment as shown in Table 7.7 (IAEA, 1966).

TABLE 7.7
FIELD EXPERIMENT ON THE EFFICIENCY OF UTILIZATION OF N FROM VARIOUS N^{15} LABELED NITROGEN FERTILIZER SOURCES[a]

Depth (cm)	N (%) derived from labeled portion of fertilizer source (60 kg N/ha)					LSD 05
	Urea	$(N^{15}H_4)_2SO_4$	$N^{15}H_4NO_3$ + $NH_4N^{15}O_3$	$N^{15}H_4NO_3$	$NH_4N^{15}O_3$	
			First harvest			
0	9	11	13	10	3.4	7.2
5	26	23	26	21	5.4	
			Grain			
0	8.8	14	17.5	9.2	8.3	4.2
5	13	17	24.8	18	6.8	

[a] Data taken at Vercelli, Italy.

The efficiency of N fertilizers of low solubility is likely to be lower than that of soluble fertilizers, depending on particle size of the materials (De Ment *et al.*, 1961). In the finely divided state, however, the plant availability of N in oxamide was found to be almost identical to that of ammonium nitrate (De Ment *et al.*, 1961).

3.4.2. *Potassium*

The efficiency of K from different K fertilizers is likely to be identical for most soluble K fertilizers. Although isotope techniques cannot be applied in the case of K within the limits of tests using yield of K and dry matter, no differences have been found between KCl and K_2SO_4. Insoluble K materials such as K silicates are poorer source of K when the material is not finely divided. But in finely ground potassium metaphosphate and fused potassium phosphates, K has been found equivalent to KCl and K_2SO_4 in the ability to supply K to the plant (Barber and Humbert, 1963; Lunt and Kwate, 1956; De Ment and Stanford, 1959; Jackson and Burton, 1958; Leaf, 1959; Caldwell and Kline, 1963).

3.4.3. *Phosphorus*

The efficiency of P fertilizers with regard to the P supply to plants depends on the extent, and for how long, the fertilizer is able to increase the P concentration of the soil solution in the vicinity of the plant roots (see Chapter 2, Section 3).

Water-soluble phosphate fertilizers such as monocalcium and ammonium phosphates differ from water-soluble K and N salts in the degree of chemical interaction between soil and fertilizers. In acid soils, highly soluble P compounds react rapidly and are partially transformed into dicalcium phosphate and complex precipitates, and are partially adsorbed to exposed surfaces containing Fe and Al. As a result, the phosphate concentration of the soil solution may be only slightly increased in the vicinity of the fertilizer particle (Coleman *et al.*, 1960; Lindsay and Stephenson, 1959; Lindsay and Moreno, 1960).

Increasing particle size of highly soluble phosphate compounds will result in less contact between soil and particle and therefore slower dissolution of the phosphate and a reduced soil-fertilizer interaction. Therefore, a relatively higher P concentration will be found in the soil solution near larger fertilizer particles. On acid soils, particularly those high in Fe and Al, granulated water-soluble phosphates will therefore be more efficient P suppliers than liquid or finely powdered sources (Terman *et al.*, 1960; Stanford and Bouldin, 1960).

On alkaline soils, where the immobilization by Al and Fe surfaces is not as important, liquid or finely powdered sources of highly soluble phosphates will be more effective in increasing the P content of the soil solution than granulated sources (Bouldin and Sample, 1959; Cole and Olsen, 1959; Hagin, 1958; Lawton *et al.*, 1956; Moreno, 1959; Rennie and McKercher, 1959; Webb *et al.*, 1961). Relatively insoluble phosphate sources such as dicalcium phosphate will always be less efficient in supplying P than water-soluble phosphate fertilizers in both acid and alkaline soils, providing comparisons are made of materials of the same particle, size, and identical method of application (Petersburghskii and Debreczeni, 1961; Oberländer and Zeller, 1962; Armiger and Fried, 1958). An apparent exception to this generalization is the comparison of superphosphate with a mixed treatment of dicalcium or even tricalcium phosphate in acid, so-called high-phosphate-fixing soils. This is not really an exception, since under these conditions all the materials will react completely with the soil, forming the same products. The rates of reaction differ, however, and at any given time one or the other treatment may appear more effective.

The availability of P from relatively insoluble processed P fertilizer will generally increase with decreasing particle size, particularly in alkaline soils, but in acid soils decreasing particle size may again lead to an increase in rate of phosphate immobilization by free Al and Fe and consequently to an early lowering of the efficiency of P supply to the plant (Petersburghskii and Debreczeni, 1961).

Conflicting observations with regard to the relative availability of P from different phosphate fertilizer materials are usually due to differences in particle size and methods of application used by various investigators and the use of yield or yield of dry matter as test criteria. The principles outlined above are largely responsible for the differences in availability of P from high-analysis compound NPK, NP, or PK fertilizers (Nelson and Terman, 1963).

In addition to solubility and particle size effects, the chemical interaction of N and K salts with the phosphate, as well as the effect of these salts on the soil, may be responsible for an effect on the availability of P (Terman *et al.*, 1963). Water moves toward the fertilizer particle owing to the osmotic effect resulting from the high solubility of the N and K components of the material or mixture, thus increasing the rate of P dissolution. This results in an increase in the P concentration in the soil solution adjacent to the fertilizer particle (Kolaian and Ohlrogge, 1959). Another reason for the stimulating effect is the intimate association of the root with the fertilizer particle (Chapter 7, Section 6).

Finally, the efficiency of P uptake from P fertilizers may be affected by the chemical form in which the phosphate is supplied into the soil solution. $H_2PO_4^-$ and HPO_4^{2-} may be taken up by plant roots, but anions of different form such as metaphosphate $(PO_3)_n$ or pyrophosphate $H_2P_2O_7^{2-}$ are not necessarily taken up by plant roots as easily as orthophosphate (Sutton and Larsen, 1964; Allmaras and Black, 1962).

4. Timing of Application

The object of proper timing of fertilizer applications is to obtain the highest yield with a minimum investment in fertilizer material and labor. Unfortunately, the conditions that are required for the most efficient use of fertilizers by the crops do not generally correspond to a reduction in the labor requirement.

At least two different considerations are involved in deciding the timing of the fertilizer application: (1) cost is reduced if the number of applications are kept as a minimum and if they are carried out during periods when labor is most available; and (2) efficiency is improved if the application of fertilizers is made at a time when the fertilizer is utilized most effectively by the crop.

Both considerations have in common the principle that an attempt is made to make nutritional conditions for crop development as close to optimal as possible. However, the first approach is, in general, a fertilization of the soil, whereas the second is a fertilization of the crop.

4.1. Minimum Number of Applications

From considerations of labor, soil tillage, and cost of fertilizer, there may be an economic advantage in applying fertilizers during a particular period of the year. However, this must be done with caution.

There are a number of factors that may have an unfavorable effect on crop yield or quality if large doses of fertilizers are applied at one particular time. Leaching, gaseous losses, and immobilization may result in an insufficient supply of nutrient to the crop. On the other hand, a high concentration of fertilizers may have an unfavorable effect on crop development owing to high salt concentration and toxicities.

When the time of application is determined by economic considerations of labor or fertilizer cost, a means of avoiding possible unfavorable effects on crop yield or quality is by choosing the appropriate fertilizer and placement method. Depending on soil conditions and the nature of the crop, large infrequent applications of fertilizers are feasible.

The criteria for judging the effect of this type of fertilizer application are primarily crop yield and crop quality.

The efficiency of fertilizers that differ physically or chemically becomes a consideration when it has been established that single applications at a particular time can produce the same yield and quality as frequent applications.

4.1.1. *Nitrogen*

The application of fertilizers in the fall, in the case of spring-seeded crops, may be of economic importance. Nitrogen applications in the fall are not possible when the climate and soil conditions are such that, owing to leaching of nitrates, the N supply for the crop of the following year becomes insufficient.

The United States can probably be divided into four regions in relation to the advisability of fall application of N (Nelson and Stanford, 1958; Nelson and Uhland, 1955). In the western part losses due to leaching are unlikely to occur. In the higher rainfall areas of the north central region, the susceptibility to leaching is greater, particularly if the applications of NH_4 fertilizers are made so early in the fall that temperature conditions are favorable for nitrification. In the remainder of the country, fall applications are risky and likely to result in heavy N losses due to nitrification and subsequent leaching. In the southeastern United States, leaching losses from fertilizer N practically prohibit fall N applications. Thus fall N applications on corn during 3 years were found on the average to be only 49% as effective as spring applications (Pearson *et al.*, 1961). Unfavorable effects of fall N applications, as compared with spring or split applications, were also found for beets (Adams, 1961) and winter wheat (Widdowson *et al.*, 1961; Prün, 1959).

Under conditions of moderate or little rainfall and on heavy soils, the unfavorable effects of fall N applications seem to disappear, e.g., in experiments with malting barley in Hungary, where calcium ammonium nitrate was plowed down in the fall, and grain yields increased to the same degree as spring or split applications (Pekáry, 1960).

The texture of the soil seems to be a major factor in the effectiveness of single N applications. In the case of heavy soils, even preplant N applications for corn have resulted in the same yields as side dressing under certain conditions (Krantz and Chandler, 1954). Similarly, applications at planting time gave the same yield of pineapples on a silty clay as split applications (Samuels and Diaz, 1958). Most of the literature, however, comparing single and split applications of N fertilizers show than many different crops give the highest yields when N is applied in at least two doses during the growing period.

The applications of N to pastures has been found to be more advantageous when applied several times during the year than as one single dose (Laughlin, 1963; Morris and Celecia, 1962; Voigtländer, 1961). Single applications of N to perennial pasture may result in early lodging, while top grasses develop at the expense of the bottom grasses (Bartosova and Bartos, 1961). Under relatively dry conditions, single N applications applied at a low rate were superior to splitting into two or three applications. At high rates of N application, the difference between single and split applications disappeared (Kunze, 1959). Nitrogen applied to flooded rice soils in the field may be most efficiently utilized when applied as a single dose shortly before primordial initiation as demonstrated by the results of 11 field experiments using N^{15}-labeled $(N^{15}H_4)_2SO_4$ as shown in Table 7.8 (IAEA, 1965).

Of the N fertilizers that are currently applied, urea seems to be the least suited for heavy single applications in the fall or early spring. The ammonia losses that arise when urea hydrolyzes and the pH around the fertilizer particle is increased, may be about 20 to 40% when applied on the surface of grass sods (Volk, 1959, 1961). The physical status of the urea fertilizer apparently markedly influences the extent of these losses, large granules showing greater losses than fine granules. Volk reported a loss of NH_3 from urea-NH_4NO_3 solutions amounting to 11.5%, whereas granular NH_4NO_3 did not show any appreciable loss of NH_3 when applied to the surface.

The loss of NH_3 from urea may have a twofold effect on the growing crop: the supply of nitrogen may seriously decrease and, at the same time, NH_3 may become toxic to roots (Low and Piper, 1961; Meyer *et al.*, 1961; Volk, 1959). Unfavorable effects of heavy single urea dressings as compared with other N fertilizers have been reported for hay (Burton and DeVane, 1952), broom grass (Laughlin, 1963), sugar cane (Parish and Feillafe, 1960), and corn (G. Benavides, 1959).

TABLE 7.8
EFFECT OF TIME AND RATE OF NITROGEN APPLICATION ON THE PER CENT NITROGEN IN THE PLANT DERIVED FROM THE FERTILIZER[a]

Geographical origin of the soil	Treatment in kg N/ha at three different times of application[b]						
	60-0-0	0-60-0	0-0-60	30-30-0	30-0-30	0-30-30	20-20-20
Korea	33	41	40	36	35	33	37
China	16	24	37	22	29	31	22
Thailand I	16	28	37	21	26	33	28
Thailand II	27	44	46	33	43	40	33
East Pakistan	9	23	37	14	26	27	24
West Pakistan	18	32	36	26	34	36	28
United Arab Republic	26	23	31	21	32	29	28
Philippines	10	23	30	14	20	24	20
Burma I	20	35	40	28	34	41	33
Burma II	16	29	37	25	30	29	24
Hungary	18	17	25	13	16	19	15

[a] Mean value of six replications.

[b] The three numbers refer to transplanting, halfway between transplanting and 2 weeks before primordial initiation, and 2 weeks before primordial initiation, respectively.

It seems likely that the demand for slowly soluble N fertilizers such as methyl urea products that provide a continuous N supply over a relatively long period will increase if the timing of N fertilizer application is going to depend on labor rather than on fertilizer efficiency factors (Kunze, 1959; Datta *et al.*, 1961; Hensel, 1960; Hayase, 1961; Jung, 1961; Jurcak, 1961).

Another promising product is a condensation product of crotonaldehyde and urea (CD urea) (Jung, 1963). Figure 7.3 illustrates the gradual N supply from CD urea of different possible sizes to pasture in four subsequent cuttings as compared with the N release from ammonium nitrate. Nitrogen from NH_4NO_3 is found mainly in the first two cuttings, whereas relatively less N from CD urea is found in the first cuttings and more in the last cuttings. The methyl and aldehyde urea condensation products can be applied to plants at much higher rates than NH_4NO_3 without toxic effects as shown by Fig. 7.4 for corn, summer wheat, and sunflowers (Jung, 1963).

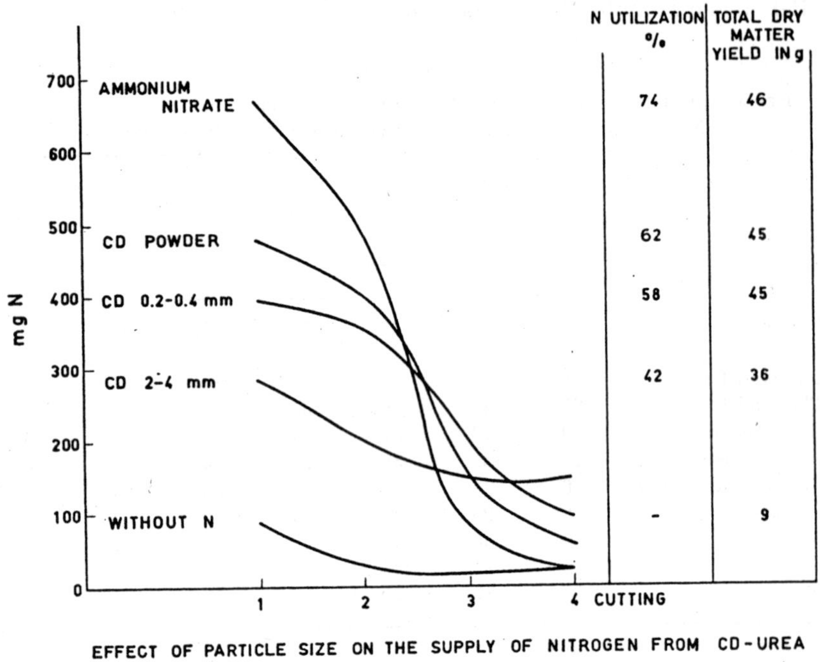

FIG. 7.3. Effect of particle size on the supply of nitrogen from CD-urea (pasture 4 cuttings). From Jung (1963).

4.1.2. *Potassium*

Single K applications as part of a mixed fertilizer are generally made for most crops prior to seeding or planting. Not only is the K ion not readily leached from the topsoil, but even when K immobilization in clay lattices takes place, it has been shown that this K will be slowly released upon depletion of the soil solution during subsequent crop growth. Potassium fixation is, therefore, not considered as a serious agricultural problem which would necessitate frequent K applications (Nelson and Stanford, 1958) but rather as a favorable soil characteristic (Marel, 1954). Still, under conditions where K leaching or removal is extensive and "K fixation" limited as for very sandy soils, single applications of K fertilizers may result in an insufficient supply to the growing crop and a consequent reduction of yield. This is particularly valid for crops that require large quantities of K such as legumes (Brown, 1961; Nelson and Stanford, 1958; Brown, 1957; Burton and Jackson, 1962; Kreoge and Youats, 1962).

In the case of mixed grass-legume associations, the competition be-

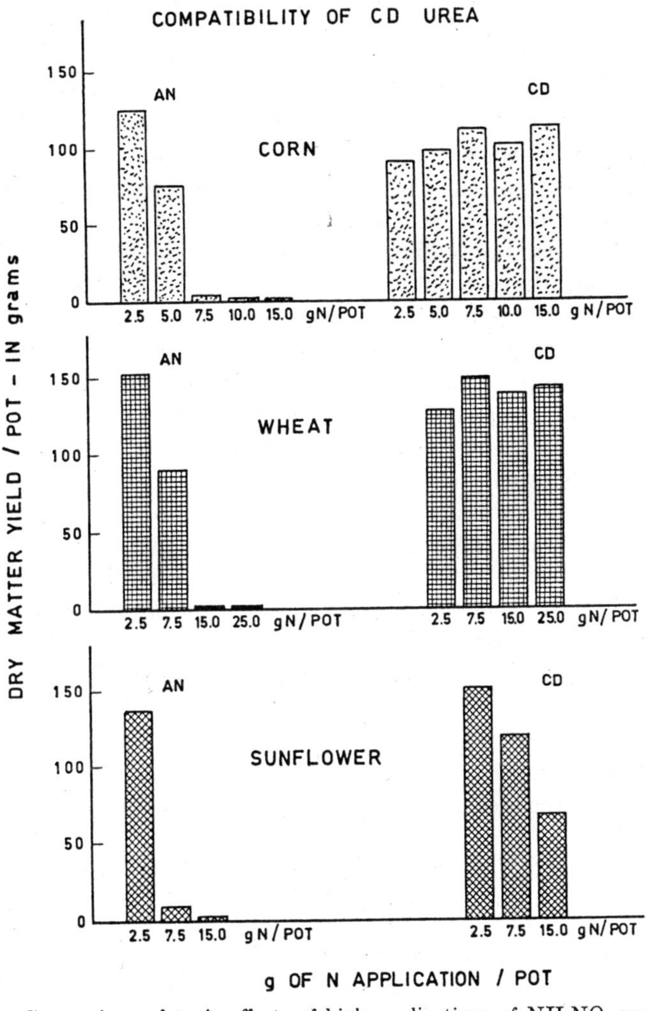

FIG. 7.4. Comparison of toxic effects of high applications of NH_4NO_3 and CD-urea on the dry matter yield of corn, wheat, and sunflowers. From Jung (1963).

tween grasses and legumes for K may result in an elimination of the legumes in the mixture (Gray *et al.*, 1953; Parsons *et al.*, 1953; Hanway *et al.*, 1953; Blaser and Brady, 1953). Frequent K applications appear to be a practical solution (Nelson and Stanford, 1958).

4.1.3. *Phosphorus*

In contrast to N and K, the feasibility of single large P applications is not related to losses from the soil, but rather to immobilization processes. Highly soluble phosphates will tend to become immobilized more rapidly than slightly soluble phosphates, particularly in acid soils (see Chapter 6). Although "fixation" occurs almost immediately, however, the fertilizer is not rapidly removed from the system M(solid) $\rightleftharpoons$ M(solution). The P immobilization results in an increase in the capacity factor M(solid) and, if rate of dissolution is not limiting, immobilization is only important when the resultant concentration of M(solution) is still too low to result in the rate of P uptake required by the crop. Annual applications, or even applications every 2, 3, or 4 years, have been shown to maintain optimal yields in a corn-soybean-wheat rotation (Barber, 1958; Bronson and Barber, 1957; Stanberry *et al.*, 1960).

The need for annual applications depends, to some extent, on the nature of the crop. Crops with a reduced root system, such as potatoes, will generally respond to annual P applications even when the soil P status is high (Terman *et al.*, 1952; Struchtemeyer *et al.*, 1955).

However, single heavy applications of phosphate fertilizers at 1- to 4-year intervals often produce yields that are identical to those obtained by more frequent application (Nelson and Stanford, 1958) although the quantity of fertilizer necessary is sometimes higher (Wakefield *et al.*, 1957; Stanford *et al.*, 1955; Hartfiel, 1958; Cheaney *et al.*, 1956).

The P application in the fall or prior to planting is usually adequate to maintain the P status of the soil at an optimum level. There seems to be little advantage from considerations of nutrient supply in applying phosphate at a more specific time (Nelson and Uhland, 1955), except in the case of crops with reduced root systems where timing is usually combined with proper placement methods (Olson and Dreier, 1956) such as drilling with the seed and subsequent row applications.

4.2. Minimum Amount of Fertilizer

If the efficiency of utilization of the fertilizer becomes of economic importance (e.g., in countries where fertilizers are expensive and labor cheap), the fertilizer practices will be directed toward applying highly soluble fertilizers at times when the crop has its greatest nutrient needs and combining this with proper methods of fertilizer placement.

The assessment of crop needs is a rather simple procedure and consists of determining the removal of nutrients as a function of time. As early as 1887, Liebscher determined the removal of nutrient by crops in order to adjust fertilizer applications according to the crop requirements

(Liebscher, 1887). Results of the more recent work of Remy (1938) are shown in Table 7.9, illustrating the relative dry matter production and nutrient uptake of N, P, and K at various stages of development of barley, rye, and wheat.

It is evident that the early rate of nutrient uptake is much more rapid than the rate of dry matter production. Figure 7.5 shows the uptake pattern of N and dry matter production as a function of the time after seedling emergence of corn. The rapid initial N accumulation is illustrated by the uptake of about 50% of the total N requirements at tasselling, a time when the dry matter production is only 20% of the final

TABLE 7.9
UPTAKE OF NUTRIENTS DURING SUCCESSIVE PERIODS OF GROWTH IN WINTER CROPS[a]

Period	Dry matter[b]			Nitrogen[b]			Phosphoric acid[b]			Potash[b]		
	I	II	III	I	II	III	I	II	III	I	II	III
Emergence to end of winter	4	3	4	16	15	15	8	8	7	11	7	11
End of winter to beginning of stalk growth	16	16	15	27	32	27	21	29	23	34	42	32
Beginning of stalk growth to blossoming	57	57	61	35	39	42	44	42	49	49	46	52
Blossoming to maturity	23	24	22	22	15	16	27	21	21	6	5	6
Sum	100	100	100	100	100	100	100	100	100	100	100	100
Total uptake in Kg per hectare	—	—	—	88	92	94	42	47	45	86	103	90

[a] Values given as per cent. From Remy (1938).
[b] I, Barley; II, rye; and III, wheat.

yield. During ripening of the grain, N is translocated from leaves, stalk, and cob to grain A similar situation exists with rice at tillering, when relative total uptake of N, P, and K exceeds the relative total dry matter production (Reyes *et al.*, 1962 or 1963). Lowland rice, however, shows a leveling off of N uptake after tillering and a second period of rapid N uptake just before maturity (Reyes *et al.*, 1952 or 1963; Singh and Murayama, 1963; Tanaka *et al.*, 1958).

From the uptake pattern of grain corps, it is evident that large quantities of nutrients, particularly N, are required during the initial stages of growth. Experiments with grain crops comparing split N applications with single doses at sowing time have conclusively shown that for low to normal levels of fertilizer the highest yields are obtained when N is applied as a split application, two or three times during the growing periods. At the same time, the quality of grain is improved by increased amounts of protein and glutein (Bockholt *et al.*, 1962; Brouwer *et al.*, 1961; Mitsui, 1960; Primrost and Rittmeyer, 1962; Widdowson *et al.*, 1961; Domanska, 1960; Hendrysiak, 1960; Kofoed, 1960; Linser and

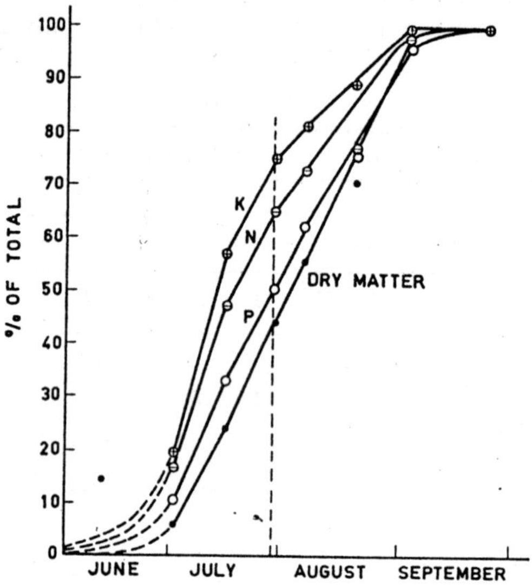

FIG. 7.5. Accumulation of dry matter. N, P, and K in corn plants during the growing season. Average of MLP, Cr, and CrP plots in a corn-corn-oats-meadow-meadow rotation. Date of silking indicated by vertical dashed line. From Hanway (1962).

Primrost, 1959; Selke, 1959; Dobben, 1957). The application of N during the early stages serves to promote the growth of the crop, whereas later applications at about ear or panicle emergence have an effect on the quality of the grain (Bockholt *et al.*, 1962; Primrost and Rittmeyer, 1962; Brouwer *et al.*, 1961; Mitsui, 1960).

The timing of the P applications in the case of cereals is often dependent on climatic conditions and generalizations cannot be made. Late applications of superphosphate may improve grain-straw ratios, particularly in wet summers when ripening is delayed (Mieschlag, 1957).

The inclusion of P with N as starter fertilizer for corn and grain sorghum has been found to be of advantage in soils of low P status, but may even be harmful in soils of high P-supplying power, particularly in years of drought and moisture stress (Olson *et al.*, 1962).

In the case of perennial grasses, frequent application is common practice, particularly with regard to K fertilization (Voigtländer, 1961; Bartosova and Bartos, 1961; Burton and Jackson, 1962). For many crops the timing of N and P applications is, however, associated with a particular method of placement, and it is often not possible to discriminate between the two effects. This is particularly the case with crops such as potatoes (Semma, 1959), beets (Adams, 1961; Hernando *et al.*, 1961), pineapples (Chen *et al.*, 1958), and tobacco (Khemchandarin *et al.*, 1958) where split applications of N and K are often made. The foregoing examples serve to illustrate that frequent applications of fertilizers, particularly N and sometimes P, may be more efficient than single applications of the same amount of fertilizer, because either higher yields are obtained with the same investment in fertilizer, or the same yield with a lower fertilizer investment.

A comparison of split applications under conditions of no yield response have erroneously led to conclusions of no effect due to split applications. In actuality, using yield response as a criteria, it is difficult to tell the difference between single and split applications of the fertilizer (See Chapter 7, Section 2.2). Examination of the crop is required to establish that split applications or single applications have not harmed the crop in yield or quality. Once the absence of adverse effects has been demonstrated, the actual determination of the efficiency of split application of various kinds of fertilizer can be determined directly by means of isotopic techniques, where suitable isotopes are available.

The use of N^{15} for large-scale field experimentation has, unfortunately, not been possible owing to the high cost of N^{15}. Only recently did large-scale field experimentation using N^{15} to study efficiency of timing of N applications in rice cultivation become possible through international cooperation (IAEA, 1965). Pot experiments using N^{15} to study the timing of N fertilizer in rice showed that 60 to 70% of the N was taken up by the rice plants from a top dressing of ammonium sulfate as compared to 30 to 40% when the application was made at planting time (Mitsui, 1960).

The efficiency of P^{32}-labeled superphosphate applied at various times during rapid vegetative development of rice was studied in a large number of locations of widely different soil and climatological conditions (IAEA, 1963; Broeshart and Brunner, 1964). The results are given in Table 7.10.

The efficiency, as judged by the percentage of the total P in the plants derived from the fertilizer, is slightly lower at later times of application, suggesting that the highest P utilization is obtained from applications at transplanting time. There is no serious loss in efficiency, however, when P applications are made at a later date, a pertinent observation particularly for those parts of the world where transportation problems may result in the late arrival of fertilizer. This illustration also serves to demonstrate that even in the absence of phosphate responses, isotope techniques successfully assess whether the timing or splitting of fertilizer applications is efficient.

TABLE 7.10
EFFECT OF SPLIT APPLICATION OF PHOSPHORUS ON THE PER CENT PHOSPHORUS IN RICE DERIVED FROM THE FERTILIZER[a]

Geographical origin of the soil	P (%) from fertilizer; treatment in kg P_2O_5/ha at three different times of application[b]						
	60-0-0	0-60-0	0-0-60	30-30-0	30-0-30	0-30-30	20-20-20
Burma (Gyogon)	7	8	6	7	8	8	8
Hungary	12	8	7	12	9	10	10
Philippines	10	10	9	11	11	11	10
Korea	19	22	19	23	24	23	23
Burma (Mandalay)	31	30	22	33	28	30	28
U.A.R.	39	27	33	34	35	33	35
W. Pakistan (Kalasho)	44	46	39	41	45	49	46
W. Pakistan (Tando Jam)	46	48	41	45	46	43	49
E. Pakistan	66	63	48	64	63	64	60
Thailand	82	74	64	82	84	81	85

[a] From Broeshart and Brunner (1964).

[b] The three numerical values refer to transplanting, halfway between transplanting and 2 weeks before primordial initiation, and 2 weeks before primordial initiation, respectively.

It may be concluded that in situations where the cost of labor is of much more economic importance than the cost of fertilizers, heavy single applications of mixed fertilizers which supply nutrient at the necessary slow rate will be all-important. On the other hand, in large areas of the world where the cost of fertilizer is prohibitive and labor is relatively cheap, the timing of fertilizer applications to feed the crop most efficiently must be the approach for the near future and until the economics of the situation change.

5. Fertilizer Placement

The object of fertilizer placement is to obtain the highest yield with a given limited amount of fertilizer. This can only be accomplished by the most efficient use of the fertilizer. Many different placements are in use such as surface broadcast, row placement, deep placement, plow-down, discing, drilling with seeds, and foliar spraying. The success of such methods in terms of efficient use of the fertilizer depends on the nature of the root system of the crop and the extent to which favorable or unfavorable soil-fertilizer interactions or both take place.

The nature of the root system includes the morphology, the effective zone of feeding, and the physiology. The morphology alone is not adequate for evaluating placement methods since considerations of extent of root proliferation do not necessarily coincide with the feeding area. Ion uptake consistently shows variations even over short distances of a given root (Handley and Overstreet, 1961, 1963; Tanada, 1962; Kramer, 1954). The physiology of the root and the physiological response of the root to the soil environment may well be the most significant factor in determining the relative efficiency of different placements. The morphology of the root, the active area of uptake, soil-fertilizer interactions, and soil physical and chemical characteristics determine the effective soil volume which affects not only the soil contribution of nutrient but also the effectiveness of a given placement.

Soil-fertilizer interactions determine the effectiveness of many materials. This effectiveness can often be either increased or decreased and may determine placement. An increased effectiveness may be due to the solubilization of the material, whereas a decrease may be due to leaching, gaseous losses, or immobilization. In certain cases, this decreased effectiveness may be so severe as to necessitate foliar applications in order to supplement the supply of a given nutrient.

Once the nature of plant soil and fertilizer factors in relation to placement methods have been evaluated, proper field tests are required to test which placement method results in the highest efficiency of fertilizer use by the crop. This can best be done by using isotope techniques where available, because a direct quantitative measurement can be made.

5.1. NATURE OF THE CROP

5.1.1. *Root Morphology*

The development of a root system differs for different plant species, but is very much influenced by the fertility level, moisture content, and soil structure. The soil structure and the related moisture content, the O_2 tension, and CO_2 tension may have a pronounced effect on the density and distribution of the plant roots, particularly if horizons of different density occur close to the soil surface (Wiersum, 1958; Troughton, 1957; Bertrand and Kohnke, 1957; Goedewagen, 1955; Roo, 1957).

When a layer of soil of low density covers a high density subsoil, plant roots have difficulty in penetrating into the subsoil. When the soil is of uniform high density, however, root penetration is much deeper (Wiersum, 1958). In dense soils, roots tend to follow cracks and fissures without necessarily penetrating the dense aggregates (Weaver, 1958; Hulshof *et al.*, 1960; Kuntze and Neuhaus, 1960). Generally, root growth is faster in loose than in dense soils (Phillips, 1959; Lugo-Lopez, 1960) and will be greatly hampered in closely packed soils (Wiersum, 1958).

The most extensive root development is generally found where the soil nutrients are most abundant (Goedewagen, 1954; Wiersum, 1962; Wilkinson and Ohlrogge, 1962; Ohlrogge, 1961), particularly in layers of humus and clay, where because of the high adsorption capacity for cations the supply of nutrients such as K, Ca, Mg, and NH_4 is more favorable than in sandy layers (Weaver, 1958; Butijn, 1955).

Undoubtedly, the composition of the soil solution has an effect on the extent of root branching (Wiersum, 1958), i.e., there are indications that NO_3^-, K^+, and Mg^{2+} stimulate branching more than $H_2PO_4^-$, Ca^{2+}, and SO_4^{2-}. Under normal conditions, reduction in root development due to high salt concentrations in the soil solution is unlikely to occur except in special cases such as in reclaimed soils, soils receiving excessive fertilizer applications, those receiving irrigation water of high salt content, or irrigated soils where removal of salts by leaching does not occur.

Under optimal conditions of crop development, unfavorable effects of texture and structure of the soil on root developments are essentially absent and the nature of the root development will be a species characteristic only. Crops can be subdivided into a number of categories (de Wit, 1953) based on the nature of the root systems.

Small grain crops produce primary roots during germination. After a few leaves are formed, secondary roots appear from the nodes above the primary roots. At the initiation of the tillering stage, the horizontal development of the primary roots in the subsoil may be about several centimeters. The secondary roots develop rapidly in the topsoil during tillering, and at the end of the tillering stage the lateral development is about 10 to 15 cm, i.e., roots are found under the entire soil surface (Goedewagen, 1942; de Wit, 1953; Weaver, 1926). Corn roots expand rapidly in all directions in the soil. This expansion is so rapid in a horizontal direction that a few weeks after planting, roots are found under the entire soil surface (Weaver, 1926). Potatoes have a root system that develops horizontally, but the extent of development is much more limited than in the case of corn (Goedewagen, 1942). Beets and mangold form tap roots that grow downward, although small lateral roots develop. Horizontal growth is rather limited and often does not exceed a few centimeters (Weaver, 1926). Peas and beans also form tap roots, but lateral root development is much more extensive than for beets (Sayre and Clarke, 1935).

Direct morphological root studies can only give qualitative information on relationships concerning the placement of fertilizers because the technique of studying root development in the field is often unsatisfactory, and the rate of uptake of nutrients differs in different parts of the root system.

One of the techniques for study morphology involves the use of glass containers. This has the disadvantage that root development takes place under conditions that are greatly different from those in the field. Another approach is sampling soil cores around the plant in a geometric pattern and determining the distribution of the weight of roots. When the soil is washed carefully from the soil cores (using a pin board), the arrangement of the roots can be observed. However, the loss of root hairs and fine roots cannot be prevented. Even when the morphology is obtained, the relative efficiency of the root samples taken in such a way is not known. An extreme example of possible error is the inclusion of large quantities of dead roots in the case of perennial grasses (Burton, 1957).

5.1.2. *Effective Feeding Zone*

The first attempt to determine the extent of the effective feeding zone of crops was by Sayre and Morris (1940) who used Li as a "tracer" to determine the active root distribution of corn. Lithium does not occur in soils in appreciable amounts and any Li taken up by the crop must come from the place where it was supplied.

This general method of tracing root activity was greatly expanded as a result of the availability of labeled phosphate solutions. Hall *et al.* (1953) injected solutions containing approximately 50 μc of P^{32} per gram of KH_2PO_4 into the soil around the plant base in a regular pattern of distance and depth, using a hollow probe and a hypodermic syringe to place the radioactive solution at the required depth. The method is attractive, because the effectiveness of roots corresponding to a particular depth and distance from the plant base is determined from a measurement of the activity of P^{32} in the plant shoots. This obviates the difficulty of root sampling and separation from soil particles and gives a quantitative measure of the actual feeding zone, irrespective of the over-all morphology.

The basic requirements for successful application of the injection method are that the labeled material remains at the place of injection, that its specific activity is the same at all injection points, and that the labeled element is relatively rapidly taken up by the plant and translocated to the shoots.

These requirements are generally fulfilled when solutions or powders are used containing sufficiently high carrier concentrations of an element that is not very mobile in the soil but is very mobile in the plant. Phosphorus salts meet these requirements very nicely and most of the root activity determinations have utilized phosphate salts labeled with P^{32}. Because of the possibility that relatively high concentrations of phosphate salts may either damage or stimulate root activity in the injected area, attempts have also been made to use carrier-free or high specific activity phosphate salts with a correction for any effects of isotopic exchange with native soil P (Nye and Foster, 1958, 1960).

Corn, cotton, peanuts, and tobacco root activities have been evaluated using the technique of soil injection with P^{32} and measuring uptake from each injection location as a function of time (Hall *et al.*, 1953). Although the uptake pattern did not disagree with morphological observations, it afforded a quantitative measure of root activity in the profile.

The root system of corn, 4 weeks after planting, showed an activity corresponding to the morphology of a hemispherical shape extending about 18 inches in depth and 24 inches in radius. Close to maturity, a lateral extension to 30 inches was found. Phosphorus placed at a depth of 3 inches contributed half of the plants' supply of P through the first 7 weeks, and over one-third throughout the growing period. Phosphorus from an 8-inch depth contributed about one-third of the total P uptake and the remaining third came from a depth of 13 to 18 inches. Phosphate applications close to the plant are particularly advantageous during germination and early growth. Supplemental applications can be placed some distance from the stalk, however, without any particular expected decrease in efficiency.

Cotton roots show a marked lateral extension during the first 3 weeks, followed by vertical development between 4 and 7 weeks. During the 9th and 11th week there is a general lateral spreading of roots in the surface 8 inches. This seems to indicate that early applications of fertilizer should be made close to or under the plant and side dressings should be made within 8 inches of the plant.

With peanuts, a tap root develops very rapidly and is active to a depth of approximately 24 inches, whereas active lateral roots develop more slowly, reaching a distance of 18 inches from the plant within 7 weeks after planting. The most active part of the root system is the tap root. Four weeks after planting, 95% of the phosphate taken up by the plant came from directly under the plant, and at the end of 7 weeks, 53% still came from this location. It was not until after the

11th week that lateral roots within a radius that did not exceed 6 inches from the plant started to contribute to P uptake. These findings stress the importance of fertilizer placement below the seed and at a depth of not over 8 inches. Although root development activity was evident as deep as 24 inches, the activity of the root at that depth with respect to phosphate uptake was very small.

Tobacco has a deep root system that proliferates very rapidly during the first 6 weeks. The phosphate in close proximity to the plant is taken up. Initially, the surface (6 to 10) inches supply most of the phosphate, but at later stages of growth the major part of phosphate is taken up from deeper layers, 10 to 18 inches. Fertilizer placement under the transplanted seeds seems to be the most advantageous method of placement.

Perennial grasses were evaluated by a similar technique (Boggie and Knight, 1958; Hunter and Knight, 1958; Burton *et al.*, 1954; Burton, 1957). After seeding, the active root systems of the various species developed quite differently, both laterally and horizontally. After long establishment, however, most species had penetrated quite deeply and the differences between species decreased.

Similar techniques were also utilized to evaluate root activities of vegetable crops (Hammes and Bartz, 1963). Carrots were found to have an active root system to depths as deep as 30 inches, depending on the texture of the soil. Snap beans were found to have a very limited active root system located in a sphere around the plant base that extended downward during the time between blossoming and harvest. The most active part of the root system for peppers and onions is in the surface layer of the soil, 80% of the absorption coming from roots in a surface layer of 8 inches.

Instead of solutions or suspensions, labeled superphosphate can be used for evaluating root activity simply by placing the fertilizer at various depths. This was done by Lawton *et al.* (1954) to study the root development of alfalfa and alfalfa-broom grass mixtures. The most active parts of the root systems were located near the surface to a depth of 3 inches.

In contrast to labeled solutions, finely powdered anion-exchange resins containing the labeled P have been used by Nishigaki *et al.* (1958) to study root distribution and root efficiency of rice as affected by intermittent drainage and flooding. The most active portion of the rice roots was found near the surface and periodic oxidation of the surface layer seemed to increase the P uptake.

Root activity distributions of tree cultures require additional care and even additional assumptions, since only part of the crop can be sampled for analysis. The use of rapidly translocated labeled nutrient, such as P^{32}-labeled phosphate, is essential to compare the activity of samples taken according to a standard procedure. This may be illustrated from root distribution studies of the African oil palm, using P^{32}- and Ca^{45}-labeled calcium phosphates (Broeshart, 1959). Rapid distribution of P^{32} was found in the entire tree after a period of 12 days. However, the distribution of Ca^{45} was very irregular and remained unchanged during the experimental period which lasted 33 days. A rapid decrease in root activity was found at a depth of 2 to 6 inches. For 1-year-old palms, the most active roots were close to the stem, but the zone of active roots moved in a lateral direction as the tree aged. In trees 2 to 4 years old, the active root zone was identical to the radius of the crown, whereas in adult plantations, where crowns are overlapping, the root activity was not limited to a particular zone. Thus the activity distribution studies with P^{32} revealed that for oil palms the fertilizers should be placed on the surface, the distance from the crown depending on the age of the tree.

A technical problem frequently encountered with root activity studies using P^{32}-labeling techniques is the high variability of the activity in the plant tissue harvested (Burton, 1957). Extreme care and adequate replication are necessary to obtain quantitative results.

Another precaution in the interpretation of the data is the assumption that the active root system with regard to P is the same as the active root system with respect to other nutrient ions. This may or may not be a reasonable assumption, and although this can be checked experimentally by using a double labeling technique, clear evidence is still lacking.

5.2. Effective Soil Volume

The distribution of the active root surface in the soil is one of the most important factors that help decide where fertilizers should be placed. Whether the plant root actually profits from the fertilizer, however, depends on the distance between the fertilizer particle and the root surface and how much of the nutrient moves to the root. If the uptake from fertilizers only takes place from the soil volume in contact with the young roots and root hairs, a low percentage of the soil in the rooting zone would be effectively involved in the supply of plant nutrient. Wiersum (1962) reviewed the estimates made by various investigators of the soil volume in contact with crop roots and found that with the exception of grasses, where the contact volume in the surface is often close to 100%, the soil volume in contact with roots and root hairs varies between 0.1% and 5%. These figures are of limited value because of the mechanisms of mass flow of ions along with the water and movement by diffusion; plant nutrients from the fertilizers may migrate into the "effective" soil volume and a higher percentage of the soil volume will thus be important for plant nutrition.

The transport of elements through the soil is also governed by the valency and charge of the ions (Bray, 1954) and the nonadsorbed anions such as NO_3 and Cl are likely to be more mobile than the cations. This has been demonstrated by Tepe and Leidenfrost (1958), using ion exchange resins in dialysis bags to extract the elements. The values found by them for the distances of ion movement per day were about 1 to 15 mm for cations, 1 to 50 mm for nitrate, and 1 to 10 mm for H_2PO_4. Ions present in the soil solutions were found to move over distances of not more than 50 mm.

Soil water movement and transpiration of plants are responsible for mass flow of water toward the plant root. The ions in the soil solution move along with this water stream. This is particularly important for nonadsorbed negative ions such as sulfate in some soils and nitrate in all soils and may explain the differences in effect of row placement of fertilizers, as observed by Prummel (1957). For N, P, and K fertilizers the greatest effect of row placement was found for phosphate and the smallest for N, illustrating that added nitrates move from much greater distance to the plant root than the much less mobile phosphate ion. The amount of ion that can reach the root surface by mass flow will be governed by the concentration of the ion in the soil solution and is likely to be much smaller for phosphate than for nitrate, K, or Ca (Barber, 1962). Barber and Humbert (1963) found that the K concentration in saturation extracts from 93 midwestern soils averaged 10 ppm K. They calculated that for a consumption of 500 lb of water per pound of dry matter mass flow would bring less than half the K requirement to the plant root.

The extent of the movement of ions around absorbing plant roots was demonstrated by Walker and Barber (1961, 1962), using an autoradiographic technique. Roots were forced to grow along a surface in a soil labeled with Rb^{86} or P^{32}. The autoradiographs, Figs. 7.6 and 7.7 show the depleted area of Rb and P around the root surfaces. The width of the depleted area is much less for P than for Rb (Vasey and Barber, 1963).

The effective soil volume is a particularly important consideration at the early stages of growth of crops such as peanuts, tobacco, or potatoes that have little lateral development of secondary roots. In these cases the placement of nitrate fertilizer is likely to be more effective than NH_4 fertilizer, and the placement of K and P fertilizers as top dressings on the surface may be rather inefficient.

Placement of fertilizers too close to the root surface may result in damage owing to the high concentration of soluble salts such as KCl or K_2SO_4 or effects due to toxic gases of NH_3 which may originate from anhydrous ammonia, ammonia solutions, or be liberated during the hydrolysis of urea. For germinating seeds that are particularly sensitive, there should be some soil between seed and fertilizer.

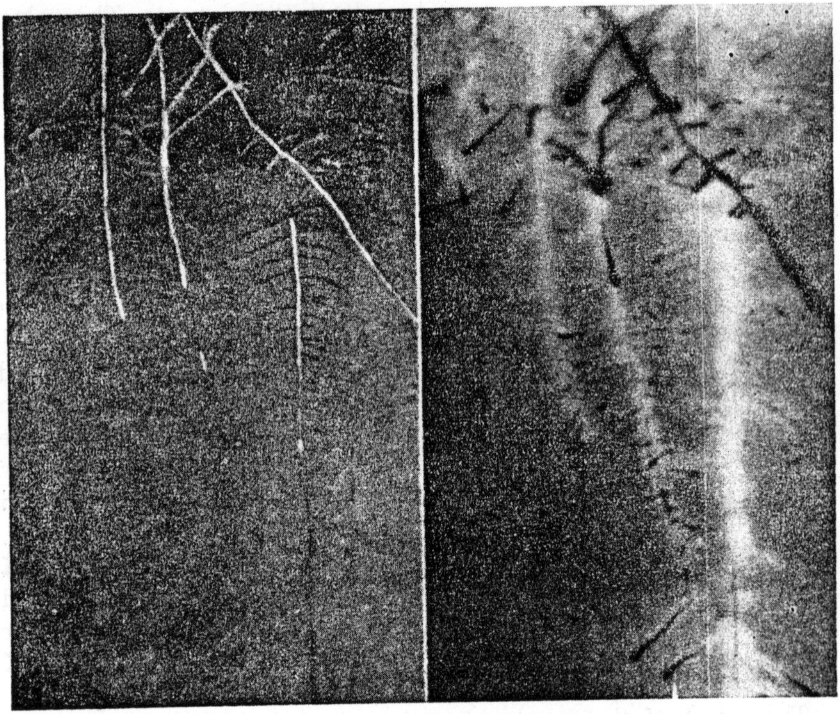

FIG. 7.6. Photograph (left) and autoradiograph (right) showing depleted areas of rubidium about corn roots. Dark areas represent areas of high Rb^{86} content. Light areas represent areas where Rb^{86} has been removed.

Although the small effective soil volume that is found for field crops might suggest that proper fertilizer placement would be the dominant factor in obtaining normal yields, it is known in practice that the differences in yield obtained when various methods of placement are compared are often small. There are at least two explanations for these conclusions from practical experience that do not seem to correspond to the picture of a small effective root volume of crops. The first again is the lack of sensitivity of yield tests. The second is the observation that a small part of the root system may be able to provide the entire nutrient re-

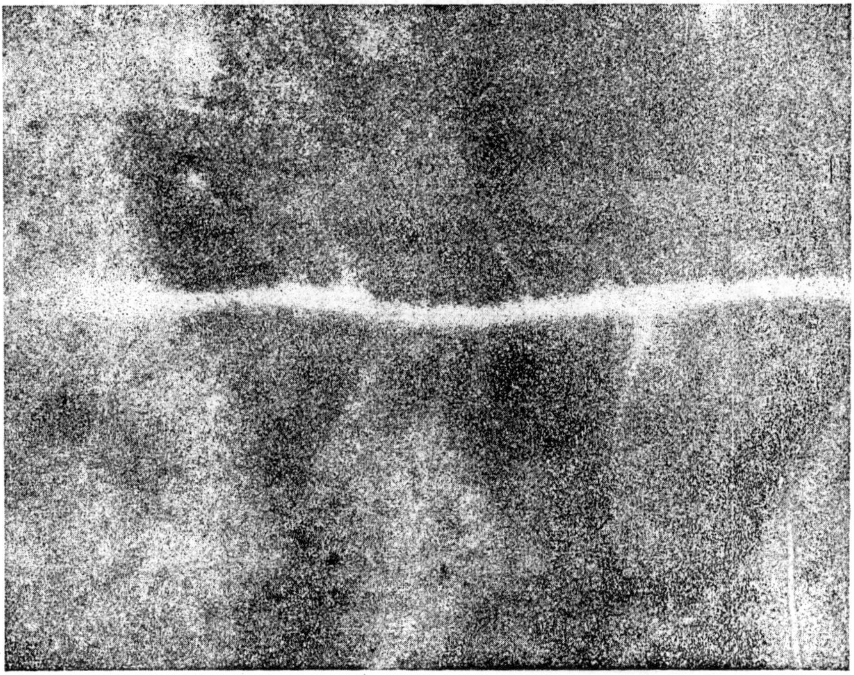

FIG. 7.7. Autoradiograph showing narrow depleted area of P^{32} about a corn root resulting from diffusion of phosphorus to root and uptake. From Barber *et al.* (1963).

quirements of the whole plant as demonstrated by Ohlrogge (n.d., 1961) and Wilkinson and Ohlrogge (1962, 1960). Thus, for nutrient uptake from the applied fertilizer, it may only be necessary that a small part of the plant roots reaches the vicinity of the fertilizer particles. This may, depending on the composition of the fertilizer, result in an active proliferation of roots and corresponding increase of nutrient uptake from the fertilized area.

5.3. N:P INTERACTIONS ASSOCIATED WITH THE FERITILIZER BAND

Once the extent of the active root surface and the effective soil volume are known, a firm basis for determining fertilizer placement is established. An additional consideration is the observation that the efficiency of utilization of a particular element may be markedly affected by the presence or absence of another element in the fertilizer band. This interaction has been clearly established and quantitatively measured for the increase in efficiency of phosphate fertilizer when ammonia nitrogen is mixed with in contrast to separate applications (Grunes, 1959; Essafi *et al.*, 1962; Rennie and Soper, 1958; Olson and Dreier, 1956; Barber and Stivers, 1959; Caldwell, 1960; Miller and Ohlrogge, 1958) (see Fig. 7.8).

Possibly N:P interaction reflects both chemical and biological factors (Grunes, 1959). The chemical effects refer to changes in solubility of fertilizer components owing to the presence of other salts, whereas the biological effects can be divided into stimulatory or inhibitory effects of one plant nutrient on the ion uptake process of another and the

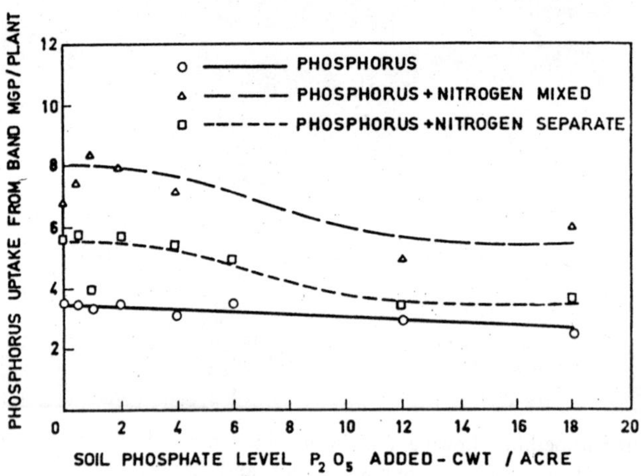

FIG. 7.8. The influence of soil phosphate level and nitrogen additions on the phosphorus absorption by corn from band-applied phosphorus fertilizer. From Miller and Ohlrogge (1958).

effect of a particular nutrient element such as N on the proliferation of roots.

5.3.1. *Chemical*

In a review of the N effect on P availability Grunes (1959) showed that one of the factors that govern the solubility of P fertilizers is the nature and solubility of associated or mixed salts in the fertilizer as related to the pH of the soil. Similar observations were made by earlier reviewers (Olson, 1953; Wild, 1950). This was primarily based on the observation that when superphosphate is added to a soil an appreciable amount of dicalcium phosphate is formed. More is formed on high pH soils, and if the associated salt results in residual acidity, dissolution of the dicalcium phosphate should be enhanced. Rennie and Soper (1958), however, clearly demonstrated that another acid-forming salt with no N present had even the opposite effect.

Ionic strength can markedly affect the solubility of associated dicalium phosphate. The solubility of dicalcium phosphate is increased in the presence of $(NH_4)_2SO_4$ without any corresponding increase in the presence of $Ca(NO_3)_2$ and a lesser increase in the presence of NH_4NO_3 (Starostka and Hill, 1955). Greenhouse tests also reflected this interaction of the different salts. These observations, however, were made on dicalcium phosphate and not superphosphate. Although residual dicalcium phosphate is formed when superphosphate is added to the soil, this is only a fraction of the P and it is only this fraction that would be affected by this chemical interaction. It is more reasonable to suppose that other effects discussed under "Biological Effects" may also explain the experimental results even with dicalcium phosphate.

5.3.2. *Biological*

The mixing of N and P fertilizer may result in a stimulation of root development that can account for a greater efficiency of P uptake by the plant from the fertilizer. In extensive studies, Miller and Ohlrogge (1958), Ohlrogge *et al.* (1957), Duncan and Ohlrogge (1958), and Wilkinson and Ohlrogge (1960) showed that banding of N and P together increased root growth. If one particular root was allowed to tap from mixed N–P sources, heavy root proliferation was induced, and the entire needs of N and P were supplied by one root. Many investigators have observed that in N-deficient conditions the roots of crops are very extensive, thin, and with many branches. Application of N in such cases results in the development of thick fibrous roots (de Wit, 1953; Cooke, 1954; Boscmarck, 1954).

The general stimulatory effect of N on root growth is further illustrated by the higher percentage of the P in the plant taken up from the fertilizer band by sugar beets and potatoes, both when the N was mixed with the phosphate in the band and when the N fertilizer was mixed with the soil. The greatest effect on uptake of P from the fertilizer was, however, when both N and P were mixed together in the band (Grunes *et al.*, 1958). When the phosphate is not banded but finely divided and mixed with the soil, there may be no stimulating effect of N on P uptake from the fertilizer (Grunes *et al.*, 1958). Ammonium ions may stimulate root development and if P is in the same vicinity, there will be an increase in the utilization of this phosphate. An interesting suggestion made by Cole *et al.* (1963) was that a higher N level in the plant increased the rate of P uptake. They demonstrated that high N-excised roots had a greater effect on rate of P uptake than a tenfold increase in phosphate concentration.

An entirely different explanation for the stimulating effect of N on P uptake from the fertilizer band is that N may change the characteristics of the plant roots to such an extent that phosphate is more rapidly taken up (Rennie and Soper, 1958). Nitrogen application usually only increases fertilizer phosphate uptake when both elements are intimately mixed in the fertilizer band. Even a distance as small as 1 inch between N and phosphate bands may fail to produce any increase in P uptake from the fertilizer (Rennie and Soper, 1958). Therefore, it appears that the ammonium ion influences the ability of the roots to absorb P.

Although absence and presence of ammonium may result in differences in adsorption capacity (Russell and Ayland, 1955), O_2 uptake, and CO_2 release (Willis and Yemm, 1955), and glutamine production (Yemm and Willis, 1956), a definite relationship between these observations and P uptake has not been shown (Grunes, 1959). Nitrogen compounds are possibly involved in the formation of ion carriers in the plant root.

A difference in the rate of uptake of ions including phosphate may therefore be inferred from differences in phosphate uptake by plants growing in solutions in which either NO_3 or NH_4 is the N source (Russell, 1959; Loughman and Russell, 1957; Russell and Ayland, 1955). Hagen *et al.* (1957) concluded from their experiments on the uptake of $H_2PO_4^-$ and HPO_4^{2-} by plant roots that the rate constant of the rate-limiting step in phosphate uptake is related to oxidative phosphorylation. It may well be, therefore, that N can affect the rate-limiting step in phosphate uptake (Grunes, 1959), but direct experimental evidence for this phenomenon is as yet unavailable.

5.4. Losses and Immobilization of Fertilizers

The losses due to N leaching are usually markedly reduced when crops have a well-developed root system (see Chapter 5, Section 4.1). Only in special cases, e.g., when crops with a limited root system such as tobacco or potatoes are growing in sandy soils under heavy rainfall conditions, are N losses likely to occur. Under such conditions, foliar sprays may be of advantage (Loughlin 1962, 1962a).

Nitrogen losses may become severe through denitrification or directly as NH_3 from NH_4 fertilizers, urea and anhydrous ammonia. Under submerged conditions in rice cultivation, losses from denitrification may be particularly severe. Ammonia fertilizers applied to the topsoil of rice fields are oxidized to nitrate by nitrifying bacteria in the surface of the soil, but are reduced to nitrous and nitric oxides and elemental N after leaching into the subsoil where the conditions for denitrification are favorable. These losses can be reduced by placing the ammonia fertilizer below the soil surface (Abichandani and Patnaik, 1958; IAEA, in press).

Nitrogen losses as NH_3 from ammonium fertilizers and urea are a function of soil pH, moisture content, and temperature (see Chapter

5, Section 4.2). Generally, application of any ammonium salt or urea at a depth of 2 inches will greatly reduce NH_3 losses (Martin and Chapman, 1951). Anhydrous ammonia applications may give rise to NH_3 losses, particularly if soils are of light texture, since relatively little NH_4^+ ion can be held in the adsorbed state. Subsequent drying of the topsoil may result in severe NH_3 losses. In fine textured soils, a high-moisture content at the time of application results in the formation of NH_4OH and relatively little NH_4^+ ion will be adsorbed. Subsequent drying of the soil may again result in severe losses of NH_3. Soil applications of anhydrous ammonia should therefore be made when the moisture content is below field capacity and should be placed deeper in sandy soils of low exchange capacity than in fine textured soils (Donald *et al.*, 1963; Mortland, 1955).

The relationship between P effectiveness in soils of different pH as a function of the particle size and water solubility of phosphate fertilizers was discussed in Section 1 of this chapter. A detailed discussion and literature review of those fertilizer properties and practices affecting P availability in soils was recently made (Seatz and Stanberry, 1963).

In acid soils immobilization of added water-soluble P may be such that there is no appreciable P increase in the soil solution. This interaction should, if possible, be prevented either by use of granulated fertilizers (large particle size) ([see Section 10 of this Chapter] Seatz and Stanberry, 1963; Terman, 1961) or by banding. In contrast, low water-soluble phosphate fertilizers are increased in effectiveness when they interact with the soil, and this interaction should therefore always be promoted by increasing the surface area reacting (decreasing the particle size) as well as by physical mixing with the soil. For alkaline soils, high water solubility of the phosphate source is highly essential, and banding of low water-soluble phosphate fertilizers may render them relatively ineffective.

Potassium fixation on alkaline soils can be reduced by band placement, particularly if the rate of application of K fertilizers is low (Barber and Humbert, 1963). On high K-fixing river clay soils, the banding of K for cereals was found to be more than three times as effective in terms of grain yield as broadcasting. For crops with less extensive root systems, the effect of banding was only 1.6 times as effective in the case of potatoes, whereas no differences from banding and broadcasting were found for sugar beets (Prummel, 1956, 1957).

5.5. Efficiency of Placement

The assessment of the best method of placement of fertilizers can only be done in the field. A comparison of different methods of placement of fertilizers in pots may give information on certain soil characteristics, such as relative extent of immobilization of phosphates when comparing different soils, or soil fertilizer interactions when comparing different fertilizers; but comparisons of different placements and the most efficient location and depth of fertilizer application can only be carried out under conditions of normal root development, i.e., in field experiments.

The information obtained from the nature of the crop (see Section 5.1 of this chapter) is a valuable guide for choosing the most likely effectiveness of different placement methods. However, root distribution studies do not reflect the nature of soil–fertilizer–root interactions. As was shown in the previous sections, the method of fertilizer application may affect the nature of root development and the interception of the fertilizer band by a few roots may be sufficient to supply the entire crop needs of plant nutrients. In addition, the soil may have a pronounced effect on the supply of nutrients from the fertilizer. Placement in a row might not be required because of the nature of the root system, but may still be a necessary placement method owing to the higher local concentration of elements in the soil solution, particularly phosphate, near a fertilizer band than in the vicinity of fine fertilizer particles mixed with the soil. Therefore, field experimentation is the only means of evaluating placement methods as affected by the soil–fertilizer–crop interactions. Furthermore, plant yield is the primary criterion for evaluating unfavorable effects on crop development due to salt effects or reduction in availability of certain plant nutrients such as microelements.

The quantitative determination of the efficiency of a placement method in terms of the utilization of the fertilizer by the crop can best be determined by means of isotope techniques. Yield and yield of nutrient are not sensitive enough criteria (see Section 1 of this chapter). There is, therefore, little advantage in applying theories (de Wit, 1953) which attempt to make generalizations with regard to the method of application of fertilizers if specific soil–fertilizer–plant interactions are not taken into consideration and the yield-uptake relationship is taken as a basis for efficiency calculations.

5.5.1. *Crop Injury*

Placement methods are important in the application of high-analysis fertilizers. Seedlings are easily damaged by too high a salt concentration. The importance of placing fertilizer in a band to the side or below the seed has been recognized and is generally practiced in many countries where agricultural production is at a high level (Nelson and Stanford, 1958). The higher yield of corn owing to band placement of the starter fertilizer about 2 inches below and 2 inches to one side of the seed row as compared with placement of fertilizer with the seed are a good example of this toxicity (MacGregor and Johnson, 1956; Miller *et al.*, 1947). This is illustrated by Table 7.11.

Injury to the germinating small grain plants when the starter fertilizer is placed with or too close to the seed row may be responsible for large reductions in plant population and yield, particularly if the moisture conditions are such that the seeds are in contact with highly concentrated salt solutions. Calcium cyanide, NH_4OH, and urea and also K salts are particularly harmful. Placement of the fertilizer at 1 to 2 inches to the side and 1 inch below the seed row eliminates these harmful effects. There are indications that for crops with tap roots and little lateral root development such as lucerne or onions, placement at 1 to 3 inches below the seed is not only more effective than placement 1 inch to the side of the seeds but also eliminates the injurious effects from more direct fertilizer contact. Broadcasting and top dressing of fertilizers do not generally result in injurious effects.

TABLE 7.11
RESPONSE OF CORN TO SIDE-BAND PLACEMENT[a]

Placement	Increase in yield: From 200 lb 5-20-20[b] (Bu/A)	Increase in yield: From 400 lb 3-12-12[c] (Bu/A)
Standard split boot	6.8	3.3
2 inches below and 2.5 inches one side	10.7	9.5

[a] From Nelson and Stanford (1958).
[b] From MacGregor and Johnson (1956).
[c] Millar *et al.* (1947).

Unfavorable effects of placement on the uptake of other elements by crops have been reported (Langin *et al.*, 1962), e.g., the most effective method of P placement for corn and grain sorghum resulted in a reduction of Zn uptake. Applications of Zn cured the harmful effects.

Excellent reviews concerning the effects of different placement methods have been published recently (Nelson and Stanford, 1958; Donald *et al.*, 1963; Seatz and Stanberry, 1963; Barber and Humbert, 1963; Nelson and Terman, 1963).

5.5.2. *Evaluation of Efficiency*

Early results obtained using P^{32}-labeled techniques for evaluating fertilizer placement methods were reviewed by Fried (1953) and Black

(1957). Prior to 1953, the knowledge of effective root distribution of crops was small, and evaluations of the efficiency of placement methods were generally limited to a comparison of empirically chosen treatments and standard practices. The prohibitively high cost of N^{15}-labeled fertilizer virtually prevented large-scale field experiments with labeled N and the present practices of N fertilizer application are still based on the results of yield comparisons. Little quantitative information is available concerning the relative efficiencies of various placement methods.

TABLE 7.12
EFFECT OF FERTILIZER PLACEMENT ON THE PER CENT PHOSPHORUS IN THE RICE PLANT DERIVED FROM THE FERTILIZER[a]

		P (%) from fertilizer; fertilizer placement (field)							
	Greenhouse			Hill		Row		
Location	*A* value (mg P/kg soil)	Surface	Hoeing in	10 cm depth	20 cm depth	10 cm depth	20 cm depth	LSD 5%
Hungary	210	3.4	2.7	1.5	0.7	0.9	1.0	0.8
Philippines	66	17	17	5.9	4.2	4.4	3.4	2.1
Burma (Gyogon)	36	11	16	6.4	3.5	4.3	3.1	1.6
Burma (Mandalay)	26	25	25	6.2	6.5	6.0	4.4	2.6
U.A.R.	19	64	59	37	38	38	36	9.8
W. Pakistan	15	48	49	5.0	3.6	4.5	3.8	5.1
Thailand (Surin)	11	36	39	22	15	25	22	2.3
E. Pakistan	10	19	24	22	16	12	6.7	8.4
Thailand (Bangkhen)	1	67	67	50	34	50	36	9.0

[a] Plants harvested 60 days after transplanting; 60 kg P_2O_5/ha; average of six replications. From Broeshart and Brunner (1964).

Although the knowledge of active root distribution of crops and root–soil–fertilizer interactions in fertilizer bands (see Section 5.3 in this chapter) has greatly increased, comparatively few field experiments have been carried out to determine the most efficient methods of placement. One of the reasons for the relatively low priority given to fertilizer efficiency studies is that in countries such as the United States agricultural production is at a high level and the cost of fertilizer is gradually becoming a smaller percentage of the total investment. Even in these areas, however, the cost of fertilizer is still substantial. Moreover, in large areas of the world it is an economic necessity to obtain the highest yield with a minimum of fertilizer, particularly for crops such as corn and rice. In order to determine general principles regarding methods of application that are valid for a wide variety of soil and climatic conditions, intercomparison of the results of field experiments of similar design, using the same fertilizer material and carried out according to standardized procedures of planting, maintenance, sampling, and analytical procedures are highly desirable.

TABLE 7.13

N (%) IN PLANTS DERIVED FROM $(^{15}NH_4)_2SO_4$ APPLIED AT TRANSPLANTING[a]

	In Rows						
Location	N,P, mixed on surface	N,P, separated on surface	N at 5 cm depth P on surface	N,P, mixed at 5 cm depth	N,P, separated at 5 cm depth	N,P, surface broadcast and harrowing	LSD 5%
Burma	9	11	24	28	23	12	5.0
E. Pakistan	13	13	22	22	20	15	7.7
Hungary	42	42	48	52	48	46	7.2
India I	34	37	47	47	44	38	6.4
India II	29	32	44	36	46	35	8.2
Italy	14	10	13	20	15	12	n.s.[b]
Korea I	34	38	47	49	50	39	7.7
Korea II	18	17	25	22	26	18	7.5
Philippines	20	18	28	28	31	19	4.8
China	18	19	28	27	27	21	6.1
Thailand	22	24	25	30	28	21	5.1
U.A.R. I	24	24	24	25	25	28	n.s.[b]
U.A.R. II	22	27	34	33	33	32	9.9
W. Pakistan	32	31	35	36	33	33	n.s.[b]

[a] Sixty-days harvest. 60 kgN/ha; average of six replications. From IAEA (in press).

[b] n.s., not significant.

A series of fertilizer placement experiments of identical design and using the same source of labeled materials, P^{32}-labeled superphosphate in the 1962 experiments and N^{15}-labeled $(NH_4)_2SO_4$ in the 1965 experiments were carried out to study the efficiency of placement of P and N fertilizer for rice. The results are summarized in Tables 7.12 (Broeshart and Brunner, 1964) and 7.13 (IAEA, in press).

Table 7.12 shows that in rice soils of widely different P status, under widely different climatic conditions, surface placement and hoeing-in of P is much more efficient than placement at certain distances or depths.

Table 7.13 shows that shallow placement at 5 cm depth generally increases the efficiency of utilization of N supplied as $(^{15}NH_4)_2SO_4$ as compared with surface placement.

6. Quantitative Assessments

From the foregoing chapters, it is evident that there is no simple answer to the question: "How much fertilizer should be applied?"

The quantity to apply will depend on the efficiency of the chosen fertilizer, and time and method of application. In principle, using isotope techniques to determine in the plant the quantity of element derived from the fertilizer, it is possible to ascertain the quantitative relationships between the efficiency of a fertilizer and the nature, time, and method of application. Once it is known that fertilizer A is X times as efficient in supplying an element to the plant as fertilizer B, it will always be possible to transform findings of field experiments obtained with fertilizer A into fertilizer B. The same principle is valid for time and method of application, e.g., a particular time or method of placement may result in an efficiency of uptake of a particular element that is Y times as efficient as another time or method of application. It is thus possible to arrive at an equation for any site where Quantity to apply equals *f* (nature of fertilizer, time of application, method of placement). Therefore, when reference is made to "quantity to apply" it is understood that this quantity refers to a particular fertilizer supplied at a particular time and applied according to a well-defined way. In this way there will be little difficulty in transforming a particular quantity of superphosphate applied on the surface of a particular soil at planting time into a quantity of dicalcium phosphate applied in rows for a split application which results in identical quantities of P taken up from the fertilizer. Once these parameters are determined, the estimate of the quantity to apply of a given fertilizer at a particular time, at a particular placement can only be a result of, or a reflection of, field experimentation.

The problem, however, is to assess the relationship between yield of crop and the quantity to apply for a particular site from laboratory or greenhouse experiments only. The most straightforward way to find out how much of a particular fertilizer is needed at a particular location is to carry out a field experiment in which various rates of application of that fertilizer are compared with respect to their influence on crop yield. The nature of the fertilizer and the relative proportion of the nutrients depend on local soil, climate, and crop characteristics. In fact, for annual crops it is not possible to predict what the succeeding crops would require. The maximum that can be expected is that when the experiment has been done, one is in a position to say what would have been the best fertilizer treatment for that year in that location. For the succeeding crop, however, the nature of the soil has changed because of fertilizer and crop residual effects and the climate conditions may be entirely different. For perennial and tree crops the conditions are likely to alter to a lesser extent in the year succeeding the field experiment, although a large variation in rainfall may be responsible for the absence of a fertilizer response that was very significant in the previous year. There will, therefore, always be a large error involved in making quantitative recommendations, even when they are based on the results of field experiments at the very site of interest.

In practice, it is not possible to carry out field experiments on a large enough scale which would enable every farmer to base his fertilizer application on field experimentation. For this reason, soil and plant analyses are a valuable means of finding out to what extent the soil conditions at a particular farm are similar to those where data from field experiments are available. It is evident, however, that the present methods of soil analysis in use by soil-testing laboratories can never replace field experimentation but rather are a means of extrapolating the results from field experimentation to similar soil, climatic, and crop conditions.

In the past, a large number of field experiments comparing different rates and types of fertilizers have been carried out and soil data on these sites have been determined using such methods as were discussed in Chapter 2.

Soil-testing laboratories base their recommendations for the quantitative application of fertilizers on accumulated experience. By means of statistical treatment, the relationship between soil and yield data has been determined for numerous locations; the effects of crop rotations and residual effects of fertilizers have been studied. By means of statistical methods which often are very involved, the yield in a certain location can be described as a function of parameters, determined by laboratory and greenhouse methods. It should be stressed, however, that the basis for such equations is field experimentation.

As long as it is not possible to determine the capacity, intensity, and rate factors of plant nutrients in the soil and to define the plant requirements when grown under natural soil conditions, soil-testing laboratories will have to continue to base their recommendations for quantitative fertilizer applications on the rather empirical yield method of soil analysis equations obtained from field and laboratory studies. The degree of refinement of these methods has, in many cases, reached its maximum possibility. Further refinement of techniques seems meaningless since it is impossible to forecast nutrient supply factors in the soil that depend on soil–climatic–crop interactions in the next growing season.

Once the methods of soil analysis have been chosen and the mathematical relation between yield and results of such methods has been established, there is little advantage for a soil-testing laboratory to change its methods, because the likelihood of a correct recommendation is proportional to the experience that has been obtained with the chosen methodology. This explains why reasonable results are obtained by soil-testing laboratories all over the world using entirely different laboratory methods. Generally, in those areas where production levels are high, the present methods of routine soil testing of major nutrient elements have been fully explored and further refinement is unlikely to improve diagnostic methods to any appreciable extent. For any further improvement in soil fertility diagnosis in relation to quantitative fertilizer recommendation, more fundamental methods would be needed to replace the present empirical approach. The final aim of soil testing, i.e., the use of universal methods that determine capacity, intensity, and rate factors in the soil, has not as yet been reached, but may be approached with further knowledge of fundamental processes in soil–plant relationships.

PART IV

Epilogue

Farmers have been using fertilizer for centuries. This fertilization has been based on empirical observations, although the discovery of nutrient element requirements of crop plants placed supplemental nutrient applications on a firmer footing. The observation need no longer be that "putting fish on a hill of corn" increased the growth or that animal bones in the soil resulted in healthier looking plants. Nevertheless, the observations are still empirical even if they have reached a peak of refinement that almost keeps more statisticians busy than soil scientists. It is because this peak of refinement has been reached that any improvement, at this stage, can only come from knowledge and understanding and not from trial-an-error experimentation.

It takes more to grow a crop than just fertilization and it must surely not be inferred that this is the only area in which this knowledge is necessary. However, we leave the plant breeder, entomologist, plant pathologist, etc., with their problems. Insofar as this book is concerned, it is in the *nutrient supply* aspect that further knowledge is required to make the next steps forward over the information-gathering stage. This knowledge is being gathered and Part II has described the progress to date in terms of the simple equation:

$$\text{M(solid)} \rightleftharpoons \text{M(solution)} \rightleftharpoons \text{M(plant root)} \rightleftharpoons \text{M(plant top)}$$

showing that the soil-plant system insofar as plant nutrition is concerned can be treated as a continuum, a sequence of physicochemical processes that follow physicochemical laws in which rates are functions of the concentration of reactants and reaction rate constants. Within this continuum, the interrelationships are such that one or more steps may be rate-limiting to the uptake of a given nutrient. The knowledge as to which step is rate-limiting is necessary to increase most efficiently the over-all rate by focusing on this rate limitation.

References

Abdulla, I., and M. S. Smith. Influence of Chelating Agents on the Concentration of Some Nutrients for Plants Growing in Soil under Acid and under Alkaline Conditions, *J. Sci. Food Agr.*, **14**: 18–109 (1963).

Abichandani, C. T. and S. Patnaik. Nitrogen Changes and Fertilizer Losses in Lowland Waterlogged Soils, *J. Indian Soc. Soil Sci.*, **6**: 87–93 (1958).

Abruña, F., and J. Figarella. Some Effects of Calcium and Phosphorus Fertilization on the Yield and Composition of a Tropical Kudzu-Grass Pasture, *J. Agr. Univ. P.R.*: 41–231 (1957).

Acharya, C. N., and S. P. Jain, Nitrifiability of Soil Organic Matter, *J. Indian Soc. Soil Sci.*, **2**: 43–48 (1954).

Acock, A. M. An Examination of MULDER's Rapid Biological Method for Estimating the Amount of Copper in Soils, *J. Australian Council Sci. Ind. Res.*, **14**: 288–300 (1941).

Adams, S. N. Fertilizers for Beet. Can They Be Applied Before Ploughing in the Autumn?, *Fertiliser Feeding Stuffs J.*, **55**: 13–14 (1961).

Aldrich, D. G., and J. R. Buchanan. Anomalies in Techniques for Preparing H Bentonites, *Soil Sci. Am. Proc.*, **22**: 281–285 (1958).

Alexander, M. *Introduction to Soil Microbiology*, John Wiley & Sons, Inc., New York, pp. 1–472, 1961.

Allerup, S., and J. Nielsen. P^{32}-Transport in Young Barley Plants Following Variations in Transportation and Water Uptake, *Physiol. Plantarum*, **15**: 172–176 (1962).

Allison, F. E. Azobacter Inoculation of Crops. I. Historical, *Soil Sci.*, **64**: 413 (1947).

Allison, F. E. The Enigma of Soil Nitrogen Balance Sheets, *Adv. Agron.*, **7**: 213–250 (1955).

Allison, F. E. Losses of Nitrogen from Soils by Chemical Mechanisms Involving Nitrous Acid and Nitrates, *Soil Sci.*, **96**: 404–409 (1963).

Allison, F. E. The Fate of Nitrogen Applied to Soil, *Advan. Agron.*, **18**: 219–258 (1966).

Allison, F. E., and L. D. Sterling. Nitrate Formation from Soil Organic Matter in Relation to Total Nitrogen and Cropping Practices, *Soil Sci.*, **67**: 239–252 (1949).

Allison, F. E., J. H. Doetsch, and E. M. Roller. Ammonium Fixation and Availability in Harpster Clay Loam, *Soil Sci.*, **72**: 187–200 (1951).

Allison, F. E., J. H. Doetsch, and E. M. Roller. Availability of Fixed Ammonium in Soils Containing Different Clay Minerals, *Soil Sci.*, **75**: 373–381 (1953).

Allison, F. E., E. M. Roller, and J. E. Adams. Soil Fertility Studies in Lysimeters Containing Lakeland Sand, *U.S. Dept. Agr. Tech. Bull. No. 1199*, pp. 1–62 (1959).

Allmaras, R. R., and C. A. Black. Relative Value of Calcium Metaphosphate and Superphosphate as Sources of Phosphorus for Plants on Different Soils, *Soil Sci. Soc. Am. Proc.*, **26**: 470–475 (1962).

Allred, S. E., and A. J. Ohlrogge. Principles of Nutrient Uptake from Fertilizer

Bands: VI. Germination and Emergence of Corn as Affected by Ammonia and Ammonium Phosphate, *Agr. J.*, **56**: 309–313 (1964).

Amer, F. Determination of P-32 Exchangeable Phosphorus in Soils, *IAEA/FAO Symp. Radioisotopes in Soil-Plant Nutrition Studies,* IAEA, Vienna, pp. 43–58 (1962).

Amer, F., D. R. Bouldin, C. A. Black, and F. R. Duke. Characterization of Soil Phosphorus by Anion Exchange Resin Adsorption and P-32 Equilibration, *Plant Soil,* **6**: 391–408 (1955).

Amphlett, C. B., and L. A. McDonald. Equilibrium Studies on Natural Ion-exchange Minerals I. Caesium and Strontium, *J. Inorg. Nucl. Chem.,* **2**: 403–414 (1956).

Amphlett, C. B., and L. A. McDonald. Equilibrium Studies on Natural Ion-exchange Minerals. II. Caesium, Sodium, and Ammonium Ions, *J. Inorg. Nucl. Chem.,* **6**: 145–152 (1958).

Anderson, M. S. and M. G. Keyes. Soluble Material of Soils in Relation to Their Classification and General Fertility, *U.S. Dept. Agr. Tech. Bull. No. 813,* pp. 79, 1942.

Andersen, A. J., and Th. Mogensen. A Comparison of Various Laboratory Methods for Determining the Phosphate Conditions in Soils, *Acta Agron. Scand.,* **12**(4): 315–329 (1962).

Andersen, A., B. Greger-Hansen, and G. Nielsen. Determination of the Phosphate Status of Soils by Means of Radioactive Phosphorus in Pot Experiments, *Acta Agr. Scand.,* **11**: 270–290 (1961).

Annet, H. E., A. R. P. Aiver, and R. N. Kayasth. Losses and Gains of Nitrogen in an Indian Soil Studied in Relation to the Seasonal Composition of Well Waters, and the Bearing of the Results on the Alleged Deterioration of Soil Fertility, *Mem. Dept. Agr. India Chem.,* **9**: 155–234 (1928).

Anon. Ernährung der Pflanzen, *Mitt. Kali-Syndikates,* Berlin, **24**: 303 (1938).

Anon. Faustzahlen für die Landwirtschaft, *Aufl. Deut. Ammoniak-Vertrieb Bochum:* 129 (1951).

Anon. Use of Leguminous Plants in Tropical Countries as Green Manure, as Cover and as Shade, Rome, *Intern. Conf. Agr.,* 1963.

Antipov-Karatayev, I. N. Application of the Isotope Method to the Study of Adsorption of Electrolytes by the Soils in Connection with Land Improvement, *Proc. First Intern. Conf. Peaceful Uses At. Energy, Geneva, P/698,* **12**: 130–137 (1956).

Arambarri, P., and O. Talibudeen. Factors Influencing the Isotopically Exchangeable Phosphate in Soils. I. The Effect of Low Concentrations of Organic Ions, *Plant Soil* **11**(4): 343–354 (1959a).

Arambarri, P., and O. Talibudeen. Factors Influencing the Isotopically Exchangeable Phosphate in Soils. II. The Effect of Base Saturation with Sodium and Calcium in Non-Calcareous Soils, *Plant Soil* **11**(4): 355–363 (1959b).

Arambarri, P., and O. Talibudeen. Factors Influencing the Isotopically Exchangeable Phosphate in Soils. III. The Effect of Temperature in Some Calcereous Soils, *Plant Soil* **11**(4): 364–376 (1959c).

Armiger, W. H., and M. Fried. The Plant Availability of Various Sources of Phosphate Rock, *Soil Sci. Soc. Am. Proc.,* **21**: 183–188 (1957).

Armiger, W. H., and M. Fried. Effect of Particle Size on Availability to Plants of Phosphorus in Phosphate Rock from Various Sources, *Agr. Food Chem.,* **6**: 539–542 (1958).

Arnaud, R. J. St., M.S. Thesis, Univ. of Sask., Canada, 1950.

Arnold, P. W. Losses of Nitrous Oxide From Soil, *J. Soil Sci.,* **5**: 116–125 (1954).

Arnon, D. I., W. E. Fratzke, and C. M. Johnson, Hydrogen Ion Concentration in Relation to Absorption of Inorganic Nutrients by Higher Plants, *Plant Physiol.,* **17**: 515–524 (1942).

Asher, C. J., and P. G. Ozanne. Cation-Exchange Capacity of Plant Roots and Its Relationship to the Uptake of Insoluble Nutrients, *J. Agr. Res.,* 755–766 (1961).

Aslander, A. Nutritional Requirements of Crop Plants, *Encyclopedia of Plant Physiology,* IV, Berlin, 1958.

Aslyng, H. C. The Lime and Phosphate Potentials of Soils, the Solubility and Availability of Phosphates, *Royal Vet. Agr. Coll. Copenhagen, Yearbook 1954,* pp. 1–50, 1954.

Ayres, A. S., and H. H. Hagihara. Effect of the Anion on the Sorption of Potassium by Some Humic and Hydro Humic Latosols, *Soils Sci.,* **75**: 1–17 (1953).

Baird, B. L., J. W. Fitts, and D. D. Mason. The Relationship of Nitrogen in Corn Leaves to Yield, *Soil Sci. Soc. Am. Proc.,* **26**(4): 378–381 (1962).

Baird, G. B., and J. E. Dawson. Determination of that Portion of Soil Boron Available to the Plant by a Modified Soxhlet-Extraction Procedure, *Soil Sci. Soc. Am. Proc.* **19**: 219–222 (1955).

Baker, J. H. The Fate of Ammonia Applied to Soils, *Dissertation Abstr.,* **20**: 1501 (1959).

Bange, G. G. J. Interactions in the Potassium and Sodium Absorption by Intact Maize Seedling, *Plant Soil,* **11**: 17–29 (1959).

Bange, G. G. J. The Carrier Theory of Ion Transport: A Reconsideration, *Acta Boton. Neerl.* **11**: 139–146 (1962).

Bange, G. G. J., and R. Overstreet. Some Observations on Absorption of Cesium by Excised Barley Roots, *Plant Physiol.* **35**: 605–609 (1960).

Bange, G. G. J., and Van Gemerden. The Initial Phase of Ion Uptake by Plant Roots, *Plant Soil* **18**: 85–98 (1963).

Bange, G. G. J., and E. Van Vliet. Translocation of Potassium and Sodium in Intact Maize Seedlings, *Plant Soil,* **15**(47): 312–328 (1961).

Barber, D. A., and H. V. Koontz. Uptake of Dinitrophenol and Its Effect on Transpiration and Calcium Accumulation in Barley Seedlings, *Plant Physiol.,* **38**(1): 60–65 (1963).

Barber, D. A., and R. Russell. The Relationship Between Metabolism and the "Exchangeability" of Ions in Plant Tissues, *J. Exptl. Botany,* 12(35): 252–260 (1961).

Barber, S. A. Relation of Fertilizer Placement to Nutrient Uptake and Crop Yield, *Agr. J.,* **50**: 535–539 (1958).

Barber, S. A. North Central Regional Potassium Studies. II. Greenhouse Experiments with Millet, *Ind. Agr. Expt. Sta. Res. Bull.* 717, 19 pp., 1961.

Barber, S. A. A Diffusion and Mass Flow Concept of Soil Nutrient Availability, *Soil Sci.,* **93**: 39–49 (1962).

Barber, S. A., and R. P. Humbert. *Advances in Knowledge of Potassium Relationships in the Soil and Plant in Fertilizer Technology and Usage,* Symposium Publ. by *Soil Sc. Soc. Amer.,* Madison Wis., 1963.

Barber, S. A., and R. K. Stivers. Phosphorus Fertilization of Field Crops in Indiana, *Purdue Univ. Agr. Expt. Sta. Lf. Res. Bull. No. 787,* 1959.

Barber, S. A., and J. M. Walker. Principles of Ion Movement Through the Soil to the Plant Root, *Trans. Fourth Comm. Intern. Soc. Soil Sci.,* New Zealand, pp. 121–124 (1962).

Barber, S. A., J. M. Walker, and E. H. Vasey. Principles of Ion Movement Through the Soil to the Plant Root, *New Zealand Intern. Soils Conf. Proc. 3, 1962.*

Barber, S. A., J. M. Walker, and E. H. Vasey. Mechanisms for the Movement of Plant Nutrients from the Soil and Fertilizer to the Plant Root, *J. Agr. Food Chem.*, **11**: 204–207 (1963).

Barbier, G., and A. Gouere. Fertilizer Action of a Synthetic Potassium Silicate, *Compt. Rend. Acad. Agr. France*, **32**: 334–338 (1946).

Barbier, G., M. Lesaint, and E. Tyszkiewicz. Investigations by Means of Isotopes in Autodiffusion Phenomena in Soil and of the Nutrition of Plants, *Ann. Inst. Natl. Rech. Agron., Ser. A*, pp. 923–959 (1954).

Bardsley, C. E., and V. J. Kilmer. Sulfur Supply of Soils and Crop Yields in Southeastern U.S. *Soil Sci. Soc. Am. Proc.* **27**: 197–199 (1963).

Barrer, R. M., and J. D. Falconer. Ion Exchange in Feldspathoids as a Solid State Reaction, *Proc. Roy. Soc. (London), Ser. A*, **236**: 227 (1956).

Barshad, I. Vermiculite and Its Relation to Biotite as Revealed by Base-Exchange Reactions, X-Ray Analysis, Differential Thermal Curves, and Water Content, *Am. Mineralogist*, **33**: 655–678 (1948).

Bartholomew, R. P. Changes in the Availability of Phosphorus in Irrigated Rice Soils, *Soil Sci.*, **31**: 209–218 (1931).

Bartosova E., and J. Bartos. The Effect of Split Application of Mineral Fertilizer on the Composition of Permanent Grassland, *Sb. Cesk. Akad. Zemedel. Ved (Rostlinna Vyroba)* **7**: 1661–1680 (1961).

Bassham, J. A. *The Path of Carbon in Photosynthesis*, Prentice-Hall, Englewood Cliffs, N.J., pp. 104, 1957.

Bates, T. E., and S. L. Tisdale. The Movement of Nitrate Nitrogen Through Columns of Coarse Textured Soil Materials, *Soil Sci. Soc. Am. Proc.* **21**: 525–528 (1957).

Batjer, L. P., and B. L. Rogers. Fertilizer Applications as Related to Nitrogen, Phosphorus, Potassium, Calcium, and Magnesium Utilization by Apple Trees, *Proc. Am. Soc. Hort. Soc.* **60**: 1–6 (1952).

Baule, H. Zu Mitscherlichs Gesetz der biologischen Beziehugen, *Landwirtsch. Jahrb. Schweig.*, **51**: 363–385 (1918).

Bauman, H., and G. Maasz. Uber den Verlauf des Nitratgehaltes unter verschiedenen Früchten im Ackerboden, *Z. Pflanzenernaehr. Düng. Bodenk.*, **79**: 155–167 (1957).

Bauwin, G. R., and E. H. Tyner. The Nature of Reductant-Soluble Phosphorus in Soils and Soil Concretions, *Soil Sci. Soc. Am. Proc.*, **21**: 250–257 (1957).

Bear, F. E. Adjusting Soil and Cropping Programs to the Nitrogen Shortage. Making the Most of Our Nitrogen Resources, *Soil Sci. Soc. Am. Proc.* **7**: 294 (1942).

Bear, F. E. *Soils and Fertilizers*, 4th ed., John Wiley & Sons, Inc., New York, pp. 420, 1953.

Bear, F. E. *Chemistry of the Soil*, 2nd ed., Reinhold Publishing Corp., New York, pp. 501, 1964.

Beaton, J. D., D. W. L. Read, and W. C. Hinman. Phosphorus Uptake by Alfalfa as Influenced by Phosphate Source and Moisture, *Can. J. Soil Sci.*, **42**: 254–265 (1962).

Beauchamp, C. E. The Mineral Composition of the Alcoholic Extract of Potato Leaves and Its Relationship to Crop Yields, *Plant Physiol.*, **15**: 485 (1940).

Becker-Dillinger, J. *Handbuch der Ernährung der gärtnerischen Kulturpflanzen,* Paul Parey, Berlin, 1937.

Becket, P. Potassium-Calcium Exchange Equilibria in Soils, *Soil Sci.,* **97**: 376–382 (1964).

Becking, J. H. On the Mechanism of Ammonium Ion Uptake by Maize Roots, *Acta Botan. Neerl.,* **5**: 1–79 (1956).

Beeson, Kenneth C. The Mineral Composition of Crops with Particular Reference to the Soils in which They Were Grown, *U.S. Dept. Agr. Misc. Publ. No. 369,* pp. 164, 1941.

Behrens, W. U. Graphische Methoden, *Fortschr. Landwirtsch.* (G): 404–406 (1928).

Behrens, W. U. Die Methoden zur Bestimmung des Kali und Phosphorsaurebedarfs, Potasse, *Landwirtsch. Genutzter Boden.* 1935.

Behrens, W. U. *Methode zur Bestimmung des Kali-und Phosphorsäurebedarfs landwirtschaftlich genutzter Böden,* Verlag Chemie, Berlin, 1935.

Behrens, W. U. Die Berechnung der Konstanten in der logarithmischen Formel von Mitscherlich, *Z. Pflanzenernaehr. Düng. Bodenk.,* **49**: 60–70 (1950).

Bel, B., H. Denizot, and G. Mathieu. Mineralization of Nitrogen in Some Soils in the Dijon Region, *Compt. Rend. Acad. Agr. France,* **37**: 537–539 (1951).

Bell, C. W., and O. Biddulph. Translocation of Calcium. Exchange Versus Mass Flow, *Plant Physiol.,* **38**(5): 610–614 (1963).

Benko, V. Phosphorus Diffusion in Soil, Pol'nohospodarstvo, **9**: 337–343 (1962). Abstract in *Soils Fertilizers,* **26**(4): 1620; *Chem. Abs.* **58**: 4994.

Berg, W. A., and G. W. Thomas. Anion Elution Patterns from Soils and Soil Clays, *Soil Sci. Soc. Am. Proc.,* **23**: 348 (1959).

Berger, K. C., and E. Truog. Boron Deficiencies as Revealed by Plant and Soil Tests, *J. Am. Soc. Agron.,* **32**: 297–301 (1940).

Berger, K. C., and E. Truog. Boron Tests and Determination for Soils and Plants, *Soil Sci.,* **57**: 25–36 (1944).

Bergman, W. Die Ermittlung der Nährstoffdürftigkeit des Bodens, *Encyclopedia of Plant Physiology IV,* Springer Verlag, Berlin-Göttingen, Heidelberg, pp. 867, 1958.

Bernstein, L., and R. H. Nieman. Apparent Free Space of Plant Roots, *Plant Physiol.,* **35**: 589–598 (1960).

Bertramson, B. R., M. Fried, and S. L. Tisdale. Sulfur Studies of Indiana Soils and Crops, *Soil Sci.* **70**: 27–41 (1950).

Bertrand, A. R., and H. Kohnke. Subsoil Conditions and their Effects on Oxygen Supply and the Growth of Corn Roots, *Soil Sci. Soc. Am. Proc.,* **21**: 135–141 (1957).

Bingham, F. T. Relation Between Phosphorus and Micronutrients in Plants, *Soil Sci. Soc. Am. Proc.,* **27**: 389–391 (1963).

Birch, H. F. Relationship Between Base Saturation and Crop Response to Phosphate in Acid Soils, *Nature,* **168**: 388–389 (1951).

Bishop, R. F., A. J. MacLean, and L. E. Lutwick. Fertility Studies on Soil Types. IV. Potassium Supply and Requirements as Shown by Greenhouse Studies and Laboratory Tests, *Can. J. Agr. Sci.* **34**: 374–384 (1954).

Bizzell, J. A. Lysimeter Experiments V. Comparative Effects of $(NH_4)_2SO_4$ and $NaNO_3$ on Removal of N from the Soil, *Mem. 252, Agr. Expt. Sta.* pp. 20, 1943.

Bizzell, J. A. Lysimeter Experiments VI. The Effects of Cropping and Fertilization on the Losses of N from the Soil, *Mem. 265, Agr. Expt. Sta.,* 1944.

Black, C. A. *Soil Plant Relationships,* John Wiley & Sons, Inc., pp. 1–332, 1957a.

Black, C. A. Use of Radioisotopes in Soil and Fertilizer Research, *Proc. Intern. Symp. Peaceful Application Nuclear Energy,* Brookhaven, Nat. Lab 1957, II D-7554, pp. 504–525, 1957b (Available as *Iowa State College Agr. Exp. Sta. Bull. No. J-3207*).

Black, C. A., and C. O. Scott. Fertilizer Evaluation. I. Fundamental Principles, *Soil Sci. Soc. Am. Proc.,* **20**: 176–179 (1956).

Bladergroen, W. *Problems in Photosynthesis,* Charles C. Thomas, Springfield, Ill., pp. 198, 1960.

Blanchard, R. W., and A. C. Caldwell. Phosphorus Uptake by Plants and Readily Extractable Phosphorus in Soils, *Agron, J.,* **56**: 218–221 (1964).

Blanck, E. Die Glimmer als Kaliquelle für die Pflanzen und ihre Verwitterung, *J. Landwirtsch.,* **60**: 97–110 (1912).

Blanck, E. Die Bedeutung des Kalis in den Feldspaten für die Pflanzen, *J. Landwirtsch.,* **61**: 1–10 (1913).

Blanck, E., and F. Scheffer. Die Neubauer-Methode und die Bestimmung des Stickstoffbedürfnisses der Böden, *Z. Pflanzenernaehr. Düng. Bodenk.,* **4**: 553 (1925).

Blaser, R. E., and N. C. Brady. Grasses and Weeds—the Potash Robbers, *Better Crops,* **37**(2): 6–10, 47–48 (1953).

Boawn, L. C., F. G. Viets, C. L. Crawford, and J. L. Nelson. Effect of Nitrogen Carrier, Nitrogen Rate, Zinc Rate and Soil pH on Zinc Uptake by Sorghum, Potatoes and Sugarbeets, *Soil Sci.,* **90**: 329–337 (1960).

Bockholt, K., P. W. Kurten, and W. Seibel. The Effect of Late Additional Nitrogen Fertilizing on the Yield and Quality of Winter Wheat, *Z. Acher Pflanzenbau,* **115**: 273–296 (1962).

Boggie, R., and A. H. Knight. Studies of the Root Development of Plants in the Field Using Radioactive Tracers. Part I: Communities Growing in a Mineral Soil, *J. Ecol.,* **46**: 621–639 (1958).

Boguslavski, E. von. Der Gefässversuch zur Bestimmung der Bodenfruchtbarkeit, *Trans. Intern. Soc. Soil Sci., Dublin,* **2**: 195–225 (1952).

Boguslavski, E. von. Das Ertragsgesetz, *Handb. Pflanzenph.,* Band IV, Springer-Verlag Berlin-Göttingen, 1958.

Bolhuis, G. G., and R. W. Stubbs. The Influence of Calcium and Other Elements on the Fructification of the Peanut, in Connection with the Absorption Capacity of Its Gynophores, *Neth. J. Agr. Sci.,* **3**: 220–237 (1955).

Bolt, G. H. Analysis of the Validity of the Gouy-Chapman Theory of the Electric Double Layer, *J. Colloid Sci.,* **10**: 206–218 (1955).

Bolt, G. H. Ion Adsorption by Clays, *Soil Sci.,* **79**: 269–276 (1955).

Bolt, G. H. Determination of the Charge Density of Silica Soils, *J. Phys. Chem.,* **61**: 1166 (1957).

Bolt, G. H. Cations in Relation to Clay Surfaces, *Trans. 7th Intern. Congr. Soil Sci.,* Proc. Com. II, pp. 321–327, 1960.

Bolt, G. H., and R. D. Miller. Compression Studies of Illite Suspensions, *Soil Sci. Soc. Am. Proc.,* **19**: 285–288 (1955).

Bolt, G. H., and M. Peech. The Application of the Gouy Theory to Soil Water System, *Soil Sci. Soc. Am. Proc.,* **17**: 210–213 (1953).

Bolt, G. H., and B. P. Warkentin. Influence of the Method of Sample Preparation on the Negative Adsorption of Anions in Montmorillonite Suspensions, *Trans. 6th Congr. Intern. Soil Sci. Soc.,* Paris, 1956.

Bolt, G. H., and B. P. Warkentin. The Negative Adsorption of Anions by Clay Suspensions, *Kolloid Z.,* **156**: 41 (1958).

Bond, G. Evidence for Fixation of Nitrogen by Root Nodules of Alder (Alnus) under Field Conditions, *New Phytologist*, **55**: 145–153 (1956).

Bond, G., W. W. Fletcher, and F. P. Ferguson. The Development and Function of the Root Nodules of *Alnus, Myrica,* and *Hippophas, Plant Soil,* **5**: 309–323 (1954).

Bondorff, K. A., and F. Steenbjerg. Studien over Jordens Fosforsgreindhold, *Tidsskr. Planteavl,* **38**: 273 (1932).

Borland, J. W., and R. F. Reitemeier. Kinetic Exchange Studies on Clays with Radioactive Calcium, *Soil Sci.,* **69**: 251–260 (1950).

Bortels, H., and C. Wetter. Microbiological Investigations on the Micro-Nutrients Manganese and Molybdenum, *Nachrbl. Deut. Pflanzenschutz dienst, Stuttgart,* **6**: 2–5 (1954).

Bosemarck, N. O. The Influence of Nitrogen on Root Development, *Physiol. Plantarum,* **7**: 497–502 (1954).

Boswell, F. C., and O. E. Anderson. Nitrogen Movement in Undisturbed Profiles of Fallowed Soil, *Agron J.,* **56**: 278–281 (1964).

Böszörményi, Z., and G. G. J. Bange. Personal Communication. Institute of Plant Physiology, Eotvos University, Budapest.

Böszörményi, Z., and E. Cseh. Relationships Between the Chloride and Iodide Uptake of Wheat Seedlings, *Nature,* **182**: 1811–1812 (1958).

Böszörményi, Z., and E. Cseh. Studies of Ion-Uptake by Using Halide Ions. I. Changes in the Relationships Between Ions Depending on Concentration, *Physiol. Plantarum,* **17**: 81–90 (1964).

Böszörményi, Z. The Ion Uptake of Excised Barley Roots with Special Reference to the Low Concentration Process, *Advancing Frontiers of Science,* **16**: 11–50 (1966).

Bould, C. Availability of Nitrogen in Composts Prepared from Waste Materials, *Empire J. Exptl. Agr.,* **16**: 103–110 (1948).

Bould, C. Seasonal Changes in the Major Nutrients of Black Currant Leaves, *J. Hort. Sci.,* **30**: 188–196 (1955).

Bouldin, D. R., and E. C. Sample. Calcium Phosphate Fertilizers: III The Effect of Surface Area on the Availability Coefficients of Dicalcium Phosphates, *Soil Sci. Soc. Am. Proc.,* **23**: 276–281 (1959).

Bouldin, D. R., J. D., and E. C. Sample. Interaction Between Dicalcium and Mono-Ammonium Phosphates Granulated Together, *J. Agr. Food Chem.,* **8**: 470–474 (1960).

Bower, C. A. Availability of Ammonium Fixed in Difficultey Exchangeable form by Soils of Semi acid Regions, *Soil Sci. Soc. Am. Proc.,* **15**: 119–132 (1951).

Bower, C. A. Cation Exchange Equilibria in Soils Affected by Sodium Salts, *Soil Sci.* **88**: 32–35 (1959).

Bower, C. A., and J. O. Goertzen. Replacement of Adsorbed Sodium in Soils by Hydrolysis of Calcium Carbonate, *Soil Sci. Soc. Am. Proc.* **22**(1): 33–35 (1958).

Bower, C. A., and E. Truog. Base Exchange Catpacity Determination as Influenced by Nature of Cation Employed and Formation of Basic Exchange Salts, *Soil Sci. Soc. Am. Proc.* **5**: 86–89 (1941).

Bower, C. A., W. R. Gardner, and J. O. Goertzen. Dynamics of Cation Exchange in Soil Columns, *Soil Sci. Soc. Am. Proc.* **21**(1): 20–24 (1957).

Bowling, D. J. F., and P. E. Weatherley. Potassium Uptake and Transport in Roots of *Ricinus communis, J. Exptl. Botany,* **15**: 413–427 (1964).

Boyd, G. E., J. Schubert, and A. W. Adamson. The Exchange Adsorption of Ions from Aqueous Solutions by Organic Zeolites. I. Ion Exchange Equilibria, *J. Am. Chem. Soc.*, **69**: 2818–2829 (1947).

Boyd, G. E., and B. A. Soldano. Self-diffusion of Cations in and through Sulfonated Polystyrene Cation-Exchange Polymer, *J. Am. Chem. Soc.*, **75**: 6091–6099 (1953).

Branton, D., and L. Jacobson. Iron Localization in Pea Plants, *Plant Physiol.*, **37**(4): 546–551 (1962).

Bray, R. H. A Field Test for Available Phosphorus in Soils, *Ill. Sta. Bull. 337*, pp. 589, 1929.

Bray, R. H. A Nutrient Mobility Concept of Soil-Plant Relationships, *Soil Sci.*, **78**: 9–27 (1954).

Bray, R. H. Confirmation of the Nutrient Mobility Concept of Soil-Plant Relationships, *Soil Sci.*, **95**: 124–130 (1963).

Bray, R. H., and L. T. Kurtz. Determination of Total, Organic and Available Forms of Phosphorus in Soils, *Soil Sci.* **59**: 39–45 (1945).

Bremner, J. M., and K. Shaw. Determination of Ammonia and Nitrate in Soil, *J. Agr. Sci.*, **46**: 320–328 (1955).

Briggs, G. E., and R. N. Robertson. Apparent Free Space, *Ann. Rev. Plant Physiol.*, **8**: 11–30 (1957).

Briggs, G. E., A. B. Hope, and R. N. Robertson. *Electrolytes and Plant Cells*. Blackwell Scientific Publications, Oxford, pp. 212, 1961.

Briggs, L. J., and A. G. McCall. An Artificial Root for Inducing Capillary Movement of Soil Moisture, *Science*, **20**: 566–569 (1904).

Briggs, L. J., and J. W. McLane. The Moisture Equivalents of Soils, *U.S. Dept. Agr. Bureau Soils Bull.*, **45**: 5–23 (1907).

Broadbent, F. E. Denitrification in Some California Soils, *Soil Sci.*, **72**: 129–137 (1951).

Broadbent, F. E. Basic Problems in Organic Matter Transformation, *Soil Sci.*, **79**: 107–114 (1955).

Broadbent, F. E., and G. R. Bradford. Cation-Exchange Grouping in the Soil Organic Fraction, *Soil Sci.*, **74**: 447–457 (1952).

Broadbent, F. E., and B. F. Stojanovic. The Effect of Partial Pressure of Oxygen on Some Soil Nitrogen Transformation, *Soil Sci. Soc. Am. Proc.*, **16**: 359–363 (1952).

Broeshart, H. The Application of Foliar Analysis in Oil Palm Cultivation, Thesis, University of Wageningen, 1955.

Broeshart, H. The Application of Foliar Analysis to Oil Palm Cultivation I. Sampling Technique, *Trop. Agr. Trin.*, **33**(3): 127–133 (1956).

Broeshart, H. The Application of Foliar Analysis to Oil Palm Cultivation. IV. The Diagnosis of Multiple Deficiencies, *Trop. Agr. Trin.*, **34**(4): 269–276 (1957).

Broeshart, H. The Application of Radioisotope Techniques for Fertilizer Placement Studies in Oil Palm Cultivation, *Neth. J. Agr. Sci.*, **7**(2): 95–109 (1959).

Broeshart, H. Cation Adsorption and Adsorption by Plants, Radioisotopes in Soil Plant Nutrition Studies, *Symp. Soil Plant Nutrition,* IAEA, Vienna, pp. 303–313, 1962.

Broeshart, H., and H. Brunner. The Efficiency of Phosphate and Nitrogen Fertilization in Rice Cultivation, *Proc. Intern. Soil Sci. Soc.*, Bucharest, Romania, 31 Aug.–9 Sep., 1964.

Broeshart, H., and G. C. Redlich. Perfection of Foliar Analysis as Applied to Tomato Cultures, *Ann. Agr.*, **12**(5): 549–560 (1961).

Broeshart, H., and H. V. Schouwenburg. Early Diagnosis of Mineral Deficiencies by Means of Plant Analysis, *Neth. J. Agr. Sci.*, **9**(2): 108–111 (1961).

Bronson, R. D., and S. A. Barber, *Bull. Purdue Agr. Expt. Sta. 127* (1957).

Brouwer, H., A. Gotz, and K. U. Heyland. The Importance of Late Applications of Nutrients for the Yield of Oats, *Z. Acher. Pflangenbau*, **113**: 117–140 (1961).

Brouwer, R. Investigations into the Occurrence of Active and Passive Components in the Ion Uptake by *Vicia Faba*, *Acta Botan. Neerl.*, **5**: 287–314 (1956).

Brouwer, R. Diffusible and Exchangeable Rubidium Ions in Pea Roots, *Acta Botan. Neerl.*, **8**: 68–76 (1959).

Brown, B. A. Increase Ladino Clover by Timing Your Potash Use, *Better Crops*, **41**: 38–43 (1957).

Brown, B. A. Fertilizer Experiments with Alfalfa (1915–1960), *Conn. Agr. Exptl. Stn. Bull. 363*, 1961.

Brown, J. C. Iron Chlorosis in Plants, *Advan. Agron.* **13**: 329–369 (1961).

Brown, W. E., and J. R. Lehr. Application of Phase Rule to the Chemical Behaviour of Monocalcium Phosphate Monohydrate in Soils, *Soil Sci. Soc. Am. Proc.*, **23**: 7–12 (1959).

Broyer, T. C. Further Observations on the Adsorption and Translocation of Inorganic Solubles Using Radioactive Isotopes with Plants, *Plant Physiol.*, **25**: 367–376 (1950).

Buck, S. F. The Use of Rainfall, Temperature and Actual Transpiration in Some Crop-Weather Investigations, *J. Agr. Sci.*, **57**: 355–365 (1961).

Burd, J. S. Chemistry of the Phosphate Ion in Soil Systems, *Soil Sci.*, **65**: 227–247 (1948).

Burd, J. S. and J. C. Martin. Water Displacement of Soils and the Soil Solution, *J. Agr. Sci.*, **13**: 265–295 (1923).

Burd, J. S. and J. C. Martin. Secular and Seasonal Changes in the Soil Solution, *Soil Sci.*, **18**: 151–167 (1924).

Burd, J. S. and J. C. Martin. Secular and Seasonal Changes in Soils, *Hilgardia*, **5**: 455–509 (1931).

Burger, H. C. Das Leitvermögen Verdünnter Mischkristallfreier Legierungen, *Phys. Z.*, **20**: 73–75 (1919).

Burgess, P. S. The Soil Solution, Extracted by Lipman's Pressure Method, Compared with 1:5 Water Extracts, *Soil Sci.*, **14**: 191–216 (1922).

Burleson, C. A., A. D. Dacus and C. J. Girard. The Effect of Phosphorus Fertilization on the Zinc Nutrition of Several Irrigated Crops, *Soil Sci. Soc. Am. Proc.*, **25**: 365–368 (1961).

Burton, G. W. Role of Tracers in Root Development Investigations, in *Atomic Energy in Agriculture*, Am. Assoc. Adv. Sci. No. 49, Washington, D.C., pp. 71–80, 1957.

Burton, G. W., and E. H. DeVane. Effect of Rate and Method of Applying Different Sources of Nitrogen upon the Yield and Chemical Composition of Bermudagrass *Cyanodon Dactylon* (L) Pers. Hay. *Agron J.*, **44**: 128–132 (1952).

Burton, G. W., and J. C. Jackson. Single vs. Split Potassium Applications for Coastal Bermudagrass, *Agron. J.* **54**: 13–14 (1962).

Burton, G. W., E. H. DeVane, and R. L. Carter. Root Penetration, Distribution and Activity in Southern Grasses Measured by Yields, Drought Symptoms and P-32 Uptake, *Agron. J.* **46**: 229–233 (1954).

Butijn, J. Root Development of Fruit Trees on Profiles with a Thin Clay Layer Underlain by Sea-Sand, *Verslag. Landbouwk. Onderzoek.*, **61**(7): 113–121 (1955).

Butler, G. W. Ion Uptake by Young Wheat Plants. II. The Apparent Free Space of Wheat Roots, *Physiol. Plantarum,* **6**: 617–635 (1953).

Butler, G. W. Uptake of Phosphate and Sulphate by Wheat Roots at Low Temperature, *Physiol. Plantarum* **12**: 917–935 (1959).

Caldwell, A. C. The Influence of Various Nitrogen Carriers on the Availability of Fertilizer Phosphorus to Plants, *Trans. 7th Intern. Congr. Soil Sci.,* **3**:517–525 (1960).

Caldwell, A. C. and J. R. Kline. Effect of Variously Soluble Potassium Fertilizers on Yield and Composition of Some Crop Plants, *Agron J.,* **55**: 542–545 (1963).

Cameron, R. E. and W. H. Fuller. Nitrogen Fixation by Some Algae in Arizona Soils, *Soil Sci. Soc. Am. Proc.,* **24**: 353–356 (1960).

Caro, J. H., and W. L. Hill. Determination of Surface Area of Dicalcium Phosphate by Isotope Exchange, *Agr. Food Chem.,* **4**(5): 436 (1956).

Carolus, R. L. The Use of Rapid Chemical Plant Nutrient Tests in Fertilizer Deficiency Diagnosis and Vegetable Crop Research, *Virginia Truck Expt. Sta. Bull.,* **98**: 1531–1556 (1938).

Carter, J. N., and F. E. Allison. The Effect of Rates of Application of Ammonium Sulfate on Gaseous Losses of Nitrogen from Soils, *Soil Sci. Soc. Am. Proc.* **25**: 484–486 (1961).

Carter, J. N., Bennett, O. L., and Pearson, R. W. Recovery of Fertilizer Nitrogen under Field Condition using Nitrogen-15, *Soil Sci. Soc. Am. Proc.,* **31**: 50–56 (1967).

Chaminade, R. Fixation de l'ion NH_4 par les colloides argilleux des sols sous forme non échangeable, *Compt. Revd. Acad. Sci. Paris,* **210**: 264–266 (1940).

Chang, Monica L., and Grant W. Thomas. A Suggested Mechanism for Sulfate Adsorption by Soils, *Soil Sci. Soc. Am. Proc.* **27**: 281–283 (1963).

Chang, S. C. and M. L. Jackson. Fractionation of Soil Phosphorus, *Soil Sci.,* **84**: 133–144 (1957).

Chang, S. C. and S. R. Juo. Available Phosphorus in Relation to Forms of Phosphates in Soils, *Soil Sci.,* **95**: 91–96 (1962).

Chao, T. T., and M. E. Harward. Nature of Acid Clays and Relationships to Ion Activities and Ion Ratios in Equilibrium Solutions, *Soil Sci.,* **93**(4): 246–253 (1962).

Chao, T. T., M. E. Harward, and S. C. Fang. Adsorption and Desorption Phenomena of Sulfate Ions in Soils, *Soil Sci. Soc. Am. Proc.,* **26**: 234 (1962).

Chao, T. T., M. E. Harward, and S. C. Fang. Soil Constituents and Properties in the Adsorption of Sulfate Ions, *Soil Sci.* **94**: 276 (1962a).

Chao, T. T., M. E. Harward, and S. C. Fang. Movement of S-35 Tagged Sulfate Through Soil Columns, *Soil Sci. Soc. Am. Proc.,* **26**: 27 (1962b).

Chao, T. T., M. E. Harward, and S. C. Fang. Cationic Effects on Sulfate Adsorption by Soils, *Soil Sci. Soc. Am. Proc.* **27**: 35–38 (1963).

Chapman, G. W. Leaf Analysis and Plant Nutrition, *Soil Sci.* **52**: 63 (1941).

Chapman, G. W. and H. M. Gray. Leaf Analysis and the Nutrition of the Oil Palm (*Elacis guineensis* Jacy), *Ann. Bot. N.S.,* **13**: 415–433 (1949).

Chapman, H. D., G. F. Liebig, and D. S. Rayner. A Lysimeter Investigation of Nitrogen Grains and Losses Under Various Systems of Cover Cropping and Fertilization and a Discussion of Error Sources, *Hilgardia,* **19**: 57 (1949).

Cheaney, R. L., R. M. Weihing, and R. N. Ford. The Effect of Various Rates and Frequencies of Application of Rock and Superphosphate on the Yield and Composition of Forage on a Lake Charles Day Loam Soil, *Soil Sci. Soc. Am. Proc.,* **20**: 66–68 (1950).

Chen, C. C., T. T. Fang, and R. S. Ling. Effects of Fertilization Growth, Nutrient Level and Yield of Pineapple, *Mem. Coll. Agr. Taiwan. Univ.*, **5**: 15–26 (1958).

Cheng, H. H. and L. T. Kurtz. Chemical Distribution of Nitrogen in Soils, *Soil Sci. Soc. Am. Proc.*, **27**: 312–316 (1963).

Cheng, K. L., and R. H. Bray, Two Specific Methods of Determining Copper in Soil and Plant Material. *Anal. Chem.*, **25**: 655 (1953).

Chernov, V. A., and N. I. Belyaeva. Rate of Displacement of Hydrogen Ions Absorbed in Soil by Aluminum Ions, *Dokl. Akad. Nauk SSSR,* **110**: 658–660, (1956).

Chernov, V. A., and N. I. Belyaeva. Rate of Displacement of Hydrogen Ions Absorbed by Clay by Means of Aluminum Ions, *Dokl. Akad. Nauk SSSR,* **111**: 849–851 (1956a).

Chesnin, L., and C. H. Yien. Imbibimetric Determination of Available Sulfates, *Soil Sci. Soc. Am. Proc.*, **15**: 149–151 (1950).

Christensen, H. R. and S. T. Jensen. Bodenuntersuchungen in Verbindung mit Statische Düngungsversuchen auf Mineralböden, *Wiss. Arch. Landwirtsch.*, **4**: 1 (1930).

Clark, F. E., W. E. Beard, and D. H. Smith. Dissimilar Nitrifying Capacities of Soils in Relation to Losses of Applied Nitrogen, *Soil Sci. Soc. Am. Proc.*, **24**: 50–54 (1960).

Clark, J. S., and M. Peech. Solubility Criteria for the Existence of Calcium and Aluminum-bound Phosphate in Soils, *Soil Sci. Soc. Am. Proc.* **19**: 171–174 (1955).

Cole, C. V., and M. L. Jackson. Colloidal Dihydroxy Dihydrogen Phosphates of Aluminum and Iron with Crystalline Character established by Electron and X-ray Diffraction. *J. Phys. Colloid Chem.*, **54**(1): 128–142 (1950).

Cole, C. V. and M. L. Jackson. Solubility Equilibrium Constant of Dihydroxy Aluminum Dihydrogen Phosphate Relating to a Mechanism of Phosphate Fixation in Soils, *Soil Sci. Soc. Am. Proc.* **15**: 84–89 (1951).

Cole, C. V., and S. R. Olsen. Phosphorus Solubility in Calcareous Soils: I. Dicalcium Phosphate Activities in Equilibrium Solutions. II. Effects of Exchangeable Phosphorus and Soil Texture on Phosphorus Solubility, *Soil Sci. Soc. Am. Proc.* **23**(2): 116–121 (1959).

Cole, C. V., S. R. Olsen, and C. O. Scott. The Nature of Phosphate Sorption by Calcium Carbonate, *Soil Sci. Soc. Am. Proc.*, **17**: 352–356 (1953).

Cole, C. V., D. L. Grunes, L. K. Porter, and S. R. Olsen. The Effects of Nitrogen on Short-Term Phosphorus Absorption and Translocation in Corn (*Zea mays*), *Soil Sci. Soc. Am. Proc.* **27**(6): 671–674 (1963).

Cole, D. W., S. P. Gessel, and E. E. Held. Tension Lysimeter Studies of Ion and Moisture Movement in Glacial Till and Coral Atoll Soils, *Soil Sci. Soc. Amer. Proc.* **25**: 321–325 (1961).

Coleman, N. T., and D. Craig. The Spontaneous Alteration of Hydrogen Clay, *Soil Sci.*, **91**(1): 14–18 (1961).

Coleman, N. T., G. W. Thomas, F. H. LeRoux, and, G. Bredell. Salt-Exchangeable and Titratable Acidity in Bentonite-sesquioxide Mixtures, *Soil Sci. Soc. Am. Proc.*, **28**: 35–37 (1964).

Coleman, R. The Value of Legumes for Soil Improvement, Bull. 336, *Miss. Agr. Expt. Sta.*, **16**: E. S. R. 82, 447 (1939).

Coleman, N. T., J. T. Thorup, and W. A. Jackson, Phosphate Sorption Reactions that Involve Exchangeable Al, *Soil Sci.*, **90**: 1–7 (1960).

Collander, R. Selective Absorption of Cations by Higher Plants, *Plant Physiol.*, **16**: 691–720 (1941).

Collins, G. H. *Commercial Fertilizers*, 5th Ed., McGraw Hill Book Co., New York, pp. 552, 1955.

Collison, R. C., H. G. Beattie, and J. D. Harlan. Lysimeter Investigations, II. Mineral and Water Relations and Final Nitrogen Balance in Legume and Nonlegume Crop Rotations for a Period of 16 years, *N.Y. State Agr. Expt. Sta. Tech. Bull. 212*, 1933.

Colwell, W. E. Intensified Cropping to Diagnose Mineral Element Deficiencies, A Method to Determine Relative Boron Content of Soils, *Soil Sci.*, **62**: 43–49 (1946).

Colwell, W. E. and N. C. Brady. The Effect of Calcium on Yield and Generality of Large-Seeded Type Peanuts, *J. Am. Soc. Agron.*, **37**: 413–428 (1945).

Conway, E. J. and M. Downey. An Outer Metabolic Region of the Yeast Cell, *Biochem. J.*, **47**: 347–355 (1950).

Cook, H. L. and G. D. Scarseth. The Effect of Cyanamid and Potash When Plowed Under With Organic Refuse on the Yield of Corn and Succeeding Crops, *J. Am. Soc. Agr.*, **33**: 282–293 (1941).

Cook, L. J. Further Results Secured in "Top Dressing" for Southeastern Pasture Lands with Phosphatic Fertilizers, *J. Dept. Agr. S. Australia* **42**: 851 (1939).

Cooke, G. W. Recent Advances in Fertilizer Placement. II. Fertilizer Placement in England, *J. Soil Sci. Food Agr.*, **5**: 429–440 (1954).

Cooke, G. W. Field Experiments on Phosphate Fertilizers, *J. Agr. Sci.*, **48**: 74–103 (1956).

Cooke, G. W., and F. V. Widdowson. Field Experiments on Phosphate Fertilizers, A Joint Investigation, *J. Agr. Sci.*, **53**: 46–63 (1959).

Cooke, M. E., R. C. Stephen, and J. S. Ward. Nitrate Toxicity Arising from the Use of Urea as a Fertilizer, *Nature* **194**: 1263–1275 (1962).

Cornfield, A. H. The Mineralization of the Nitrogen of Soils During Incubation. Influence of pH, Total Nitrogen and Organic Carbon Contents, *J. Sci. Food Agr.*, **3**: 343–349 (1952).

Cornfield, A. H. The Mineralization of Nitrogen in a Soil Acidified with Sulfur Aluminum, Sulfate or Ferrous Sulfate, *J. Sci. Food Agr.* **4**: 298–301 (1953).

Cowie, D. B., R. B. Roberts, and I. Z. Roberts. Potassium Metabolism in *Escherichia coli*. I. Permeability to Sodium and Potassium Ions, *J. Cellular Comp. Physiol.*, **34**: 243–257 (1949).

Crafts, A. S. *Translocation in Plants.* Holt, Rinehart & Winston, New York, pp. 152, 1961.

Crafts, A. S., and T. C. Broyer. Migration of Salts and Water into Xylem of the Roots of Higher Plants, *Am. J. Botany* **25**: 529–535 (1938).

Cremers, A., and H. Laudelout. Conductivité Electrique des Gels Argileux et Anisometrie de leurs Elements, submitted to *J. Chem. Phys.*

Cremers, A., and H. Laudelout. Surface Conductivity in Clay Gels, Submitted to *J. Phys. Chem.*

Cremers, A. E., and Laudelout, H. Surface Mobility of Cations in Clays, *Soil Sci. Soc. Am. Proc.*, **30**: 570–576 (1966).

Cressman, H. K. and J. F. Davis. Sources of Sulfur for Crop Plants in Michigan and Effect of Sulfur on Fertilization on Plant Growth and Composition, *Agron. J.* **54**: 341–344 (1966).

Crooke, W. M. Effect of Heavy-Metal Toxicity on the Cation-Exchange Capacity of Plant Roots, *Soil Sci.,* **86**(5): 231–240 (1958).

Crooke, W. M. and A. H. Knight. An Evaluation of Published Data on the Mineral Composition of Plants in the Light of the Cation-Exchange Capacities of Their Roots, *Soil Sci.,* **93**(6): 365–373 (1962).

Crooke, W. M., A. H. Knight, and I. R. MacDonald. Cation Exchange Properties and Pectin Content of Storage-Tissue Disks, *Plant Soil* **13**(1): 55–67 (1960).

Crooke, W. M., A. H. Knight, and I. R. MacDonald. Cation-Exchange Capacity and Pectin Gradients in Leek Root Segments, *Plant Soil,* **13**(2): 123–127 (1960a).

Cseh, E., and Z. Böszörményi. Further Investigations Concerning the Initial Stage of Anion Uptake, *Acta Botan. Acad. Sci.* Hung., **7**(3–4): 221–227 (1961).

Culot, J. P., and H. Laudelout. Retrogradation et utilisation des engrais phosphatés dans les sols du Congo Belge, *Pedologie,* **8**: 162–168 -Gand. (1957).

Culot, J. P., A. V. Wambeke, and J. Croegaert. Contribution to The Study of Mineral Deficiencies of Arabian Coffee Trees at Kivu, *INEAC Serie Sci.,* No. **73**: 105 (1958).

Currie, J. A. Importance of Aeration in Providing the Right Conditions for Plant Growth, *J. Sci. Food Agr.,* **13**: 380–385 (1963).

Czubchik, A. A. and Z. I. Konashevick. Effectiveness of Azotobacterin on Cherno-Podzolic Soil, Abstract in *Soils and Fertilizers,* **25**: 378, Abstract No. 2726 (1962).

Dakshinamurti. Studies on the Conductivity of Clay Systems, *Soil Sci.,* **90**: 302–305 (1960).

Damsgaard-Sorensen, P. Studies on the Soil's Phosphoric-acid Content. IV. The Organically Combined Phosphorus, *Tidsskr. Planteavl.* **50**: 653–675 (1946).

Daragan-Sushchova, A. Yu. Some Data on the Dynamics of Biological Forms of Nitrogen in Soils of the Vologod Region, *Sb. Rabot. Tsentr. Muzeya Pochvoved.,* **3**: 293–303 (1960).

Das, S. An Improved Method for Determination of Available Phosphoric Acid of Soils, *Soil Sci.,* **30**: 33 (1933).

Datta, N. P., and N. N. Goswami. Effect of Organic Matter and Moisture Levels on the Uptake and Utilization of Soil and Fertilizer Phosphorus by Wheat, *J. Indian Soc. Soil Sci.,* **10**: 263–276 (1962).

Datta, S. N., V. Iswaran and A. K. Rishi. Ureaform, a New Fertilizer, *Current Sci. (India),* **30**: 182–183 (1961).

Davies, B. E., and R. I. Davies. A Simple Centrifugation Method for Obtaining Small Samples of Soil Solution, *Nature,* **198** (4876): 216–217 (1963).

Davis, L. E. Sorption of Phosphates by Non-calcareous Hawaiian Soils, *Soil Sci.,* **40**: 129–159 (1935).

Davis, L. E. Significance of Donnan Equilibria for Soil Colloidal Systems, *Soil Sci.,* **54**: 199–219 (1942).

Davis, L. E. Theories of Base-Exchange Equilibriums, *Soil Sci.,* **59**: 379–395 (1945).

Davis, L. E. Ionic Exchange and Statistical Thermodynamics, I. Equilibria in Simple Exchange Systems, *J. Colloid Sci.,* **5**(1): 71–79 (1950).

Davis, L. E. The Instability of Neutralized H, Al-Bentonites, *Soil Sci. Soc. Am. Proc.,* **25**(1): 25–27 (1961).

Davis, L. E., R. Turner, and L. D. Whittig. Some Studies of the Autotransformation of H-Bentonite to Al-Bentonite, *Soil Sci. Soc. Am. Proc.,* **26**(5): 441–443 (1962).

Day, P. R., and W. M. Forsythe. Hydronamic Dispersion of Solutes in the Soil Moisture Stream, *Soil Sci. Soc. Am. Proc.,* **21**: 477–480 (1957).

Dean, L. A. An Attempted Fractionation of the Soil Phosphorus, *J. Agr. Sci.*, **28**: 234 (1938).

Dean, L. A. Fixation of Soil Phosphorus, *Advan. Agron.*, **1**: 391–411 (1949).

Dean, L. A. Yield of Phosphorus Curves, *Soil Sci. Soc. Am. Proc.*, **18**: 462–466 (1954).

Dean, L. A. Die Zersetzung der Zitronensäure durch den Boden, *Soil Sci.*, **28**: 281 (1929).

Dean, L. A., and O. C. Magistad. The Determination of Replaceable Hydrogen in Manganese Dioxide-Free Hawaiian Pineapple Soils, *J. Am. Soc. Agron.*, **23**: 10 (1931).

Dean, L. A., W. L. Nelson, A. J. MacKenzie, W. H. Armiger, and W. L. Hill. Application of Tracer Technique to Studies of Phosphatic Fertilizer Utilization by Crops: I. Greenhouse Experiments, *Soil Sci. Soc. Am. Proc.*, **12**: 107–112 (1948).

Deger, E. Grenzzahlen humider Tropenböden Mittelamerikas, *Z. Pflanzenernaehr. Düng., Bodenk.*, **A 35**: 1–6 (1934).

de Groot, S. R. *Thermodynamics of Irreversible Processes*. North-Holland Publ. Co., Amsterdam, pp. 242, 1952.

de Haan, F. A. M. The Interaction of Certain Inorganic Anions with Clays and Soils. Thesis, Agricultural University of Wageningen, Netherlands, pp. 167, 1965.

de Haan, F. A. M., and G. H. Bolt. Determination of Anion Adsorption by Clays, *Soil Sci. Soc. Am. Proc.*, **27**: 636–640 (1963).

De Ment, J. D., and G. Stanford. Potassium Availability of Fused Potassium Phosphate, *Agron. J.*, **51**: 282–285 (1959).

De Ment, J. D., C. M. Hunt and G. Stanford. Hydrolysis, Nitrification and Nitrogen Availability of Oxamide as Influenced by Granule Size, *J. Agr. Food Chem.*, **9**: 453–456 (1961).

Demolon, A. Die Bestimmung des Düngemittelbedarfes der Böden, *Potasse*, **7**: 100 (1933).

Demolon, A., and G. Barbier. La Concentration Critique d'Equilibre et l' Appréciation de besoin des Sols en Acide Phosphorique, *2nd Intern. Soil Conf. Leningrad-Moscow*, Bd. **4**: 86 (1932).

Denbigh, K. G. *The Thermodynamics of the Steady State*. Methuen & Co. Ltd., London, pp. 103, 1951.

Diest, A. van, H. W. Jespersen, R. F. White, and C. A. Black. Test of Two Methods for Measuring a labile Fraction of Inorganic Phosphorus in Soils, *Soil Sci. Soc. Am. Proc.* **24**(6): 498–502 (1960).

Dijkshoorn, W. Neth. The Rate of Uptake of Chloride, Phosphate, and Sulphate in Perennial Ryegrass, *J. Agr. Sci.*, **7**: 194–201 (1959).

Dixon, M., and E. C. Webb. *Enzymes*, Academic Press, Inc., New York, pp. 782, 1958.

Dobben, W. H. van. Split Nitrogen Manuring of Cereals, *Landbouwk Tydschr.*, **69**: 34–37 (1957).

Doll, E. C. Effects of Fall-Applied Nitrogen Fertilizer and Winter Rainfall on Yield of Wheat, *Agron. J.* **54**: 471–473 (1962).

Domanska, N. Nitrogen Topdressing of Several Spring Wheat Varieties, Dono. Ros. Zborz. **1**: 33–46 (1960).

Donald, C. M. and C. H. Williams. Fertility and Productivity of a Podzolic Soil as Influenced by Subterranean Clover (*Irifolium subterraneum* L.) and Super-Phosphate, *Australian J. Agr. Res.*, **5**: 664 (1954).

Donald, L., H. J. Stangel, and J. T. Pesek. Advances in Knowledge of Nitrogen Fertilization in the USA since 1950, *Iowa Agr. Home Econ. Expt. Sta. J.*, 4465—Proj. 1148/1189, 1963.

Doss, B. D., O. L. Bennett, and D. A. Ashley. Evapotranspiration by Irrigated Corn, *Agron. J.*, **54**: 497–498 (1962).

Doughty, J. L. The Fixation of Phosphate by a Peat Soil, *Soil Sci.*, **29**: 23 (1930).

Drake, M., J. Vengris, and W. Colby. Cation-Exchange Capacity of Plant Roots, *Soil Sci.*, **72**(2): 139–147 (1951).

Dreibelbis, F. R. Soil Type and Land Use Effects on Percolation of Soil Water Through Monolith Lysimeters, *Soil Sci. Soc. Am. Proc.*, **18**: 358–363 (1957).

Droner, D. P., *J. Roy. Soc. W. Aust.* **43**: 81–82 (1960).

Drouineau, G., and G. Lefevre. (Mineralize Nitrogen in Soil). Premiere contribution a l'étude de l'azote mineralisable dans les sols, *Ann. Agron.*, **19**: 518–536 (1949).

Drouineau, G., and G. Lefevre. Nouvelles conceptions sur la dynamique de l'azote dans les sols, *Compt. Rend. Acad. Agr. France,* **35**: 328–330 (1949a).

Duchaufour, P. The Mineralization of Nitrogen in Forest Humus, *Compt. Rend. Acad. France,* **37**: 567–571 (1951).

Dumenil, L., and R. P. Nicholson. Nitrogen Carry-Over? More Than You Think, *Iowa Farm Sci.*, **6**(9): 134–136 (1952).

Duncan, W. G., and A. J. Ohlrogge. Principles of Nutrient Uptake from Fertilizer Bands II. Root Development in the Band, *Agron. J.*, **50**: 605–608 (1958).

Dunton, E. M., R. B. Hall, and M. E. Taylor. The Influence of Certain Fertilizer Materials on the Soil Reaction and Nutrient Level in the Potato Row During the Growing Season, *Soil Sci. Soc. Am. Proc.*, **18**: 47–53 (1954).

Dutt, G. R. Prediction of the Concentration of Solutes in Soil Solutions for Soil Systems Containing Gypsum and Exchangeable Ca and Mg, *Soil Sci. Soc. Am. Proc.*, **26**(4): 341–343 (1962).

Dutt, R., and P. F. Low. Diffusion of Alkali Chlorides in Clay-Water Systems, *Soil Sci.*, **93**(4): 233–240 (1962).

Dutt, G. R., and P. F. Low. Relationship Between the Activation Energies for Deuterium Oxide Diffusion and Exchangeable Ion Conductance in Clay Systems, *Soil Sci.*, **98**: 195–203 (1962).

Dyer, B. On the Analytical Determination of Probably Available "Mineral" Plant Food in Soils, *J. Chem. Soc.* (see Biedermann Zbl.) **23**: 799 (1894).

Eagle, D. J. Responses of Winter Wheat to Nitrogen and Soil Nitrogen Status, *J. Sci. Food Agr.* **14**: 391–394 (1963).

Eaton, F. M. The Behavior of B in the Soils, *U.S. Dept. Agr. Tech. Bull.* **696**: 1–57 (1939).

Eaton, F. M., R. B. Harding, and T. J. Ganje. Soil Solution Extractions at Tenth-Bar Moisture Percentages, *Soil Sci.* **90**(4): 253–258 (1960).

Eaton, S. V. and J. H. Eaton. Sulfur in Rainwater, *Plant Physiol.*, **1**: 77–87 (1926).

Eden, T. The Nutrition of a Tropical Crop as Exemplified by Tea, *Rept. 13th Intern. Hort. Congr.*, 1952.

Eeckman, J. P. and H. Laudelout. Chemical Stability of Hydrogen-Montmorillonite Suspension, *Kolloid Z.* **178**: 99–107 (1961).

Egner, H. Method att Gestamma lattlolig fosforsyra i akerjord, *Medd. Centralanst. Forsokssta Jordbuks*, Stockholm, No. 425, 1932.

Egner, H., H. Riehm, and W. R. Domingo. Untersuchungen über die chemische

Bodenanalyse als Grundlage für die Beurteilung des Nährstoffzustandes der Böden II. Chemische Extraktionsmethoden zur Phosphor- und Kaliumbestimmung, *Kungl. Lantbrukshogskolans Ann.* **26**: 199–215 (1960).

Eid, M. T., C. A. Black, O. Kempthorne, and J. H. Zoellner. Significance of Soil Organic Phosphorus to Plant Growth, *Iowa Agr. Expt. Sta. Res. Bull.* **406**: 747–776 (1954).

Eid, M. T., A. M. Keleg, and M. A. Samie. 'A'-Value as a Method for the Determination of Available Soil Phosphorous, *Agr. Res. Rev. (Cairo)*, **40**(4): 1–20 (1962).

Elgabaly, M. M. On the Mechanism of Anion Uptake by Plant Roots. III. Effect of the Dominant Cation in the Root on Chloride Uptake by Excised Barley Roots, *Plant Soil,* **16**: 157–164 (1962).

Ellis, R., J. F. Davis, and D. L. Thurlow. Zinc Availability in Calcareous Michigan Soils as Influenced by Phosphorus Level and Temperature, *Soil Sci. Soc. Am. Proc.,* **28**: 83–86 (1964).

Emmert, E. M. Plant Tissue Tests as a Guide to Fertilizer Treatment of Tomatoes, *Kentucky Agr. Expt. Sta. Bull. 430*, 1942.

Eno, Ch. F., and H. W. Reuzer. The Availability of Potassium in Certain Minerals to *Aspergillus niger, Soil Sci. Soc. Am. Proc.* **15**: 155–159 (1951).

Eno, Ch. F., and H. W. Reuzer. Potassium Availability from Biotite, Muscovite Greens and Microline as Determined by Growth of *Aspergillus niger, Soil Sci.* **80**: 199–209 (1955).

Ensminger, L. E. Some Factors Affecting the Adsorption of Sulfate by Alabama Soils, *Soil Sci. Soc. Am. Proc.,* **18**: 259–264 (1954).

Epstein, E. Ion Absorption by Plant Roots, *Proc. Ann. Oak Ridge* Summer Symp. 4th Symposium, pp. 418–434 (U.S. Atomic Energy Commission, TID-5115, 1953a).

Epstein, E. Mechanism of Ion Absorption by Roots, *Nature* **171**: 83 (1953).

Epstein, E. Passive Permeation and Active Transport of Ions in Plant Roots, *Plant Physiol.,* **30**(6): 529–535 (1955).

Epstein, E. Passive Passage and Active Transport of Ions in Plant Roots, *Conf. Radioactive Isotopes Agr.* (USAEC Report No. TID-7512), pp. 297–301, 1956.

Epstein, E. Calcium-Lithium Competition in Absorption by Plant Roots, *Nature* **185**: 705–706 (1960a).

Epstein, E. Spaces, Barriers and Ion Carriers: Ion Absorption by Plants, *J. Botany,* **47**(5): 393–399 (1960b).

Epstein, E. Dual Pattern of Ion Absorption by Plant Cells and by Plants, *Nature* **212**: 1324–1327 (1966).

Epstein, E. and C. E. Hagen. A Kinetic Study of the Absorption of Alkali Cations by Barley Roots, *Plant Physiol.,* **27**: 457–474 (1952).

Epstein, E., and J. E. Leggett. The Absorption of Alkaline Earth Cations by Barley Roots: Kinetics and Mechanism, *Am. J. Botany,* **41**(10): 785–791 (1954).

Epstein, E., D. W. Rains, and W. E. Schmidt. Course of Cation Absorption by Plant Tissue. *Science* **136**: 1051–1052 (1962).

Epstein, E., D. W. Rains, and D. E. Elzam. Resolution of Dual Mechanisms of Potassium Absorption by Barley Roots, *Proc. Natl. Acad. Sci. U.S.,* **49**: 684–692 (1963).

Erdei, P. The Effect of Split Application of Fertilizers on the Rhythm of Grass Growth on Pastures, *Magy. Mezögazd,* **15**: 17–18 (1960).

Eriksson, E. Cation Exchange Equilibria on Clay Minerals, *Soil Sci.,* **74**: 103–113 (1952).

Eriksson, E. Composition of Atmospheric Precipitation. I. Nitrogen Compounds II. Sulfur, Chlorine, Iodine Compounds, *Tellus,* **4**: 145–270, 280–303 (1952a).

Essafi, A., J. Cline, and A. Mathieu. Utilization of P-32 in Studying Plant Assimilation of Phosphoric Acid According to Its Location in the Soil and Whether or Not It Is Combined with Nitrogen, *IAEA/FAO Symp. Radioisotopes in Soil-Plant Nutrition Studies,* IAEA Vienna, 1962.

Evans, S. D. and S. A. Barber. The Effect of Cation-Exchange Capacity Clay Content, and Fixation on Rubidium 86 Diffusion in Soil and Kaolinite Systems, *Soil Sci. Soc. Am. Proc.* **28**: 53–55 (1964).

Evans, S. D., and S. A. Barber. The Effect of Rubidium 86 Diffusion on the Uptake of Rubidium 86 by Corn, *Soil Sci. Soc. Am. Proc.,* **28**: 56–57 (1964a).

Fang, S. C., R. D. Nance, T. Chao, and M. E. Harward. A Chromatographic Approach to the Determinations of Sulfate Adsorption and exchange of Less Retentive Soils, *Soil Sci.,* **94**: 14–18 (1962).

FAO. Long Term Trend of World Fertilizer Consumption, *FAO Monthly Bull. Agri. Econ. Sta.,* **11**: 1–3 (1962).

Faucher, J. A., and H. C. Thomas. Adsorption Studies on Clay Minerals. IV. The System Montmorillonite-Cesium-Potassium. *J. Chem. Phys,* **22**(2): 258–261 (1954).

Fawzy, H., R. Overstreet, and L. Jacobson. The Influence of Hydrogen Ion Concentration on Cation Absorption by Barley Roots, *Plant Physiol.,* **29**: 234–237 (1954).

Ferrière, J. J. F. de, R. Blanchet, C. Milot, and T. Camen. Effect of the Nature of Clay on the Dynamics of Soil Potassium in some Experimental Fields, *Am. Inst. Nat. Res. Agr., Ser. A,* pp. 163–175 (1960).

Finck, A. Methoden zur Bestimmung des für Hafer verfügbaren Mangans, *Z. Pflanzenernaehr. Düng. Bodenk.* **67**(112): 198–211 (1954).

Finck, A. and E. Schlichting. Möglichkeiten und Grenzen der chemischen Bodenuntersuchung zur Ermittlung des Düngerbedarfs, *Z. Acker- Pflanzenbau,* **99**: 267–272 (1955).

Fine, L. O. et al. The Influence of Nitrogen and Potassium on the Availability of Fertilizer Phosphorus, *S. Dakota State Coll. Agr. Expt. Sta., North Centr. Reg. Publ. No. 67, Bull.* 453, 1955.

Fitts, J. W. Research Plus Extension, Higher Farming Profits, Summer *Plant Food Rev.,* 1959.

Fitts, J. W., W. V. Bartholomew, and H. Heidel. Correlation Between Nitrifiable Nitrogen and Yield Response of Corn to Nitrogen Fertilization on Iowa Soils, *Soil Sci. Soc. Am. Proc.* **17**: 119–122 (1953).

Fitts, J. W., W. V. Bartholomew, and H. Heidel. Predicting Nitrogen Fertilizer Needs of Iowa Soils. I. Evaluation and Control of Factors in Nitrate Production and Analysis, *Soil Sci. Soc. Amer. Proc.,* **19**: 69–77 (1955).

Fitts, J. W., J. J. Hanway, L. T. Kardos, W. T. McGeorge, L. A. Dean, J. A. Reed, and D. D. Mason, (Soil Test Work Group) also W. L. Nelson, R. O. Parks, J. A. Rigney, and D. D. Mason. Soil Tests Compared with Field, Greenhouse and Laboratory Results, *N. Carolina Agr. Expt. Sta. Tech. Bull. 121,* 1956.

Flaig, W., and H. Sochting. Einfluss organischer Stoffe auf die Aufnahme anorganischer Ionen, *Agrochimica,* **6**(3): 251–264 (1962).

Florell, C. Calcium Mitochondria and Anion Uptake, *Physiol. Plantarum,* **10**: 781–790 (1957).

Fordham, A. W. Forms of Phosphate in Calcium Chloride Extracts of Soil, *Nature,* **195**(4841): 627–628 (1962).

Foy, C. D. and J. C. Brown. Toxic Factors in Acid Soils. I. Characterization of Aluminum Toxicity in Cotton, *Soil Sci. Soc. Am. Proc.*, **27**: 403–407 (1963).

Foy, C. D. and J. C. Brown. Toxic Factor in Acid Soils. II. Differential Aluminum Tolerance in Plant Species, *Soil Sci. Soc. Am. Proc.*, **28**: 27–32 (1964).

Fraps, G. S. Availability of Phosphoric Acid of the Soil, *J. Am. Chem. Soc.*, **28**: 824 (1902).

Fraps, G. S. Active Phosphoric Acid and Its Relation to the Needs of the Soil for Phosphoric Acid in Pot Experiments, *Texas Agr. Expt. Sta., Bull. 126,* 1909.

Frear, W., and E. S. Erb. Condition of Fertilizer Potash Residues in Hagerstown Silty-Loam Soil, *J. Agr. Res.*, **15**: 59–81 (1918).

Frear, D. E. *Agricultural Chemistry,* Vol. II, Practical Applications, D. Van Nostrand Co., New York, pp. 573, 1951.

Freney, J. R. Determination of Water Soluble Sulfate in Soils, *Soil Sci.*, **86**: 241–244 (1958).

Freney, J. R., N. J. Barrow, and K. Spencer. A Review of Certain Aspects of Sulfur as a Soil Constituent and Plant Nutrient. *Plant Soil,* **17**: 295–308 (1962).

Fribourg, H. A. and I. J. Johnson. Dry Matter and Nitrogen Yields of Legume Tops and Roots in the Fall of the Seeding Years, *Agron J.*, **47**: 73 (1955).

Fricke, H. A Mathematical Treatment of the Electric Conductivity and Capacity of Disperse Systems, *Phys. Rev.*, **24**: 575–587 (1924).

Fried, M. Contamination in Orthophosphates Irradiated in a Neutron Pile, *Science,* 3: 492–493 (1950).

Fried, M. The Feeding Power of Plants for Phosphate, *Soil Sci. Soc. Am. Proc.*, **17**: 357–359 (1953).

Fried, M. Progress in Agronomic Research Through the Use of Radioisotopes, in *Role of Atomic Energy in Agricultural Research,* Proc. 4th Ann. Oak Ridge Summer Symp. 1952, Pub. TID-5115: 452–480 (1953a).

Fried, M. Quantitative Evaluation of Processed and Natural Phosphates, *J. Agr. Food Chem.*, **2**: 271–244 (1954).

Fried, M. Tagged Atoms Tell Where Nutrients Go, *What's New in Crops and Soils,* **8**: 3, (1955).

Fried, M. Measurement of Plant Nutrient Supply of Soils by Radioactive Isotopes, in *Atomic Energy and Agriculture,* pp. 1–18, 1957.

Fried, M. Exchangeable Sodium in Calcareous Soils, *Soil Sci. Soc. Am. Proc.*, **23**(2): 179–180 (1959).

Fried, M. The Use of Isotopes in Basic Soil-plant Relationships Research, *Trans. 7th Intern. Congr. Soil Sci.*, **2**: 197–205 (1960).

Fried, M. 'E', 'L', and 'A' Values, *Proc. 8th Intern. Congr. Soil Sci.,* Bucharest, Rumania (in press).

Fried, M., and L. A. Dean. A Concept Concerning the Measurement of Available Soil Nutrient, *Res. Rept. No. 197,* Issued by the Division of Soil Management and Irrigation, U.S. Dept. Agr., 1950.

Fried, M., and L. A. Dean. A Concept Concerning the Measurement of Available Soil Nutrient, *Soil Sci.*, **73**: 263–271 (1953).

Fried, M., and L. A. Dean. Phosphate Retention by Iron and Aluminum in Cation Exchange Systems, *Soil Sci. Soc. Am. Proc.*, **19**(2): 143–147 (1955).

Fried, M. and A. H. Franklin. Rapid Preparation of Fresh Frozen Sections of Barley Roots, *Nature,* **189**: 413–414 (1961).

Fried, M., and G. Hawkes. K-41 Enriched Potassium, K-42 and Rb-86 Tracer Studies in the Soil Plant System, (Presented at Soil Sci. Soc. Am. Meeting at Cincinnati, Ohio 1959).

Fried, M., and W. Heald. Radioisotopes in Soils: Soil-Plant Relationships, in *Symp. on Radioisotopes in the Biosphere,* Centre for Coordination Studies, Univ. of Minn., pp. 47–60, 1960.

Fried, M., and A. J. Mackenzie. Rock Phosphate Studies With Neutron Irradiated Rock Phosphate, *Soil Sci. Soc. Am. Proc.,* **14**: 226–231 (1950).

Fried, M., and J. C. Noggle. Multiple Site Uptake of Individual Cations by Roots as Affected by Hydrogen Ion, *Plant Physiol.,* **33**: 139–144 (1958).

Fried, M., and R. E. Shapiro. Phosphate Supply Pattern of Various Soils, *Soil Sci. Soc. Am. Proc.,* **20**(4):471–475 (1956).

Fried, M., and R. E. Shapiro. Soil-Plant Relations in Phosphorus Uptake, *Soil Sci.,* **90**: 69–76 (1960).

Fried, M. and R. E. Shapiro. Soil-Plant Relationships in Ion Uptake, *Ann. Rev. Plant Physiol.,* **12**: 91–112 (1961).

Fried, M., C. E. Hagen, J. F. Salz del Rio, and J. E. Leggett. Kinetics of Phosphate Uptake in the Soil-Plant System, *Soil Sci.,* **84**: 427–437 (1957).

Fried, M., F. E. Allison, and C. H. M. van Bavel. Investigations with Isotopes in Soil-Plant Relationships Research, in *Radiation Biology and Medicine, Selected Reviews in Life Sciences,* Addison Wesley Publishing Co., Inc., Reading, Mass. (Atoms for peace, Geneva), pp. 607–632, 1958a.

Fried, M., J. C. Noggle, and C. E. Hagen. The Relationship Between Adsorption and Absorption of Cations, *Soil Sci. Soc. Am. Proc.,* **22**: 495–499 (1958b).

Fried, M., G. Hawkes, and W. Z. Mackie. Rubidium-Potassium Relations in the Soil Plant System, *Soil Sci. Soc. Am. Proc.,* **23**: 360–362 (1959).

Fried, M., H. E. Oberländer, and J. C. Noggle. Kinetics of Rubidium Absorption and Translocation by Barley, *Plant Physiol.,* **36**: 183–191 (1961).

Fried, M., F. Zsoldos, P. B. Vose, and I. L. Shatokhin. Characterizing the NO_3^- and NH_4^+ Uptake Process of Rice Roots by Use of N^{15} Labelled NH_4NO_3, *Physiol. Plantarum,* **18**: 313–320 (1965).

Frink, C. R., and M. Peech. Hydrolysis and Exchange Reactions of the Aluminum Ion in Hectorite and Montmorillonite Suspensions, *Soil Sci. Soc. Am. Proc.,* **27**(5): 527–530 (1963).

Fripiat, J. J., and M. C. Gastuche. *Etude physico-chimique des surfaces des argiles,* Publication de l'Institut National pour l'Etude Agronomique due Congo Belge (INEAC), Série Scientifique, No. 54, 60 pp., 1952.

Gadet, R., L. Soubiès, and F. Fourcassie. Investigations on the Toxic Effects and Evolution of Biuret in Soil, *Am. Agr. Paris,* **10**: 609–660 (1959).

Gahan, P. B., and Rajan, A. K. Autoradiography of $^{35}SO_4$ in Plant Tissue, *Exptl. Cell Res.,* **38**: 204–207 (1964).

Gaines, G. L., and H. C. Thomas. Adsorption Studies on Clay Minerals. II. A Formulation of the Thermodynamics of Exchange Adsorption, *J. Chem. Phys.,* **21**: 714–718 (1953).

Gaines, G. L., and H. C. Thomas. Adsorption Studies on Clay Minerals. V. Montmorillonite-Cesium-Strontium at Several Temperatures, *J. Chem. Phys.* **23**: 2322–2326 (1955).

Gapon, E. N. On the Theory of Exchange Adsorption in Soils, *J. Gen. Chem.* (USSR), **3**: 144–163 (1933).

Gardner, A., R. S. Whitney, and A. Kezer. Slick Spot Soils in Western Colorado, *Colorado Agr. Expt. Sta. Techn. Bull.,* **20**: 1–13 (1937).

Gardner, W. R., and M. S. Mayhugh. Solutions and Tests of the Diffusion Equation for the Movement of Water in Soil, *Soil Sci. Soc. Am. Proc.*, **22**: 197–201 (1958).

Garman, W. L. Potassium Release Characteristics of Several Soils from Ohio and New York, *Soil Sci. Soc. Am. Proc.*, **21**: 52–58 (1957).

Gasser, J. K. R. Transformation, Leaching and Uptake of Fertilizer Nitrogen Applied in Autumn and Spring to Winter Wheat on a Heavy Soil, *J. Sci. Food Agr.*, **12**: 375–380 (1961).

Gasser, J. K. R. Transformation, Leaching and Uptake of Fertilizer Nitrogen Applied to Winter and to Spring Wheat Grown on a Light Soil, *J. Sci. Food Agr.*, **13**: 367–375 (1962).

Gasser, J. K. R. Effects of Long-Continued Treatment on the Mineral Nitrogen Content and Mineralizable Nitrogen of Soil from Selected Plots of the Broadbulk Experiment on Continuous Wheat, Rothamsted, *Plant Soil*, **17**: 209–220 (1962a).

Gastuche, M. C. *et al.*, *Soil Sci.* **98**: 281 (1964).

Gedroiz, K. K. *Chemische Bodenanalyse*, Gebr. Bomtraeger, Berlin, 1926.

Gedroiz, K. K. *Handbuch der Bodenlehre* (Blanck), Bd. VIII, pp. 106–148, Springer, Berlin, 1933.

Gehring, A. Die Bestimmung der im Boden im leichtlöslichen Zustande vorkommenden Nährstoffe, *Handbuch der Bodenlehre von Blanck*, Bd. VIII, pp. 106–148, Springer Verlag, Berlin, 1931.

Gerbon, M. L'Engrais Organique, *Feuilles Agricoles*, **12**(85): 333–337 (1954).

Gerlach, M. Ueber das Verhalten der wasserlöslichen Phosphorsäure gegen adsorbierende Bestandteile des Bodens, *Landwirtsch Versuchsstat.*, **46**: 201 (1896).

Gerretsen, F. C. En Onderzoek Naar de Oorzaken der Veenkoloniale Haverziekte, *Landbouwk. Tijdschr.*, **54**: 373–383 (1942).

Gerretsen, F. C. Nitrogen Balance and pH Changes in Some Sandy Soils as Influenced by Microbes and Plants, *Verslag. Landbouwk. Onderzoek*, **54**: 16 (1949).

Gerretsen, F. C. Microbiological tranformation of Nitrogen and Its Influence on Nitrogen Availability in the Soil, Trans. *4th Intern. Congr. Soil Sci., Amsterdam*, **2**: 114–117 (1950).

Gerretsen, F. C., and H. de Hoop. Nitrogen Losses During Nitrification in Solutions and in Acid Sandy Soils, *Can. J. Microbiol.*, **3**: 359–380 (1957).

Ghani, M. O. Fractionation of Soil Phosphorus. I. Method of Extraction, *Indian J. Agr. Sci.*, **13**: 29 (1943).

Ghani, M. O., and M. A. Islam. Phosphate Fixation in Acid Soils and Its Mechanism, *Soil Sci.*, **62**: 293–306 (1946).

Giesecke, F. Das Stickstoffkapital des Bodens und seine Bestimmung, *Handbuch der Bodenlehre*, Bd. VIII, pp. 421–452, Springer Verlag, Berlin, 1931.

Gil Benavides, A. A. A Comparison of Four Nitrogen Fertilizers for Maize in the Canca Valley, Colombia, *Acta Agron. Colombia*, **9**: 153–168 (1959).

Gilbert, M., and H. Laudelout. Ion Exchange Reactions of Hydrogen Ions in Clays, *Soil Sci.*, **100**: 157–162 (1965).

Gilligan, G. M. The Effect of Degree of Base Saturation of Soils upon the Fixation of Phosphate and Potassium and the Availability of Phosphorus. *Delaware Expt. Sta., Bull. No. 215*, pp. 1–20, 1938.

Ginniken, P. J. H. van. Mededelingen Inst. Rat. Suikerprod. XIII, 1943.

Gliemeroth, G. The Movement of Nitrogen in a Loess Loam During Winter, as Depending on Form, Amount, Time and Method of Fertilizer Application in Autumn, *Z. Pflanzenernaehr. Düng. Bodenk.*, **85**: 20–31 (1959).

Glover, J. The Nutrition of Maize in Sand Culture. II. The Uptake of Nitrogen and Phosphorus and Its Relevance to Plant Analysis, *J. Agr. Sci.,* **43**: 160–165 (1953).

Goedewagen, M. A. J. *Het Wortelstelsel der Landbouw Gewassen,* The Hague, 1942.

Goedewagen, M. A. J. Wortelgroei in Gronden Bestaande uit een Bovengrond van Klei en een Ondergrond van Zand, *Verslag. Landbouwk., Ondergock.,* **61**: 7 (1955).

Goldschmidt, V. M. *Geochemistry,* Clarendon Press, Oxford, pp. 730, 1954.

Goodall, D. W. and F. G. Gregory. Chemical Composition of Plants as an Index of Their Nutritional Status, *Imp. Bur. Hort. Plantation Crops, Tech. Com.,* **17**: 167 (1947).

Goring, C. A. Control of Nitrification of Ammonium Fertilizers and Urea by 2-Chloro-6-(trichloromethyl) Pyridine, *Soil Sci.,* **93**: 433–439 (1962).

Goring, C. A. I., and F. E. Clark. Influence of Crop Growth on Mineralization of Nitrogen in the Soil, *Soil Sci. Soc. Am. Proc.,* **13**: 261–266 (1949).

Gough, N. A., and J. D. Beaton. Influence of Phosphorus Source and Soil Moisture on the Solubility of Phosphorus, *J. Sci. Food Agr.,* **14**: 224–228 (1963).

Goy, S. Ueber das Wesen der Bodenazidität von Mineralböden, *Landwirtsch. Forstwirt. Ztg. Georgine,* pp. 18, 19, 20, 1928.

Graham, E. R. and W. A. Albrecht. Potassium Bearing Minerals as Soil Treatments, *Missouri Agr. Exptl. Sta. Res. Bull. 510,* 1952.

Graham-Bryce, I. J. Effect of Moisture Content of Soil Type on Self Diffusion of Rb^{86} in Soils, *J. Agr. Sci.,* **60**: 239–244 (1963).

Gray, B., M. Drake, and W. G. Colby. Potassium Competition in Grass-Legume Associations as a Function of Root Cation Exchange Capacity, *Soil Sci. Soc. Am. Proc.,* **17**: 235–239 (1953).

Greene, H. Paddy Soils and Rice Production, *Nature* (London), **186**: 511–513 (1960).

Gregor, P., J. I. Bregman, F. Gutoff, R. Broadley, D. Baldwin, and C. G. Overberger. Studies on Ion-Exchange Resins, Capacity of Sulfuric Acid Cation-Exchange Resins, *J. Colloid Sci.,* **6**(1): 20–32 (1951).

Grim, R. E. *Clay Mineralogy,* McGraw-Hill Book Co., Inc., New York, pp. 384, 1953.

Grissinger, E., and C. D. Jeffries. Influence of Continuous Cropping on the Fixation and Release of K in Three Pennsylvania Soils, *Soil Sci. Soc. Am. Proc.,* **21**: 409–412 (1957).

Grunes, D. L. Effect of Nitrogen on the Availability of Soil and Fertilizer Phosphorus to Plants, *Advan. Agron.,* **11**: 369–396 (1959).

Grunes, D. L., and H. Jenny. Two-Phase Studies on Availability of Iron in Calcareous Soils. II. Decomposition of Colloidal Iron Hydroxide by Ion Exchangers, *Agrochimica,* **55**(4): 279–287 (1960).

Grunes, D. L., H. J. Haas, and S. H. Shih. Effect of Long-Time Dryland Cropping on Available Phosphorus of Cheyenne Fine Sandy Loam, *Soil Sci.,* **80**: 127–138 (1955).

Grunes, D. L., F. G., Viets, and S. H. Shih. Proportionate Uptake of Soil and Fertilizer Phosphorus by Plants as Affected by Nitrogen Fertilization, *Soil Sci. Soc. Am. Proc.,* **22**: 43–48 (1958a).

Grunes, D. L., H. R. Haise and L. O. Fine. Proportionate Uptake of Soil and Fertilizer Phosphorus by Plants as Affected by Nitrogen Fertilization. II. Field Experiments with Sugar Beets and Potatoes, *Soil Sci. Soc. Am. Proc.,* **22**: 49–52 (1958b).

Grunes, D. L., H. R. Haise, F. Turner, and J. Alessi. Relationship between Yield Response to Applied Fertilizers and Laboratory Measures of Nitrogen and Phosphorus Availability. *Soil Sci. Soc. Am. Proc.,* **27**: (6): 675–679 (1963).

Gunnarson, O., and L. Frederiksson. A Method For Determining the Amount of "Plant Available" Phosphorus in Soil by Means of P-32, *Radioisotope Techniques,* **1**: 427–431 (1951).

Haddock, J. L., R. L. Hausonbruller, and C. O. Stanberry. Studies with Radioactive Phosphorus in Soils of the Western States, 1950–53, *U.S. Dept. Agr., Prod. Res. Rept. 12,* 1957.

Hagen, C. E., and H. T. Hopkins. Ionic Species in Orthophosphate Absorption by Barley Roots, *Plant Physiol.,* **30**: 193–199 (1955).

Hagen, C. E., and H. T. Hopkins, Personal Communication.

Hagen, C. E., J. E. Leggett, and P. C. Jackson. The Sites of Orthophosphate Uptake by Barley Plants, *Proc. Natl. Acad. Sci. U.S.,* **43**: 496–506 (1957).

Hagen, C. E., J. E. Leggett, and P. C. Jackson. Beltsville USDA Agr. Res. Serv., Beltsville, Md., Personal Communication.

Hagenzieker, F. Soil-Nitrogen Studies at Urambo, Tanganyika Territory, East Africa, *Plant Soil,* **9**: 97–113 (1957).

Hagin, J. Availability of Dicalcium Phosphate to Plants When Applied in Various Forms, *Plant Soil,* **10**: 101–113 (1958).

Hagin, J. On the Shape of the Yield Curve, *Plant Soil,* **12**: 285–296 (1960).

Hagin, J., and J. Berkovita. Efficiency of Phosphate Fertilizers of Varying Water Solubility, *Can. J. Soil Sci.,* **41**: 68–80 (1961).

Hagin, J., J. Hillinger, and A. Olmert. Comparison of Several Ways of Measuring Soil Phosphorus Availability, *J. Agr. Sci.,* **60**: 245 (1963).

Hahn, G. Vergleichende Versuche zur Bestimmung des Phosphorsäurebedürfnisses der Ackerböden, *Boten. Arch.,* **20**: 223 (1927).

Hall, N. S., W. V. Chandler, C. H. M. Bavel, P. H. Reid, and J. H. Anderson. A Tracer Technique to Measure Growth and Activity of Plant Root Systems, *N. Carolina Expt. Sta. Tech. Bull. 101,* 1953.

Halstead, R. L., and H. B. Heeney. Exchangeable and Water-Soluble Potassium in Soils and Degree of Saturation in Relation to Tomato Yields, *Can. J. Soil Sci.,* **39**: 129–135 (1959).

Hammes, J. K., and J. F. Bartz. Root Distribution and Development of Vegetable Crops as Measured by Radioactive Phosphorus Injection Techniques, *Agron. J.,* **55**: 329–333 (1963).

Handley, R., and R. Overstreet. Effect of Various Cations upon Absorption of Carrier-free Cesium, *Plant Physiol.,* **36**(1): 66–69 (1961).

Handley, R., and R. Overstreet. Uptake of Calcium and Chlorine in Roots of *Zea mays, Plant Physiol.,* **36**: 761–769 (1961a).

Handley, R., and R. Overstreet. Uptake of Strontium by *Zea mays, Plant Physiol.,* **38**: 180–184 (1963).

Hanson, J. B., and J. S. Kahn. The Kinetics of Potassium Accumulation by Corn Roots as a Function of Cell Maturity, *Plant Physiol.,* **32**: 497–498 (1957).

Hanway, J. J. Corn Growth and Composition in Relation to Soil Fertility I, *Argon. J.,* **54**: 204–207 (1962a).

Hanway, J. J. Corn Growth and Composition in Relation to Soil Fertility. II. Uptake of N.P and K and their Distribution in Different Plant Parts During the Growing Season. III. Percentages of N, P and K in Different Plant Parts in Relation to Stage of Growth, *Agron. J.,* **54**: 217–222; 222–229 (1962a).

Hanway, J. J., G. Stanford, and H. R. Meldrum. Effectiveness and Recovery of Phosphorus and Potassium Fertilizers Topdressed on Meadows, *Soil Sci. Soc. Am. Proc.*, **17**: 378–382 (1953).

Harmsen, G. W. and D. J. Lindenbergh. Investigations on the Nitrogen Nutrition of Plants. I. A New Method for the Determination of Nitrogen Requirement of Soil, *Plant Soil*, **2**: 1–29 (1949).

Harmsen, G. W., and D. A. van Schreven. Mineralization of Organic Nitrogen in Soil, *Adv. Agron.*, **7**: 300–398 (1955).

Harris, G. P., and T. M. Morrison. Fixation of N-15 by Excised Nodules of Coriaria Arborea Lindsay, *Nature (London)*, **182**: 182 (1958).

Harrold, L. L., and F. R. Dreibelbis. Agricultural Hydrology as Evaluated by Monolith Lysimeters, *U.S. Dept. Agr. Tech. Bull. 1050*, pp. 149, 1951.

Hartfiel, W. The Enrichment of Fodder Crops with Phosphorus for an Improved Mineral Nutrition of Domestic Animals, *Phosphorsäure*, **18**: 129–139 (1958).

Harward, M. E., and N. T. Coleman. Some Properties of H- and Al-clays and Exchange Resins, *Soil Sci.*, **78**: 181–188 (1954).

Harward, M. E., T. T. Chao, and S. C. Fang. The Sulfur Status and Sulfur Supplying Power of Oregon Soils, *Agron. J.*, **54**: 101–106 (1962).

Haseman, J. F., J. R. Lehr, and J. P. Smith. Minerological Character of Some Iron and Aluminum Phosphates Containing Potassium and Ammonium, *Soil Sci. Soc. Am. Proc.*, **15**: 76–84 (1951).

Hasenbäumer, J. Die Feststellung der Nährstoffbedürftigkeit des Bodens I. Die chemische Bodenanalyse, *Handbuch Pflanzenernährg. Düngerlehre* (Honcamp), Bd. I, pp. 771–808, Springer Verlag, Berlin, 1931.

Hassbroek, F. J., J. C. Noggle, and A. L. Fleming. High-Resolution Autoradiography Without Loss of Water-Soluble Ions, *Nature*, **195**: 615–616 (1962).

Hausenbruller, R. L., and W. H. Weavor. A Comparison Between Greenhouse and Field Procedures in Phosphate Fertilizer Testing, *Soil Sci.*, **20**: 298–301 (1960).

Hauser, G. F. Die nichtaustauschbare Festlegung des Kalis im Boden, Dissertation, Univ. of Wageningen, pp. 1–171, 1941.

Hawkes, G. R., and M. Fried. Effect of Preparation, Method and Water Solubility of Nitric Phosphate on Uptake by Greenhouse Culture, *Agr. Food Chem.*, **5**: 844–848 (1957).

Hayase, T. The Fertilizing Effect of Reaction Products of Urea Formaldehyde, *J. Sci. Soil Manure, (Tokyo)*, **32**: 628–640 (1961).

Heald, R. Characterization of Exchange Reactions of Strontium or Calcium on Four Clays, *Soil Sci. Soc. Am. Proc.*, **24**(2): 103–106 (1960).

Hegarty, M. P. Nitrogen Uptake and Nitrogen Transformation in Plants, *C/Wealth Burs. Past. and Field Crops Bull.*, **46**: 83–97 (1962).

Heintze, S. G. Readily Soluble Manganese of Soils and Marsh Spot of Peas, *J. Agr. Sci.*, **28**: 175–186 (1938).

Heintze, S. G. Studies on Cation-Exchange Capacities of Roots, *Plant Soil*, **13**(4): 365–383 (1961).

Helder, R. J. Absorption and Distribution of Labelled Rubidium Ions in Young Intact Barley Plants. *Proc. 2nd Intern. Conf. Peaceful Uses At. Energy*, Geneva, 1958, P/1472 Netherlands, **27**: 42–47 (1958).

Helfferich, F. *Ion Exchange*. McGraw Hill Book Co., Inc., 624 pp., 1962.

Helmy, A. K., M. N. Hassan, and S. Taher. Absorption of Na, K, and Ca by Barley Roots at Constant Ionic Strength, *Plant Soil*, **18**(1): 133–139 (1963).

Hemwall, J. B. The Role of Soil Clay Minerals in Phosphorus Fixation, *Soil Sci.*, **83**: 101–108 (1957).

Hendrick, J., and H. D. Welsh. The Substances Removed by the Drainage from a Scottish Soil Trans., *1st Intern. Congr. Soil Sci.,* Commission II, pp. 358–366, 1928.

Hendrysiak, J. The Effect of Repeated Spring Applications of Nitrate of Grain Yields of Rye in Sandy Soils, *Prace Zah Upraw. Rol. Podor.,* **2**: 123–128 (1960).

Henriksen, A. A Chemical and Biological Determination of Copper in Soil, *Nature (London)*, **178**: 499–500 (1956).

Hensel, D. R. Rate of Release of Nitrogen and Phosphorus from Urea-Formaldehyde Products and Urea-Formaldehyde Coated Phosphorus., *Dissertation Abstr.,* **21**: 1005 (1960).

Henze, R. Comparison of Yields from the Göttingen E-Fields and the Nutrient Potential and Supply of the Soil, *I. Pflanzenbau Düng. Bodenk.,* **103**: 9–21 (1963).

Henzell, E. F., and D. O. Norris. Processes by Which Nitrogen is Added to the Soil/Plant System, *Commonwealth Bureau of Pastures and Field Crops, Bull. 46,* Hurley, Berkshire, U.K., 1962.

Hernandez, S. C. Studies on Soil Fertility in Hyderabad, India, *J. Soil Sci. Soc. Philippines,* **8**: 19–22 (1956).

Hernando, V., L. Jimeno, and A. Guerra. Effect of Time of Application of Nitrate on the Yield of Sugarbeet, *Anales. Edafol. Agrobiol.,* **20**: 477–486 (1961).

Herrmann, R. Zusammenfassender Bericht über die Ueberprüfung der Schnellmethoden zur Bestimmung des pflanzenaufnehmbaren Kaliums und der Apparatur von Riehm-Lange, *A. Pflanzenernähr. Dueng. Bodenk.,* **29**(74): 252–274 (1943).

Herrmann, R. Handb. d. landw. Versuchs- und Untersuchungsmethodik, Bd. I, Berlin-Neumann, 1949.

Heslep, J. M., and C. A. Black. Diffusion of Fertilizer Phosphorus in Soils, *Soil Sci.,* **78**: 384–401 (1954).

Hester, J. B. Magnesium Deficiency in Soils, *Farm Chem.,* **121**: 20–21 (1958).

Hester, J. B. and F. A. Shelton. *Know Your Plant and Soil Requirements,* Dept. Agr. Res., Campbell Soup Co., Res. Mon. 3, 1949.

Hewit, E. J. Long Ashton Annual Report 1947, pp. 82.

Heytler, P. G. Uncoupling of Oxidative Phosphorylation by Carbonyl Cyanide Phenylhydrazones. I. Some Characteristics of m-Cl-CCP-Action on Mitochondria and Chloroplasts, *Biochem.,* **2**(2): 357–361 (1963).

Hibbard, P. L. Comparison of the Soil Solution by Displacement Method and the Water Extract of Alkali Soils, *Soil Sci.,* **16**: 465–471 (1923).

Hill, R. *Photosynthesis,* Methuen, London, pp. 175, 1958.

Hitchcock, D. I. The Formal Identity of Langmuir's Adsorption Equation with the Law of Mass Action *J. Am. Chem. Soc.,* **48**: 2870, 1926.

Hoagland, D. R. Lecturers on the Inorganic Nutrition of Plants, *Chronica Botanica,* Waltham, Mass, pp. 226, 1944.

Hoagland, D. R. and T. C. Broyer. Hydrogen Ion Effects and the Accumulation of Salt by Barley Roots as Influenced by Metabolism, *Am. J. Botany,* **27**: 173–185, 1940.

Hoagland, D. R. and J. C. Martin. Absorption of Potassium by Plants in Relation to Replaceable, Non-Replaceable and Soil Solution Potassium, *Soil Sci.,* **36**: 1, 1933.

Hoagland, D. R., J. C. Martin, and G. R. Stewart. Relation of the Soil Solution to the Soil Extract, *J. Agr. Res.*, **20**: 381–395, 1920.

Hockensmith, R. D., R. Gardner and J. Goodwin. Comparison of Methods for Estimating Available Phosphorus in Alkaline Calcareous Soils, *Colo. Agr. Expt. Sta. Tech. Bull. 2*, pp. 3–24, 1933.

Hoff, D. J. and H. J. Mederski. The Chemical Estimation of Plant "Available Soil Manganese." (Ohio Agr. Expt. Sta., Wooster) *Soil Sci. Soc. Am. Proc.*, **22**: 123–132 (1958).

Hofman, E. Studien über die Löslichkeit der Bodenphosphorsäure und -kieselsäure als Grundlage für die Bestimmung des Phosphorsäurebedürfniesses, *Landwirtsch. Tahrb. Schweiz*, **72**: 791, 1930.

Holben, F. J. Soil Respiration in Relation to Crop Yields, *Pa. Agr. Expt. Sta. Bull.*, **273**: 26–29, 1932.

Hood, J. T., N. C. Brady, and D. J. Lathwell. The Relationship of Water-Soluble and Exchangeable Potassium to Yield and Potassium Uptake by Ladino Clover, *Soil Sci. Soc. Am. Proc.* **20**: 228–231, 1956.

Hood, J. T. and L. E. Ensminger. Fertilizer Placement Critical for Cotton, *Alabama Agr. Expt. Sta. Highlights of Agr. Res.*, **6**(1): 1959.

Hope, A. B., and P. G. Stevens. Electric Potential Differences in Bean Roots and Their Relation to Salt Uptake, *Australian J. Sci. Res. Ser. B.* (Biol. Sc.) **5**: 335–343, 1952.

Horne, W. R. and O. T. Denmead. A Modified Colorimetric Method for the Field Determination of Soil Nitrate, *J. Australian Inst. Agr. Sci.*, **21**: 34–36, 1955.

Hsu, P. M., and M. L. Jackson. Inorganic Phosphate Transformation by Chemical Weathering in Soils as Influenced by pH, *Soil Sci.*, **20**: 16–24, 1960.

Huffaker, R. C. and A. Wallace. Sodium Absorption by Different Plant Species at Different Potassium Levels, *Soil Sci.*, **87**: 130–134, 1958.

Huffman, E. O. Reactions of Phosphate in Soils: Recent Research by T. V. A., *Fert. Soc. Proc.*, **71**: 49, 1962.

Hulshof, H. J., L. J. J. Van der Kloes, and A. F. C. M. Schellekens. Beworteling van Appelboomen en Bodemstructuur *Mededel. Dir. Tuinb.*, **23**: 33–42, 1960.

Humbert, R. P. Potash Fertilization in the Hawaiian Sugar Industry. 5th Potassium Symposium, Madrid, *Intern. Potash Inst. Proc.*, pp. 319–399, 1958.

Hunter, A. S., E. N. Hoffman, and J. A. Yungen. Residual Effects of Phosphorus Fertilizer on an Eastern Oregon Soil, *Soil Sci. Soc. Am. Proc.*, **25**: 3, 1961.

Hunter, R. F. and A. H. Knight. Studies of the Root Development of Plants in the Field Using Radioactive Tracers. Part II: Communities Growing in Deep Peat, *J. Ecol.*, **46**: 621–639, 1958.

Husted, R. F., and F. Low. Ion Diffusion in Bentonite, *Soil Sci.*, **77**(5): 343–353 (1954).

Hutton, J. T., and T. I. Leslie. *Aust. J. Agr.* **9**: 492 (1958).

Hylmö, B. Transpiration and Ion Absorption, *Physiol. Plantarum*, **6**: 333–405 (1953).

Hylmö, B. Passive Components in the Ion Absorption of the Plant. I. The Zonal Ion and Water Absorption in Brouwer's Experiments, *Physiol. Plantarum*, **8**: 433–449 (1955).

Hylmö, B. Passive Components in the Ion Absorption of the Plant. II. The Zonal Water Flow, Ion Passage, and Pore Size in Roots of *Vicia Faba*, *Physiol. Plantarum*, **11**: 382–400 (1958).

IAEA. 1964. Annual Report of 1963 Laboratory Activities, *IAEA Tech. Rpt. Ser. No. 25*, pp. 33–47.

IAEA. Annual Report of 1964 Laboratory Activities, *IAEA Tech. Rpt. Ser. No. 41*, pp. 17–25, 1965.

IAEA. Annual Report of 1965 Laboratory Activities, *IAEA Tech. Rept. Ser. No. 55*. in press.

Impey, R. L., and W. W. Jones. Effects of Biuret on Nitrogen Status of Washington Navel and Valencia Orange Leaves, *Proc. Am. Soc. Hort. Sci.*, **76**: 186–192 (1960).

Imre, L. Kinetic-Radioactive Investigations on the Active Surface of Crystalline Powders, *Trans. Faraday Soc.*, **33**: 571 (1937).

Ingelsten, B., and B. Hylmö. Apparent Free Space and Surface Film Determined by a Centrifugation Method, *Physiol. Plantarum*, **14**: 157–170 (1961).

Iowa State University of Science and Technology, *Co-op. Ext. Serv. Pamphlet 277*, 1960.

Jackson, J. C., and G. W. Burton. An Evaluation of Granite Meal as a Source of Potassium for Coastal Bermuda Grass, *Agron. J.* **50**: 307–308 (1958).

Jackson, M. L. Aluminum Bonding in Soils: A Unifying Principle in Soil Science, *Soil Sci. Soc. Am. Proc.*, **27**(1): 1–10 (1963).

Jackson, P. C., and H. R. Adams. Cation-Anion Balance during Potassium and Sodium Absorption by Barley Roots, *J. Gen. Physiol.*, **46**: 369–386 (1963).

Jackson, P. C., and C. E. Hagen. Products of Orthophosphate Absorption by Barley Roots, *Plant Physiol.*, **35**: 326–332 (1960).

Jackson, P. C., S. B. Hendricks, and B. M. Vasta. Phosphorylation by Barley Roots Mitochondria and Phosphate Absorption by Barley Roots, *Plant Physiol.*, **37**: 8–17 (1962).

Jackson, R. D., D. A. Rose, and H. L. Penman. Circulation of Water in Soil Under a Temperature Gradient, *Nature*, **205**: 314–316 (1965).

Jacob, K. D. Fertilizer Technology and Resources in the U.S., Academic Press, Inc., 454, (1953).

Jacobson, H. G. M., C. L. W. Swanson, and E. Smith. Effect of Various Fertilizer Cations and Anions on Soil Reaction, Leaching, Nitrification of Urea, and Related Characteristics in an Uncropped Soil: A Report on Windsor Lysimeter Series E., *Soil Sci.*, **65**(6): 437–460 (1948).

Jacobson, L., R. Overstreet, K. M. King, and R. Handley. A Study of Potassium Absorption by Barley Roots, *Plant Physiol.*, **25**: 639–647 (1950).

Jacobson, L., R. Overstreet, R. M. Carlson, and J. A. Chastain. The Effect of pH and Temperature on the Absorption of Potassium and Bromide by Barley Roots, *Plant Physiol.*, **32**: 658–662 (1957).

Jacobson, L., R. J. Hannapel, and D. P. Moore. Non-Metabolic Uptake of Ions by Barley Roots, *Plant Physiol.*, **33**(4): 278–282 (1958).

Jacobson, L., R. J. Hannapel, D. P. Moore, and M. Schaedle. Influence of Calcium on Selectivity of Ion Adsorption Process, *Plant Physiol.* **36**(1): 58–61 (1961).

Jacoby, B. Function of Bean Roots and Stems in Sodium Retention, *Plant Physiol.*, **39**: 445–449 (1964).

Janes, B. E. Composition of Florida-Grown Vegetables II, III, *Fla. Agr. Expt. Sta. Bull. 455*, p. 44, 1949 and *Bull. 488*, p. 32, 1951.

Jansson, S. L. Balance Sheet and Residual Effects of Fertilizer Nitrogen in a 6-Year Study with N^{15}, *Soil Sci.*, **95**: 31–37 (1963).

Jansson, S. L., M. J. Hallam, and W. V. Bartholomew. Preferential Utilization of Ammonium over Nitrate by Microorganisms in the Decomposition of Oat Straw, *Plant Soil*, **6**: 282–290 (1955).

Jenny, H. Simple Kinetic Theory of Ionic Exchange. I. Ions of Equal Valency, *J. Phys. Chem.*, **40**: 501–517 (1936).

Jenny, H. Reflections on the Soil Acidity Merry-Go-Round, *Soil Sci. Soc. Am. Proc.*, **25**: 428–432 (1961).

Jensen, G. Active and Passive Components in Ion Uptake Processes. Experiments with Intact and Excised Tomato Root Systems, *Physiol. Plantarum* **15**(2): 363–368 (1962).

Jensen, H. L. A Strain of *Nitrosomonas Europaea* from Farmyard Manure, *Tidsskr. Planteavl.*, **54**: 62 (1950).

Jensen, H. L. Microbiological Determination of Micronutrients in Soil, *Suomen Kennitil, Sar. A.*, **24**: 197–204 (1951).

Joachim, A. W. R. Drainage and Leaching Trials at Peradeniya, 1927–1930, *Trop. Agriculturist*, **74**: 323–330 (1930).

Joffe, J. S., and A. K. Levine. Fixation of Potassium in Relation to Exchange Capacity of Soils. I. Release of Fixed Potassium, *Soil Sci.*, **62**: 411–420 (1946).

Johannson, O. *K. Lantbr. Nögsk. Ann.* **25**: 57 (1959).

Johnson, R. E., and W. A. Jackson. Calcium Uptake and Transport by Wheat Seedlings as Affected by Aluminum, *Soil Sci. Soc. Am. Proc.*, **28**: 381–386 (1964).

Jolivet, E., and M. Elias. Mineralization of Organic Nitrogen in a Breton Soil After Ten Years of Different Treatments, *Compt. Rend. Acad. Agr. France*, **237**: 528–530 (1953).

Jones, E. J. Loss of Elemental Nitrogen from Soils Under Anaerobic Conditions, *Soil Sci.*, **71**: 193–196 (1951).

Jones, L. H. P., and G. W. Leeper. Available Manganese Oxides in Neutral and Alkaline Soils, *Plant Soil*, **3**: 154–159 (1951).

Joret, G., and J. Herbert. Contribution à la Détermination du besoin des Sols en Acide Phosphorique, *Ann. Inst. Natl. Rech. Agron., Ser. A*, **6**: 233–299 (1955).

Jung, H. Soil Fertility Losses under Missouri Conditions, *Missouri Agr. Expt. Sta. Bull. 324*, 1933.

Jung, J. The Effect of Crotonylidane Urea (CD Urea) as Nitrogen Fertilizers, *Plant Soil*, **15**: 284–290 (1961).

Jung, J. Slow Acting Nitrogen Fertilizers of Synthetic Origin, *Landwirtsch. Forsch. Sudest.*, **17**: 148–160 (1963).

Jurcak, F. The Effect of Some Synthetic Residues of Polyamide Character on the Yield and Quality of Cultivated Crops. II. The Effect of Ageing of Umalur on the Nitrogen Content and Yield of Oats, *Sb. Vysoke Skoly Zemedel. Brne., Rada A*, pp. 389–394 (1961).

Kahn, J. S., and J. B. Hanson. The Effect of Calcium on Potassium Accumulation in Corn and Soybean Roots, *Plant Physiol.*, **32**: 312–316 (1957).

Kaila, A. Effect of Incubation and Liming on the Phosphorus Fractions in Soil, *J. Sci. Agr. Soc. Finland*, **33**: 185–193 (1961).

Kaila, A. Organic Phosphorus in Finnish Soils, *Soil Sci.*, **95**(1): 38–44 (1963).

Kalinin, K. V., and B. Debreczeni. Studies with P-32 on the Utilization of Difficult Soluble Phosphate Fertilizers, *Agr. Okem. Talajt An.*, **10**: 223–232 (1961).

Kalthofen, H. Ueber den Begriff der aufnehmbaren Nährstoffmenge, *Z. Acker. Pflanzenbau*, **102**(2): 165–182 (1956).

Kamprath, E. J., W. L. Nelson, and J. W. Fitts. Sulfur Removed from Soils by Field Crops, *Agron. J.*, **49**: 289 (1957).

Kamprath, E. J., W. V. Chandler, and B. A. Krantz. Winter Cover Crops. Their

Effects on Corn Yields and Soil Properties. *N.C. Agr. Expt. Sta. Tech. Bull. 129*, p. 47, (1958).

Kanonova, M. M. *Soil Organic Matter. Its Nature, Its Role in Soil Formation and in Soil Fertility*. Translated from Russian. Pergamon Press, New York, London, 450 pp., 1961.

Kanswar, J. S. Influence of Organic Matter on Copper Fixation in Soil, *J. Indian Soc. Soil Sci.*, **2**: 73–80 (1954).

Kapp, L. C. Extracting a Submerged Soil Solution, *Arkansas Agr. Expt. Sta. Bull., 351*, p. 28 (1938).

Karaguishieva, D. The Question of the Role of Azotobacter in Accumulation of Nitrogen in Soil, Abstract in *Soil Fertilizers*, **26**: 113 (Abstr. No. 803) (1963).

Karlson, N. Potassium in the Soil, *Kungl. Lantsbr. Acad. Fidskr.*, **91**: 297–329 (1952).

Karraker, P. E., C. E. Bortner, and E. N. Fergus. Nitrogen Balance in Lysimeters as Affected by Growing Kentucky Bluegrass and Certain Legumes Separately and Together, *Kentucky Agr. Expt. Sta. Bull. 557*, p. 16 (1950).

Keay, J., and A. Wild. The Kinetics of Cation Exchange in Vermiculite, *Soil Sci.*, **92**: 54–60 (1961).

Keller, P., and H. Deuel. Kationenaustauschkapazität und Pektingehalt von Pflanzenwurzeln, *Z. Pflanzenernähr. Düng. Bodenk.*, **79**(124): 119–131 (1957).

Keller, P., and H. Deuel. Kationenaustauschgleichgewichte an abgetöteten Pflanzenwurzeln, *Intern. Soc. Soil Sci., Trans. II and IV*, Comm. Hamburg, **2**: 164–168 (1958).

Kelly, W. P., W. H. Dore, and S. M. Brown. The Nature of the Base Exchange Material, *Soil Sci.*, **31**: 25 (1931).

Kemper, W. D., and van Schalk, J. C. Diffusion of Salts in Clay–Water Systems, *Soil Sci. Soc. Am. Proc.*, **30**: 534–540 (1966).

Kerr, H. W. The Identification and Composition of the Soil Alumino-Silicate Active in Base Exchange and Soil Acidity, *Soil Sci.*, **26**: 385–398 (1928).

Kertscher, F. Ueber die Ergebnisse der Bodenuntersuchungen in der Deutschen Demokratischen Republik, *Z. Landwirtsch. Versuchs Untersuchungsw.*, **1**: 54–81 (1955).

Khan, J. S., and J. B. Hanson. The Effect of Calcium on Potassium Accumulation in Corn and Soybean Roots, *Plant Physiol.*, **32**: 312–316 (1957).

Khemchandarin, H. T., Krishnan, A. S. and Anantharaman, Nitrogen Manuring of Cigar Tobacco with Ammonium Sulphate and Time and Method of its Application, *Indian J. Agr. Sci.*, **3**: 15–27 (1958).

Kiessling, L. E., Untersuchungen über den Kalihaushalt eines langjährig einseitig gedüngten Versuchsfeldbodens mittels der Aspergillusmethode nach Niklas-Poschenrieder-Trischler, *Z. Pflanzenernaehr., Düng. Bodenk. A*, **19**: 241 (1931).

Kiessling, L. E., and A. Schmidt. Die Beeinflussung des Wachstums von Aspergillus Niger durch organische Substanzen, *Wiss. Arch. Landw. irtsch.*, **9**: 293 (1932).

Kihlman-Falk, E. Components in the Uptake and Transport of High Accumulative Ions in Wheat, *Physiol. Plantarum*, **14**: 417–438 (1961).

Kilmer, V. J., and D. C. Nearpass. The Determination of Available Sulfur in Soils, *Soil Sci. Soc. Am. Proc.*, **24**: 337–340 (1960).

Kilmer, V. J., O. E. Hays, and R. J. Muckenhirn. Plant Nutrient and Water Losses from Fayette Silt Loam as Measured by Monolith Lysimeters, *J. Am. Soc. Agr.*, **36**: 249–263 (1944).

Kirssanoff, A. T. Schnellmethode zur Feststellung der Phosphorsäurebedürftigkeit von Böden, *Z. Pflanzenernaehr. Düng. Bodenk., A,* **25**: 382 (1932).

Kirssanoff, A. T. Die chemische Bestimmung der Kalibedürftigkeit, *Z. Pflanzenernaehr. Düng. Bodenk., A,* **34**: 196 (1934).

Kittrick, J. A., and M. L. Jackson. Electron Microscope Observations of the Formation of Aluminum Phosphate Crystals with Kaolinite as the Source of Aluminum, *Science,* **120**(3117): 508–509 (1954).

Kittrick, J. A., and M. L. Jackson. Application of Solubility Product Principles on the Variscite-Kaolinite System, *Soil Sci. Soc. Am. Proc.,* **19**(4): 455–457 (1955).

Kittrick, J. A., and M. L. Jackson. Rate of Phosphate Reaction with Soil Minerals and Electron Microscope Observations on the Reaction Mechanism, *Soil Sci. Soc. Am. Proc.,* **19**(3): 292–295 (1955a).

Kittrick, J. A., and M. L. Jackson. Common Ion Effect on Phosphate Solubility, *Soil Sci.,* **79**(6): 415–421 (1955b).

Kittrick, J. A., and M. L. Jackson. Electron-Microscope Observations of the Reaction of Phosphate with Minerals, Leading to a Unified Theory of Phosphate Fixation in Soils, *J. Soil Sci.,* **7**(1): 81–89 (1956).

Klute, A., and J. Letey. The Dependence of Ionic Diffusion on the Moisture Content of Non-Absorbing Porous Media, *Soil Sci. Soc. Am. Proc.,* **22**: 213–215 (1958).

Knickmann, E. Der Morgan—Test in Frankreich, *Landwirtsch. Forsch. Sonderh.,* **6**: 47–50 (1955).

Knight, A. H., and E. G. Williams. An Evaluation of Phosphate Fertilizers with the Aid of P-32, *Sixieme Congr. Sci. Sol,* Paris, 1956.

Kofoed, A. D. Different Times of Application of Nitrate of Lime to Winter Sewn Crops, *Tidsskr. Planteavl,* **64**: 51–76 (1960).

Kohnke, H. Testing the Fertility of Alberta Soils by the Neubauer and Lemmermann Methods, *Sci. Agr.,* **17**: 312 (1937).

Kolaian, J. H., and A. J. Ohlrogge. Principles of Nutrient Uptake from Fertilizer Bands: IV. Accumulation of Water Around the Bands, *Agron. J.,* **51**: 106–108 (1959).

Kolthoff, I. M. Adsorption on Ionic Lattices, *J. Phys. Chem.,* **40**: 1027–1040 (1936).

Kolthoff, I. M., and C. Rosenblum. The Absorbent Properties and the Specific Surface of Lead Sulfate, *J. Am. Chem. Soc.,* **55**: 2656–2672 (1933).

König, J. *Die Ermittlung des Düngerbedarfs des Bodens,* Verlag P. Parey, Berlin, 1929.

König, J., and J. Hasenbäumer. Die Ermittlung des Düngungsbedürfnisses des Bodens, *Z. Pflanzenernähr. Düng. Bodenk.,* **3**: 497–532 (1924).

Kramer, P. J. Translocation of Radioactive Isotopes from Various Regions of Roots of Barley Seedlings, *Plant Physiol.,* **29**: 342–348 (1954).

Kramer, P. J. Outer Space in Plants, Some Possible Implications of the Concept, *Science,* **125**: 633–635 (1957).

Krantz, B. A., and W. V. Chandler. Fertilizer Corn for Higher Yields, *N. Carolina Agr. Expt. Sta. Bull. 366,* rev. 71 pages, 1954.

Krause, H. H., and S. A. Wilde. Uptake of Potassium by Red Pine Seedlings and Losses Through Leaching from Fertilizers of Various Solubility, *Soil Sci. Soc. Am. Proc.,* **24**: 513–515 (1960).

Kresge, C. B., and D. P. Satchell. Gaseous Loss of Ammonia from Nitrogen Fertilizers Applied to Soils, *Agron. J.,* **52**: 104–107 (1960).

Kresge, C. B., and S. R. Yousts. Effect of Various Rates and Frequencies of Potassium Application on Yield and Chemical Composition of Alfalfa and Alfalfa-Orchard Grass, *Agron. J.*, **54**: 313–316 (1962).

Krishnamoorthy, C. and R. Overstreet. An Experimental Evaluation of Ion-Exchange Relationships, *Soil Sci.*, **69**: 41–53 (1950).

Krishnamoorthy, C., L. E. Davis, and R. Overstreet. Ionic Exchange Equations Derived from Statistical Thermodynamics, *Science,* **108**: 439–440 (1948).

Kroontje, W., and W. R. Kehr. Legume Top and Root Yields in the Year of Seeding and Subsequent Barley Yields, *Agron. J.*, **48**: 127 (1956).

Krügel, C., C. Dreyspring, and W. Heinz. A New Suction Apparatus for the Complete Separation of the Soil Solution from the Soil Itself, *Superphosphate,* **8**: 101–108 (1935).

Kudrin, S. A. Potassium Silicate of the Soil as a Source of this Element for Plants, *Agrobiologiya,* **1**: 90–99 (1955).

Kühn, St. Untersuchungen zur Bestimmung des leicht aufnehmbaren Kali- und Phosphorsäurevorrats des Bodens, *Kisel Közlem,* **38**: 189 (1935).

Kunze, H. The Nitrogen Solubility in Different Urea Acetaldehyde Condensates as Affected by the Soil Reaction, *Z. Pflanzenernaehr. Düng. und Bodenk.*, **86**: 120–123 (1959a).

Kunze, H. The Solubility Characteristics of Difficultly Soluble Urea-acetaldehyde Condensates (Urea Z) in Respect to Their Use as Slowly Available Nitrogen Fertilizers in Soil, *Z. Pflanzenernähr. Düng. und Bodenk.*, **86**: 131–141 (1959b).

Kunze, H. The Effect of Urea Acetaldehyde Condensation on the Nutrient Uptake by Plants, *Z. Pflanzenernaehr. Düng. und Bodenk.*, **86**: 206–214 (1959c).

Kunze, G., and C. D. Jeffries. X-ray Characteristics of Clay Minerals as Related to K Fixation, *Soil Sci. Soc. Am. Proc.*, **17**: 242–244 (1953).

Kuntze, H., and H. Neuhaus. Die landeskulturelle Bedeutung der Pflanzenwurzel, *Der Kulturtechniker,* **98**: 60–77 (1960).

Kurtz, T., E. E. De Turk, and R. H. Bray. Phosphate Adsorption by Illinois Soils, *Soil Sci.*, **61**(2): 111–124 (1946).

Kurtz, L. T., L. D. Owens, and R. D. Hauck. Influence of Moisture on the Effectiveness of Winter Applied Nitrogen Fertilizers, *Soil Sci. Soc. Am. Proc.*, **25**(2): 40–43 (1961).

Kylin, A., and B. Hylmö. Uptake and Transport of Sulphate in Wheat, Active and Passive Components, *Physiol. Plantarum,* **10**: 407–484 (1957).

Lagatu, H., and L. Maume. Le diagnostic foliaire appliqué au contrôle de l'alimentation d'une vigne de coteau avec ou sans fumure, *Compt. Rend. Acad. Agr. France,* **14**(22): 762–776 (1928).

Lagatu, H., and L. Maume. Le diagnostic foliaire de la pomme de terre, *Ann. Ecole Natl. Agr. Montpellier, Pr.* **20**: 219 (1930).

Lagerwerff, J. V., and G. H. Bolt. Theoretical and Experimental Analysis of Gapon's Equation for Ion Exchange, *Soil Sci.*, **87**: 217–222 (1959).

Lagerwerff, J. V., and M. Peech. Relation between Exchange Adsorption and Accumulation of Calcium and Rubidium by Excised Barley Roots, *Soil Sci.*, **91**: 84–93 (1961).

Lahav, N., and G. H. Bolt. Self-Diffusion of Ca-45 into Certain Carbonates, *Soil Sci.*, **97**: 293–299 (1963).

Lai, T. M., and M. M. Mortland. Diffusion of Ions in Bentonite and Vermiculite, *Soil Sci. Soc. Am. Proc.*, **25**: 353–357 (1961).

Lai, T. M., and M. M. Mortland. *Self-diffusion of Exchangeable Cations in Bentonite,*

Clays and Clay Minerals, Pergamon Press, New York, Vol. 9, pp. 229–247, 1962.

Laidler, K. J. *The Chemical Kinetics of Enzyme Action,* Clarendon Press, Oxford, p. 419, 1958.

Lamm, C. G. Factors Affecting the Ratio in the Plants of a Nutrient Derived from the Fertilizer and the Soil, in *Radioisotopes in Soil-Plant Nutrient Studies,* IAEA, Vienna, 1962.

Langin, E. J., R. C. Ward, R. A. Olson, and M. F. Roades. Factors Responsible for Poor Response of Corn and Grain Sorghum to Phosphorus Fertilization. II. Lime and P Placement Effects on P-Zn Relations, *Soil Sci. Soc. Am. Proc.,* **26**: 574–578 (1962).

Larsen, J. E., G. F. Warren, and R. Langston. Studies of Phosphorus Availability in Organic Soils, *Soil Sci. Soc. Am. Proc.,* **22**(4): 336–339 (1958).

Larsen, S. Studies on the Uptake of Phosphorus in Plants with Radio-Phosphorus as an Indicator, K. Veterinaerog Landbohøgskole, Copenhagen, Denmark, 1960.

Larsen, S. The Use of P-32 in Studies on the Uptake of Phosphorus by Plants, *Plant Soil,* **4**: 1–10 (1952).

Larsen, S., and M. N. Court. The Chemical Potentials of Phosphate Ions in Soil Solutions, *Trans. 7th Intern. Cong. Soil Sci.,* Madison, Wis., **2**: 413–421 (1960).

Larsen, S., and D. Gunary. Diffusion in Soil Crumbs, in Plant Nutrient Supply and Movement, *IAEA Tech. Rpt. Ser. No. 48,* pp. 78–85, 1965.

Larsen, S., and C. D. Sutton. The Influence of Soil Volume on the Absorption of Soil Phosphorus by Plants and on the Determination of Labile Soil Phosphorus, *Plant Soil,* **18**: 77–84 (1963).

Lathwell, D. J., N. Sanchez, D. J. Lish, and M. Peech. Availability of Soil Phosphorus as Determined by Several Chemical Methods, *Agron. J.,* **50**: 366–369 (1958).

Laties, G. G. The Generation of Latent-Ion-Transport Capacity, *Proc. Natl. Acad. Sci. U.S.,* **45**(2): 163–172 (1959).

Laties, G. G. Active Transport of Salt into Plant Tissue, Ann. Rev. *Plant Physiol.,* **10**: 87 (1959a).

Laudelout, H., and J. P. Eeckman. La stabilité chimique des suspensions d'argile saturée par lion hydrogène, *Compt. Rend. Assoc. Int. Sci. Sol. Hambourg,* **2**: 194–199 (1958).

Laudelout, M. Principes de l'utilization des engrais mineraux au Conge Belge, *Agricultura,* Ser. 4, **7**: 451–470 (1959).

Loughlin, W. M. Broomegrass Response to Rate and Source of Nitrogen Applied in Fall and Spring in Alaska, *Agron. J.,* **55**: 60–62 (1963).

Lavy, T. L., and S. A. Barber. Movement of Molybdenum in the Soil and Its Effect on Availability to the Plant, *Soil Sci. Soc. Am. Proc.,* **28**: 93–97 (1964).

Lawton, K., C. Apostolakis, R. L. Cook, and W. L. Hill. Influence of Particle Size, Water Solubility and Placement of Fertilizers on the Nutrient Value of P in Mixed Fertilizer, *Soil Sci.,* **82**: 465–476 (1956).

Lawton, K., M. B. Tesar, and B. Kawin. The Effect of Rate and Placement of Superphosphate on the Yield and Phosphorus Absorption of Legume Hay, *Soil Sci. Soc. Am. Proc.,* **68**: 428–432 (1954).

Leaf, A. L. Release of Potassium from Feldspathic Rock and Minerals, *Soil Sci.,* **87**: 11–12 (1959).

Lees, H. The Soil Percolation Technique, *Plant Soil,* **1**: 221–238 (1949).

Lees, H., and J. W. Porteous. Effect of Quinhydrone on Soil Nitrification, *Nature*, **165**: 533 (1950).

Legg, J. D., and F. E. Allison. Recovery of N^{15} Tagged Nitrogen from Ammonium Fixing Soils, *Soil Sci. Soc. Am. Proc.*, **23**: 131–134 (1959).

Leggett, J. E., and E. Epstein. Kinetics of Sulfate Absorption by Barley Roots. *Plant Physiol.*, **31**: 222–226 (1956).

Leggett, J. E., and L. H. Stolzy. Anaerobiosis and Sodium Accumulation. *Nature*, **192**: 991–992 (1961).

Lehr, J. R., and H. F. Stephenson. Nature of the Reactions of Monocalcium Phosphate Monohydrate in Soils. I. The Solution that Reacts with the Soil. II. Dissolution and Precipitation Reactions Involving Iron, Aluminum, Manganese and Calcium, *Soil Sci. Soc. Am. Proc.*, **23**: 12–22 (1959).

Lehr, J. R., W. E. Brown, and E. H. Brown. Chemical Behavior of Monocalcium Phosphate Monohydrate in Soils, *Soil Sci. Soc. Am. Proc.*, **23**: 3–7 (1959).

Lehr, J. J., and J. E. van Wesemael. The Influence of Neutral Salts on the Solubility of Soil Phosphate, with Special Reference to the Effect of the Nitrates of Sodium and Calcium, *J. Soil Sci.*, **3**: 125–135 (1952).

Lemmermann, O. Neues über die Lactatmethode, *Z. Pflanzenernähr. Düng. Bodenk.*, **37**(82): 75–79 (1946).

Lemon, E. R., and C. L. Wiegand. Soil Aeration and Plant Root Relations, I. Theory, II. Root Respiration, *Agron. J.*, **54**: 167–174 (1962).

Leonard, C. D., and I. Stewart. Soil Application of Zinc for Citrus on Acid Sandy Soil, *Proc. Florida State Hort. Soc.*, **71**: 99–105 (1958).

Leonard, C. D., and I. Stewart. Soil Application of Manganese for Citrus, *Proc. Florida State Hort. Soc.*, **72**: 38–44 (1959).

Lesch, W. Die Bestimmung der zitronensäurelöslichen Phosphorussäure des Bodens auf Kolorimetrischem Wege, *Z. Pflanzeneraehr. Düng. Bodenk. A*, **21**: 222 (1931).

Letey, J. and A. Klute. Apparent Mobility of Potassium and Chloride Ions in Soil and Clay Pastes, *Soil Sci.*, **90**(4): 259–265 (1960).

Levitt, J. The Significance of "Apparent Free Space" in Ion Absorption, *Physiol. Plantarum*, **10**: 882–888 (1957).

Lewis, D. G., and J. P. Quirk, *Diffusion of Phosphate in Soil.* In *Trans. Intern. Soc. Soil. Sci.*, Commissions IV and V, Wellington, N.Z., pp. 132–138, 1963.

Lewis, T. E., and F. E. Broadbent. Soil Organic Matter-Metal Complexes IV: Nature and Properties of Exchange Sites, *Soil Sci.*, **91**: 393–399 (1961).

Liebscher, G. Der Verlauf der Nährstoffaufnahme und Bedeutung für die Düngerlehre, *J. Landw.*, **35**: 335–518 (1897).

Linder, R. C., and C. P. Harley. Nutrient Interrelations in Lime-Induced Chlorosis, *Plant Physiol.*, **19**: 420–439 (1944).

Lindsay, W. L., and E. C. Moreno. Phosphate Phase Equilibrium in Soils, *Soil Sci. Soc. Am. Proc.*, **24**: 177–182 (1960).

Lindsay, W. L., and H. F. Stephenson. Nature of the Reactions of Monocalcium Phosphate Monohydrate in Soils. I. The Solution that Reacts with the Soil. II. Dissolution and Precipitation Reactions Involving Ion, Aluminum, Manganese, and Calcium, *Soil Sci. Soc. Am. Proc.*, **23**: 7–12, 12–18 (1959).

Lindsay, W. L., J. R. Lehr, and H. F. Stephenson. Nature of the Reactions of Monocalcium Phosphate Monohydrate in Soils. III. Studies with Metastable Triple Point Solution. *Soil Sci. Soc. Am. Proc.*, **23**: 342–345 (1959).

Lindsay, W. L., A. W. Frazier, and H. F. Stephenson. Identification of Reactic

Products from Phosphate Fertilizer in Soils, *Soil Sci. Soc. Am. Proc.*, **26**: 446–452 (1962).

Linser, H., and R. Primrost. Nitrogen Fertilizing with High Split Applications. III. Field Experiments with Winter Rye, *Z. Pflanzenernaehr. Düng. und Bodenk.*, **86**: 97–110 (1959).

Lipman, C. B. A New Method of Extracting the Soil Solution, *Univ. Calif. Pub. Agri. Sci.*, **3**(7): 131–134 (1918).

Lisanti, L. E., and U. Marckwordt. Untersuchungen über Anreicherung and Verteilung von Rubidium in Gerstenkeimpflanzen, *Atompraxis*, **9**(3): 1–4 (1963).

Little, R. C. Sulfur in Soils. I. Determination of Readily Soluble Sulfates in Soil, *J. Sci. Food Agr.*, **4**: 336 (1953).

Little, R. C. Sulfur in Soils. II. Determination of the Total Sulfur Content of Soil, *J. Sci. Food Agr.*, **8**: 271 (1957).

Little, R. C. Sulfur in Soils. III. A Study of the Readily Soluble Sulfate Content and of the Total Sulfur Content of Soil, *J. Sci. Food Agr.*, **9**: 273–281 (1958).

Long, O. H., and L. F. Seatz. Correlation of Soil Test for Available Phosphorus and Potassium with Crop Yield Response to Fertilization, *Soil Sci. Soc. Am. Proc.*, **17**: 258–262 (1953).

Lopez-Gonzales, J. de Dios, and H. Jenny. Modes of Entry of Strontium into Plant Roots, *Science* **128**: 90–91 (1958).

Lopushinsky, W. Effect of Water Movement on Ion Movement into the Xylem of Tomato Roots, *Plant Physiol.*, **39**: 494–501 (1964).

Loughlin, W. M. Soil and Foliar Application of Nutrients Affect Potato Yield, Dry Matter and Foliar Necrosis, *Am. Potato J.*, **39**: 125–134 (1962).

Loughlin, W. M. Influence of Soil and Spray Applications of Phosphorus on Potato Yield, Dry Matter Content and Chemical Composition, *Am. Potato J.*, **39**: 343–347 (1962a).

Loughman, B. C., and R. S. Russell. The Absorption and Utilization of Phosphate by Young Barley Plants. IV. The Initial Stages of Phosphate Metabolism in Roots, *J. Exptl. Botany*, **8**: 280–293 (1957).

Low, A. J., and F. J. Piper. Urea as a Fertilizer, Laboratory and Pot Culture Studies, *J. Agr. Sci.*, **57**: 249–255 (1961).

Low, P. F. The Role of Aluminum in the Titration of Bentonite, *Soil Sci. Soc. Am. Proc.*, **19**: 135–139 (1955).

Low, P. F. The Apparent Mobilities of Exchangeable Alkali Metal Cations in Bentonite-Water Systems, *Soil Sci. Soc. Amer. Proc.*, **22**: 395–398 (1958).

Low, P. F. Effect of Quasi-Crystalline Water on Rate Processes Involved in Plant Nutrition, *Soil Sci.*, **93**(1): 6–15 (1962).

Lucas, R. E. Chemical and Physical Behaviour of Copper in Organic Soils, *Soil Sci.*, **66**: 119–129 (1948).

Lugo-Lopez, M. A. Pore Size and Bulk Density as Mechanical Soil Factors Impeding Root Development, *J. Agr. Univ. Puerto Rico*, **44**: 40–44 (1960).

Luken, H. Saline Soils under Dryland Agriculture in South-eastern Saskatchewan (Canada) and Possibilities for Their Improvement. Part I. Distribution and Composition of Water-Soluble Salts in Soils in Relation to Physiographic Features and Plant Growth, *Plant Soil*, **18**(1): 1–25 (1962).

Lundegårdh, H. The Triple-Analysis Method of Testing Soil Fertility and Probably Crop Reaction to Fertilization, *Soil Sci.*, **45**: 447 (1938).

Lundegårdh, H. Leaf Analysis as a Guide to Soil Fertility, *Nature* (London), **151**: 320 (1943).

Lundegårdh, H. *Die Blattanalyse,* Verlag Fischer, Jena, 1945.

Lunt, O. R., and B. Kwate. Potassium Frit as a Special Purpose Fertilizer, *Soil. Sci.,* **82**: 3–8 (1956).

Lüttge, U., and Weigl J. Mikroautoradiographische Untersuchungen der Aufnahme und des Transportes von $^{35}SO_4^{--}$ und $^{45}Ca^{++}$ in Keimwurzeln von Zea Mays L. und Pisum Sativum L., *Planta,* **58**: 113–126 (1962).

Lüttge, U., and Laties, G. G. Selective Inhibtion of Absorption and Long Distance Transport in Relation to the Dual Mechanisms of Ion Absorption in Maize Seedlings, *Plant Physiol.,* **42**: 181–185 (1967).

Lüttge, U., and Laties, G. G. Dual Mechanisms of Ion Absorption in Relation to Long Distance Transport in Plants, *Plant Physiol.,* **41**: 1531–1539 (1966).

Lüttge, U., and J. Weigl. Mikroautoradiographische Untersuchungen der Aufnahme und des Transportes von $S^{35}O_4^{--}$ und Ca^{45++} in Keimwurzeln von *Zea mays* L. und *Pisum sativum* L., *Planta,* **58**: 113–126 (1958).

Lutz, J. A., G. L. Terman, and J. L. Anthony. Rate and Placement of Phosphorus for Small Grains, *Agron. J.,* **53**: 303–305 (1961).

Lycklama, J. C. The Absorption of Ammonium and Nitrate by Perennial Rye-Grass, *Acta Botan. Neerl.,* **12**: 361–423 (1963).

Lyon, T. L., and J. A. Bizzell. Nitrogen Accumulation in Soil as Influenced by the Cropping System, *J. Am. Soc. Agron.,* **25**: 266 (1933).

Lyon, T. L., and J. A. Bizzell. A Comparison of Several Legumes With Respect to Nitrogen Excretion, *J. Am. Soc. Agron.,* **26**: 651 (1934).

Lyon, T. L., J. A. Bizzell, B. D. Wilson, and E. W. Leland. Lysimeter Experiments IV, *N.Y. State Agr. Expt. Sta. Mem. 134,* 1930.

MacGregor, J. M. and V. H. Johnson. Fertilizer Placement for Corn in Minnesota, *Better Crops,* **40**(5): 44–45 (1956).

Machado, S. A. Cation Exchange Capacity of Roots as a Basic Principle in Growing Coffee in Association with Other Plants, *Cerrcafe,* **8**: 277–285 (1957).

MacIntire, W. H., W. M. Shaw, and B. Robinson. Differential Behaviour of Potassium Metaphosphate, *Soil Sci.,* **75**: 69–80 (1953).

MacIntire, W. H., J. B. Young, W. M. Shaw, and B. Robinson. Nitrogen Recoveries From Applications of Ammonium Chloride, Phosphate and Sulfate and Outgo of Complementary Ions in Rain Water Leaching Through a Six-Foot Soil-Subsoil Column, *Soil Sci. Soc. Am. Proc.,* **16**: 301–306 (1962).

Mackie, W. Z., and M. Fried. Relative Distribution of Potassium and Rubidium 86 Within Corn Plants Grown in the Field, *Soil Sci.,* **80**: 309–312 (1955).

Mackenzie, A. J., and J. W. Borland. Non-Orthophosphate Contaminant of Neutron Irradiated Rock Phosphate, *Anal. Chem.,* **24**: 176–179 (1952).

MacLean, A. A., J. J. Doyle, and F. G. Hemlyn. Fertility Studies on Some New Brunswick Soils. I. Soil Phosphorus Supply as Shown by Greenhouse and Chemical Tests, *Can. J. Agr. Sci.,* **35**: 388–396 (1955).

MacLean, A. A., J. J. Doyle, and F. G. Hemlyn. Fertility Studies on Some New Brunswick Soils. II. Soil Potassium Supply as Shown by Greenhouse and Chemical Tests, *Can. J. Soil Sci.,* **37**: 29–33 (1957).

MacVicar, R., W. L. Garman, and R. Wall. Studies on Nitrogen Fertilizer Utilization Using N^{15}, *Soil Sci. Soc. Am. Proc.,* **15**: 265–268 (1951).

Magistad, O. C., and R. F. Reitemeier. Soil Solution Concentration at the Wilting Point and Their Correlation with Plant Growth, *Soil Sci.,* **55**(5): 351–360 (1943).

Magistad, O. C., A. D. Ayers, and C. H. Wadleigh. The Effect of Salt Concentration,

Kind of Salt and Climate on Plant Growth in Sand Cultures, *Plant Physiol.*, **118**(2): 151–166 (1942).

Magistad, O. C., R. F. Reitemeier, and L. V. Wilcox. Determination of Soluble Salts in Soils, *Soil Sci.*, **59**: 65–75 (1945).

Malavolta, E., *ABC da Adubacão,* Editôra Agronomica "Ceres" Ltda., Sao Paulo, p. 85, 1956.

Malcolm, J. L. Nitrogen Experiments with Potatoes and Tomatoes on Marl Soils, *Soil Sci. Soc. Florida Proc.*, **15**: 91–100 (1955).

Malcolm, J. L. Effect of Nitrogen, Phosphorus and Potassium Fertilizer on Fruit Yield and Composition of Tomato Leaves, *J. Agr. Food Chem.*, **7**: 415–418 (1959).

Marckwordt, U. Untersuchungen über Kationenbeziehungen bei der Anreicherung durch Gerstenwurzeln, *Landwirtsch. Forsch.*, **16**(1): 1–7 (1963).

Marel, H. W. van der. Tropical Soils in Relation to Plant Nutrition, *Soil Sci.*, **64**: 445–451 (1947).

Marel, H. W. van der. Potassium Fixation in Dutch Soils: Mineralogical Analyses, *Soil Sci.*, **78**: 163–180 (1954).

Marschner, H. von. Aktive und passive Ionenverlagerung durch die Pflanzenwurzel in den Spross unter dem Einfluss der Transpiration, *Landwirtsch. Forsch.*, **18**: 91–99 (1964).

Marschner, H. von. Einfluss von Calcium auf die Natriumaufnahme und die Kaliumabgabe isolierter Gerstenwurzeln, *Z. Pflanzenernaehr. Düng. Bodenk.*, **107**: 19–32 (1964a).

Martin, A. E. Nitrogen Transformation in Soil Excluding De-nitrification, *Bull. 46,* Commonwealth Bur. Pastures and Field Crops, Hurley Berkshire, England, 1962.

Martin, A. E., and G. W. Skyring. Losses of Nitrogen from the Soil Plant System, *Bull. 46,* Commonwealth Bur. Pastures and Field Crops, Hurley, Berkshire, England, 1962.

Martin, H., and H. Laudelout. Influence de l'acidité des suspensions argileuses sur la fixation des phosphates, *Pedologie,* **9**: 46–53 (1957).

Martin, H., and H. Laudelout. La fixation des phosphates par les suspensions d'argile acide, *Agricultura 2 Ser.*, **9**(2): 317–331 (1961).

Martin, H., and H. Laudelout. Thermodynamique de l'échange des cations alcalins dans les argiles, *J. Chim. Phys.*, **60**: 1086–1099 (1963).

Martin, J. C. Effect of Crop Growth on the Replaceable Bases in Some Californian Soils, *Soil Sci.*, **27**: 123 (1929).

Martin, J. P., and H. D. Chapman. Volatilization of Ammonia from Surface Fertilized Soils, *Soil Sci.*, **71**: 25–34 (1951).

Masoni, G. Solubility of Manganese Compounds in the Soil, *Staz. Sper. Agr. Italia,* **49**: 132–149 (1916).

Mattingly, G. E. G., and O. Talibudeen. Isotopic Exchange of Phosphate in Soils. Review of Experimental Techniques and Results of Rothamsted, 1952–1960, *Ann. Rept. Rothamsted Expt. Sta. for 1960,* pp. 246–265, 1960.

Mattingly, G. E. G., R. D. Russell, and B. M. Jephcott. Experiments on Cumulative Dressings of Fertilizers on Calcareous Soils in Southwest England. II. Phosphorus Uptake by Ryegrass in the Greenhouse, *J. Sci. Food Agr.*, 14: 629–637 (1963).

Mattson, S. The Laws of Soil Colloidal Behavior. VI. Amphoteric Behavior, *Soil Sci.*, **32**: 343–365 (1931).

McAuliffe, C. D., N. S. Hall, L. A. Dean, and S. B. Hendricks. Exchange Reactions Between Phosphates and Soils: Hydroxylic Surfaces of Soil Minerals, *Soil Sci. Soc. Am. Proc.*, **12**: 119–123 (1948).

McConaghy, S., and J. W. B. Stewart. Availability of Soil and Fertilizer Phosphates to Growing Crops, *J. Sci. Food Agr.*, **14**(5): 329–341 (1963).

McConaghy, J., Stewart, W. B., and Malek, M. Soil Phosphate Status as Measured by Isotopic-Exchange and other Techniques, *Trans. ISSS. Com. II and IV Aberdeen:* 151–160 (1966).

McElroy, W. D. (Ed.). *A Symposium on Light and Life,* The Johns Hopkins Press, Baltimore, Md., p. 924, 1961.

McGeorge, W. T. Modified Neubauer Method for Soil Cultures, *Soil Sci.*, **62**: 61–70 (1946).

McGeorge, W. T., and E. L. Breazeale. Application of the Neubauer Technique and Applied Potential to the Study of Immobilization of Iron in Plants, *Soil Sci.*, **82**: 329–336 (1956).

MacKay, D. C., and W. A. De Long. Coordinated Soil-Plant Analysis. III. Exchange Equilibria in Soil Suspensions as Possible Indicators of Potassium Availability, *Can. J. Agr. Sci.*, **35**: 181–188 (1955).

McKell, C. M., and W. A. Williams. Lysimeter Study of Sulfur Fertilization of an Annual Range Soil, *Calif. Agr.*, **15**(4): 4–5 (1961).

McLean, E. O., and R. H. Simon. Potassium Status of Some Ohio Soils as Revealed by Greenhouse and Laboratory Studies, *Soil Sci.*, **85**: 324–332 (1958).

McLean, E. O., D. Adams, and R. E. Franklin. Cation Exchange Capacities of Plant Roots as Related to Their Nitrogen Contents, *Soil Sci. Soc. Am. Proc.*, **20**: 345–347 (1956).

Mehlich, A., E. Truog, and E. B. Fred. The Aspergillus Niger Method of Measuring Available Potassium of Soil, *Soil Sci.*, **35**: 259 (1933).

Mehta, N. C. et al. Determination of Organic Phosphorus in Soils, *Soil Sci. Soc. Am. Proc.*, **18**: 443–449 (1954).

Mellado, L., J. Puerta, and F. Caballero. Absorción de fósforo por la judía. Influencia del nitrógeno y de la dosis y distribución del abono, *IAEA/FOA Symp. Radioisotopes in Soil-Plant Nutrition Studies,* IAEA, Vienna, 1962.

Memec, C. A., and A. Koppovà. A New Rapid Procedure for the Determination of the Nitrogen Needs of the Soil by Chemical Analysis, *Z. Pflanzenernaehr. Düng. Bodenk. A,* **23**: 140–148 (1932).

Menzel, R. G., and W. R. Heald. Distribution of Potassium, Rubidium, Caesium, Calcium and Rubidium Within Plants Grown in Nutrient Solutions, *Soil Sci.*, **80**: 281–293 (1955).

Menzel, R. G., and M. L. Jackson. Mechanism of Sorption of Hydroxy Cupric Ion by Clays, *Soil Sci. Soc. Am. Proc.*, **15**: 122–124 (1950).

Merriam, C. N., and H. C. Thomas. Adsorption Studies on Clay Minerals. VI. Alkali Ions on Attapulgite, *J. Chem. Phys.*, **24**: 993–995 (1956).

Merriam, C. N., W. Southworth, and H. C. Thomas. Ion Exchange Mechanism and Isotherms from Deep Bed Performance, *J. Chem. Phys.*, **20**(12): 1842–1846 (1952).

Metzger, J. Significance of Adsorption, or Surface Fixation of Phosphorus by Some Soils of the Prairie Group, *J. Am. Soc. Agron.*, **32**: 513–525 (1940).

Meyer, J., and E. Pampfer. Nitrogen Content of Rain Water Collected in the Humid Central Congo Basin, *Nature,* (London), **184**: 717 (1959).

Meyer, R. D., R. A. Olsen, and H. F. Rhoades. Ammonia Losses from Fertilized Nebraska Soils, *Agron. J.,* **53**: 241–244 (1961).

Michaelis, L., and M. L. Menten. Die Kinetik der Invertinwirkung, *Biochem. Z.,* **49**: 333–369 (1913).

Mieschlag, F. Late Applications of Superphosphate, *Mitt. Dent. Landwirtsch. Ges.,* **72**: 662–663 (1957).

Millar, C. E., L. S. Robertson, R. L. Cook, C. M. Hansen, W. C. Hulbert, and G. A. Cummings. Proc. 23rd An. Meeting Natl. Joint Comm. on Fertilizer Application, Illinois, 1947.

Millar, C. E. *Soil Fertility,* John Wiley & Sons, New York, 1955.

Millar, C. E., L. M. Turk, and H. D. Foth. *Fundamentals of Soil Science,* J. Wiley & Sons, London, 1958.

Miller, J. R., and J. H. Axley. Correlation of Chemical Soil Tests for Available Phosphorus with Crop Responses, Including a Proposed Method, *Soil Sci.,* **82**: 117–127 (1956).

Miller, M. H., and A. J. Ohlrogge. Principles of Nutrient Uptake from Fertilizer Bands. I. Effect of Placement of Nitrogen Fertilizer on the Uptake of Band-Placed Phosphorus at Different Soil Phosphorus Levels, *Agron. J.,* **50**: 95–97 (1958).

Mistry, K. B. Quantitative Evaluation of the Residual Values of Phosphatic Fertilizers in Soils, *IAEA/FAO Symp. Radioisotopes in Soil-Plant Nutrition Studies,* IAEA, Vienna, pp. 363–370, 1962.

Mistry, K. B. Relative Efficiencies of Phosphatic Fertilizers for Wheat Through Radioactive-Tracer Technique. II. Studies on the Black and Alluvial Soils of India, *IAEA/FAO Symp. Radioisotopes in Soil-Plant Nutrition Studies,* IAEA, Vienna, pp. 427–442, 1962a.

Mitchell, B. D., V. C. Farmer, and W. J. McHardy. Amorphous Inorganic Materials in Soils, *Advan. Agron.,* **16**: 327–383 (1964).

Mitscherlich, E. A. Eine chemische Bodenanalyse für pflanzenphysiologische Forschungen, *Landwirtsch. Jahrb.,* **36**: 309 (1907).

Mitscherlich, E. A. Eine chemische Bodenanalyse für pflanzenphysiologische Forschungen, *Landwirtsch. Jahrb.,* **38**: 534–541 (1909a).

Mitscherlich, E. A. Grundsätze des Minimum und das Gesetz des abnehmenden Bodenertrages, *Landwirtsch. Jahrb.,* **38**: 537–552 (1909b).

Mitscherlich, E. A. Zum Gesetz vom Minimum. Eine Antwort an Th. Pfeiffer und seine Mitarbeiter, *Landwirtsch. Vers. Sta.,* **77**: 413–428 (1912).

Mitsui, S. *Inorganic Nutrition Fertilization and Soil Amelioration for Lowland Rice,* Yokendo Ltd., Tokyo, 1960.

Mitsui, S., and K. Ishizuka. Dynamic Studies on the Nutrient Uptake by Crop Plant: XXIX. On the Mechanisms of Uptake and Translocation of $H_2P^{32}O_4$ in the Roots of Wheat Seedlings. *Soil Plant Food (Japan),* **61**(1): 7–15 (1960).

Mitsui, S., and M. Ueda. Cation Exchange Capacity of Crop Roots and Ion Uptake. Part 2. The Effect of Cation Exchange Capacity of Soil and Plant Roots on the Uptake of Some Cations. Particularly of Magnesium, *Soil Sci. Plant Nutr. (Tokyo),* **9**(2): 43–48 (1963a).

Mitsui, S., and M. Ueda. Cation Exchange Capacity of Crop Roots in Relation with Ion Uptake. Part 1. Method of Determining the Cation Exchange Capacity and Intensity of Plant Roots, *Soil Sci. Plant Nutr. (Tokyo),* **9**(1): 6–12 (1963).

Mokady, R. S., and Low, P. F. Electrochemical Determination of Diffusion Coefficients in Clay–Water Systems, *Soil Sci. Soc. Am. Proc.,* **30**: 438–442 (1966).

Mooers, C. A. An Evaluation of the Neubauer, the Cunninghamella and Aspergillus Niger Methods for Determination of the Fertilizer Needs of a Soil, *Soil Sci.,* **46**: 211–227 (1938).

Moore, A. W. Nitrogen Fixation in Latasolic Soil under Grass, *Plant Soil,* **29**: 127–138 (1963).

Moore, C. C. A Study of the Available Mineral Plant Food in Soils, *J. Am. Chem. Soc.,* **24**: 79 (1902).

Moore, D. G. Nitrogen Movement in Soil as Affected by Time of Application of Nitrogen and Fixation, *Dissertation Abst.,* **21**: 414–415 (1960).

Moreno, E. C. Probability Theory Applied to Fertilizer Granule-Size Effects, *Soil Sci. Soc. Am. Proc.,* **23**: 326–327 (1959).

Morgan, J. F. The Soil Solution Obtained by the Soil Pressure Method, *Michigan Agr. Expt. Sta. Techn. Bull.,* **28**: 1–38 (1916).

Morgan, M. F. The Universal Soil Testing System, *Trans. 3rd Intern. Congr. Soil Sci.,* **1**: 103 (1935).

Morgan, M. F. Soil and Plant Tissue Tests for Minor Elements Constituents, *Soil Sci. Soc. Am. Proc.,* **1**: 255–257 (1937).

Morgan, M. F. Chemical Soil Diagnosis by the Universal Soil Testing System, *Conn. Agr. Expt. Stat. Bull. 450,* 1941.

Morgan, M. F., and H. G. M. Jacobson. Soil and Crop Interrelations of Various Nitrogenous Fertilizers, Windsor Lysimeter Series B, *Conn. Agr. Expt. Sta. Bull. 458,* 1942.

Morgan, M. F., H. G. M. Jacobson, and S. B. LeCompte, Jr. Drainage Water Losses from a Sandy Soil as Affected by Cropping and Cover Crops, *Conn. Agr. Expt. Sta. Bull. 466,* pp. 731–759, (1942a).

Morgan, M. F., H. G. M. Jacobson, and O. E. Street. The Neutralization of Acid Forming Nitrogenous Fertilizers in Relation to Nitrogen Availability and Soil Grass, *Soil Sci.,* **54**: 127–148 (1942b).

Morris, H. D., and J. F. Celecia. Effect of Time of Fertilizer Application on Yield and Nutrient Uptake of Coastal Bermuda Grass on Cecil Sandy Loam, *Agron. J.,* **54**: 335–338 (1962).

Mortland, M. M. Adsorption of Ammonia by Clays and Muck, *Soil Sci.,* **80**: 11–18 (1955).

Moser, U. S., W. H. Sutherland, and C. A. Black. Evaluation of Laboratory Indexes of Absorption of Soil Phosphorus by Plants, *Plant Soil,* **10**(4): 356–374 (1959).

Moss, P. Some Aspects of the Cation Status of Soil Moisture. Part I. The Ratio Law and Soil Moisture Content, *Plant Soil,* **18**(1): 99–132 (1963).

Mouat, M. C. H., and T. W. Walker. Competition of Nutrients between Grasses and White Clover. II. Effect of Root Cation-Exchange Capacity and Rate of Emergence of Associated Species, *Plant Soil,* **11**(1): 41–52 (1959).

Mulder, E. G. Molybdenum in Relation to Growth of Higher Plants and Micro-organisms, *Plant Soil,* **5**: 368–415 (1954).

Munson, R. D., and J. T. Peaek. The Effects of Corn Residue, Nitrogen and Incubation on Nitrogen Release and Subsequent Nitrogen Uptake by Oats, *Soil Sci. Soc. Am. Proc.,* **22**: 542–543 (1958).

Nagelschmidt, G., and H. L. Nixon. Formation of Apatite from Superphosphate in the Soil, *Nature,* **154**: 428–429 (1944).

Nearpass, D. C. Estimation of Available Zinc in Soils from Yield-of-Zinc Curves, *Soil Sci. Soc. Am. Proc.*, **20**: 482–488 (1956).

Nearpass, D. C., M. Fried, and V. J. Kilmer. Greenhouse Measurement of Available Sulfur Using Radioactive Sulfur, *Soil Sci. Soc. Am. Proc.*, **25**(4): 287–289 (1961).

Nehring, K. Die Bestimmung des austauschbaren Kaliums als Mittel zur Beurteilung der Kalibedürftigkeit der Böden, *Bodenkultur Pflanzenernähr.*, **30**(75): 36–50 (1943).

Nelson, L. B. The Mineral Nutrition of Corn as Related to Its Growth and Culture, *Adv. Agron.*, **8**: 321–375 (1956).

Nelson, L. B., and R. E. Uhland. Factors that Influence Loss of Fall-Applied Fertilizers and their Possible Importance in Different Sections of the United States, *Soil Sci. Soc. Am. Proc.*, **19**: 492–496 (1955).

Nelson, L. E. A Comparison of Several Methods for Evaluating the Potassium Status of Some Mississippi Soils, *Soil Sci. Soc. Am. Proc.*, **23**: 313–316 (1959).

Nelson, W. L., and G. Stanford. Changing Concepts of Plant Nutrient Behaviour and Fertilizer Use, *Advan. Agron.*, **10**: 13–141 (1958).

Nelson, W. L. and G. L. Terman. Nature, Behaviour and Use of Multinutrient (Mixed) Fertilizers, in *Fertilizer Technology and Usage*, Soil Sci. Soc. Am., Madison, Wis., 1963.

Nemec, A. Ueber eine Methode zur Bestimmung des Phosphorsäurebedürfnisses des Bodens, *Compt. Rend. Acad. Sci. Paris*, **183**: 314 (1926).

Nemec, A. Amelioration of Degraded Forest Soil by Diabase Powder in the Forest District of Choltice, Sb. *Cesk. Akad. Zemedel.*, **25**: 55–64 (1952).

Neubauer, H. Die Keimpflanzenmethode nach Neubauer und Schneider, *Handb. Pflanzenernaehr. Düng.* (Honcamp), Springer-Verlag, Berlin, 1931.

Neubauer, H., and W. Schneider. Die Nährstoffaufnahme der Keimpflanzen und ihre Anwendung auf die Bestimmung des Nährstoffgehaltes der Böden, *Z. Pflanzenernaehr. Düng. Bodenk.*, **2A**: 329–341 (1923).

Newbould, P., and R. S. Russell. Isotopic Equilibration of Calcium-45 With Labile Soil Calcium, *Plant Soil*, **28**: 239–257 (1963).

Nicolić, S., Telenić, and Pantović, M. Experiments and Observations on Potassium Fertilizing, Jugoslav. leucite, *Z. Pflanzenernaehr. Düng. und Bodenk.*, **84**: 279–282 (1959).

Niklas, H., and O. Toursel. Die Bodenuntersuchung mittels Aspergillus Niger, *Bodenkultur Pflanzenernähr.*, **18**: 79–107 (1940).

Niklas, H., H. Poschenrieder, and G. Vilsmeier. Weitere Ergebnisse zur Bestimmung der Kalibedürftigkeit der Böden mittels Aspergillus Niger, *Wiss. Arch. Landwirtsch. A*, **52**: 152 (1930).

Niklas, H., G. Vilsmeier, and F. Kohl. Die Bestimmung der Phosphorsäurebedürftigkeit der Böden mittels Aspergillus Niger, *Z. Pflanzenernaehr. Düng. Bodenk. A*, **32**: 50 (1933).

Nicholas, D. J. D. The Application of Rapid Chemical Tests to the Diagnosis of Mineral Deficiencies in Horticultural Crops, *J. Hort. Sci.*, **24**: 2 (1948).

Nicholas, D. J. D. The Application of Rapid Chemical Tests to the Diagnosis of Mineral Deficiencies in Horticultural Plants. III. Comparisons of Tissue Tests and Total Analyses for K, Mg, Ca, P and N in Potatoes and Cauliflower, *J. Hort. Sci.*, **30**: 260–267 (1955).

Nishigaki, S., M. Shibuya, T. Koyama, and I. Hanaoka. Study on the Improvement of Fertilizer Dressing and Soil Management in the Rice Field. Field Experiment on Rice, Using P-32 Labelled Fertilizer, *Soil Plant Food (Tokyo)*, **4**(3): 117–126 (1958).

Noggle, J. C., and M. Fried. A Kinetic Analysis of Phosphate Absorption by Excised Roots of Millet, Barley, and Alfalfa, *Soil Sci. Soc. Am. Proc.*, **24**: 33–35 (1960).

Noggle, J. C., C. T. De Wit, and A. L. Fleming. Interrelations of Calcium and Rubidium Absorption by Excised Roots of Barley and Plantain, *Soil Sci. Soc. Am. Proc.*, **28**: 97–100 (1964).

Nommik, H. Investigations on Denitrification in Soil, *Acta Agr. Scand.*, 62: 195–228 (1956).

Nommik, H. Effect of the Addition of Organic Materials and Lime on the Yield and Nitrogen Nutrition of Oats, *Acta Agr. Scand.*, **2**: 211–226 (1961).

Nommik, H., and K. O. Nilsson. Fixation of Ammonia by the Organic Fraction of the Soil, *Acta Agr. Scand.*, 3(4): 371–389 (1963).

Norman, A. G. The Nitrogen Nutrition of Soybeans. I. Effect of Inoculation and Nitrogen Fertilizer on the Yield and Composition of Beans on Marshall Silt Loam, *Soil Sci. Soc. Am. Proc.*, **8**: 226 (1943).

Norman, A. G. Recent Advances in Soil Microbiology, *Soil Sci. Soc. Am. Proc.*, **11**: 9–15 (1947).

Norris, D. O. The Biology of Nitrogen Fixation, *Commonwealth Bur. Pasture Field Crops. Bull. 46,* Hurley, Berkshire, England, 1962.

Nydal, F. Determination of Manganese Persulfate Method, *Anal. Chim. Acta,* **3**: 144–157 (1949).

Nye, P. H., and W. N. M. Foster. A Study on the Mechanism of Soil Phosphate Uptake in Relation to Plant Species, *Plant Soil,* **9**(4): 338–352 (1958).

Nye, P. H., and W. N. M. Foster. The Use of Radioisotopes to Study Plant Feeding Zones in Natural Soil, 7th Intern. Congr. of Soil Sci., Madison, Wisc., 1960.

Oberländer, H. E., and A. Zeller. Die Bestimmung des pflanzenverfügbaren Phosphates in österreichischen Böden mit Hilfe von radioaktivem Phosphor, *Bodenkultur,* Ausg. A, **12**(3): 190–197 (1961).

Oberländer, H. E., and A. Zeller. Die Düngerwirkung von Hyperphos auf österreichischen Böden I. Mitteilung: Ermittlung der kurzfristigen Wirkung mittels radioaktivem Phosphor, *Bodenkultur,* Ausg. A, **13**: 3–4 (1962a).

Oberländer, H. E., and A. Zeller. Die Düngerwirkung von Hyperphos of österreichischen Böden, *Bodenkultur,* **13**: 240–248 (1962b).

Obershain, S. S., and P. T. Gish. Effect of Green Manure Crops on Certain Properties of Berks. Silt Loam, *Va. Agr. Expt. Sta. Tech. Bull. 73,* p. 12, 1941.

Ochse, J. J., M. J. Soule, N. J. Dykman, and C. Wehlbury. *Tropical and Subtropical Agriculture,* The Macmillan Co., New York 1961.

Oertli, J. J. A Test of a Mechanism of Salt and Water Absorption Using Evidence from Guttation Studies, *Agrochimica,* **8**(1): 37–63 (1963).

Ogata, T., and H. Aibara. Effects of Biuret on the Metabolism of the Germinating Plant 2, *J. Sci. Soil Tech.*, **31**: 193–196 (1960).

Ohlrogge, A. J. How Roots Tap a Fertilizer Band, reprinted from Summer-Fall *Plant Food Rev.*, 1958.

Ohlrogge, A. J. Some Soil-Root-Plant Relationships, *Soil Sci.*, **93**: 30–38 (1961).

Ohlrogge, A. J., M. H. Miller, and W. G. Duncan. 1957 Certain Principles for Getting Effective Nutrient Use from Fertilizer Bands, *Better Crops Plant Food,* **41**(6): 26–30 (1957).

Okruszko, M., G. E. Willcox, and G. F. Warren. Evaluation of Laboratory Tests on Organic Soils for Prediction of Phosphorus Availability, *Soil Sci. Soc. Am. Proc.*, **26**(1): 71–74 (1962).

Olesen, J., and U. Bent. Experiments on Time of Application of Anhydrous Ammonia, *Beretn. Faellesfors. Odense*, **1**: 162–166 (1959).

Olmstead, L. B. Some Moisture Relations of Soils from the Erosion Experiment Stations, *U.S. Dept. Agr. Tech. Bull., 652* (1937).

Olsen, S. R. Measurement of Surface Phosphate on Hydroxylapatite and Phosphate Rock with Radiophosphorus, *J. Phys. Chem.*, **56**: 630–632 (1952a).

Olsen, S. R. The Measurement of Phosphorus on the Surface of Soil Particles and its Relationship to Plant Available Phosphorus, *Proc. Conf. on the Use of Radioisotopes in Plant and Animal Research*, Manhattan, Kans., USAEC/TID 5098: 59–67, (1952b).

Olsen, S. R. Inorganic Phosphorus in Alkaline and Calcareous Soils. (Soil and Fertilizer Phosphorus in Crop Nutrition), *Agronomy J.*, **4**: 89–122 (1953).

Olsen, S. R., and F. S. Watanabe. A Method to Determine a Phosphorus Adsorption Maximum of Soils as Measured by the Langmuir Isotherm, *Soil Sci. Soc. Am. Proc.*, **21**: 144–149 (1957).

Olsen, S. R., and F. S. Watanabe. Diffusion of Phosphorus as Related to Soil Texture and Plant Uptake, *Soil Sci. Soc. Am. Proc.*, **27**: 648–653 (1963).

Olsen, S. R., F. S. Watanabe, H. R. Cooper, W. E. Lardon, and L. B. Nelson. Residual Phosphorus Availability in Long-Time Rotations on Calcareous Soils, *Soil Sci.* **78**: 141–151 (1954a).

Olsen, S. R., C. V. Cole, F. S. Watanabe, and L. A. Dean. Estimation of Available Phosphorus in Soils by Extraction with Sodium Bicarbonate, *USDA Circ. 939*, 1954b.

Olsen, S. R., F. S. Watanabe, and C. V. Cole. Soil Properties Affecting the Solubility of Calcium Phosphates, *Soil Sci.*, **90**(1): 44–50 (1960).

Olsen, S. R., W. D. Kemper, and R. D. Jackson. Phosphate Diffusion to Plant Roots, *Soil Sci. Soc. Am. Proc.*, **26**: 222–227 (1962).

Olson, R. A., and A. F. Dreier. Nitrogen—A Key Factor in Fertilizer Phosphorus Efficiency, *Soil Sci. Soc. Am. Proc.*, **20**: 509–514 (1956).

Olson, R. A., and A. F. Dreier. Fertilizer Placement for Small Grains in Relation to Crop Stand and Nutrient Efficiency in Nebraska, *Soil Sci. Soc. Am. Proc.*, **20**: 19–24 (1956a).

Olson, R. A., A. F. Dreier, C. A. Hoover, and H. F. Rhoades. Factors Responsible for Poor Response of Corn and Grain Sorghum to Phosphorus Fertilization. I. Soil Phosphorus Level and Climatic Factors, *Soil Sci. Soc. Am. Proc.*, **26**: 571–578 (1962).

Opitz, K. Vergleichende Untersuchungen über die Ergebnisse von chemischen Bodenanalysen und Vegetationsversuchen, *Landwirtsch. Jahrb. Schweiz*, **36**: 909 (1907).

Oppenheim, I. D. *Tropische und subtropische Kulturen*, Berlin, 1932.

Osterhout, W. J. V. The Absorption of Electrolytes in Large Plant Cells, *Botan. Rev.*, **2**: 283–315 (1936).

Overbeek, J. Th. G. The Donnan Equilibrium Progress, *Biophys. Chem.*, **6**: 58–84 (1956).

Overein, L. N., and Moe, P. G. Factors Affecting Urea Hydrolysis and Ammonia Volatilization in Soil, *Soil Sci. Soc. Am. Proc.*, **31**: 57–61 (1967).

Overstreet, R., L. Jacobson, and R. Handley. The Effect of Calcium on the Absorption of Potassium by Barley Roots, *Plant Physiol.*, **27**: 583–590 (1952).

Owens, L. D. Nitrogen Movement and Transformations in Soils as Evaluated by a Lysimeter Study Utilizing Isotopic Nitrogen and a Field Study, *Dissertation Abstr.*, **19**: 2709 (1959).

Owens, L. D. Nitrogen Movement and Transformations in Soils as Evaluated by a Lysimeter Study Utilizing Isotopic Nitrogen, *Soil Sci. Soc. Am. Proc.*, **24**: 372–376 (1960).

Page, A. L., F. T. Bingham, T. J. Canje, and M. J. Garber. Availability and Fixation of Added Potassium in Two California Soils when Cropped to Cotton, *Soil Sci. Soc. Am. Proc.*, **27**: 323–326 (1963).

Paneth, F., and W. Vorwerk. Uber eine Methode zur Bestimmung der Oberfläche absorbierender Pulver, *Z. Physik. Chem.*, **101**: 445–479 (1922).

Parish, D. M., and S. M. Feillafé. A Comparison of Urea with Ammonium Sulphate as a Nitrogen Source for Sugar Cane, *Trop. Agr.* (*London*), **37**: 223–225 (1960).

Parker, F. W. Methods of Studying the Concentration and Composition of the Soil Solution, *Soil Sci.*, **12**: 209–232 (1921).

Parker, F. W. The Absorption of Phosphate by Pasteur-Chamberland Filters, *Soil Sci.*, **20**: 149–158 (1925).

Parker, F. W. Soil Phosphorus Studies. III. Plant Growth and the Absorption of Phosphorus from Culture Solutions of Different Phosphate Concentrations, *Soil Sci.*, **24**: 129–246 (1927).

Parsons, J. L., M. Drake, and W. G. Colby. Yield and Vegetative and Chemical Composition of Forage Crops as Affected by Soil Treatment, *Soil Sci. Soc. Am. Proc.*, **17**: 42–46 (1953).

Patil, A. S., K. M. King, and M. H. Miller. Self-diffusion of Rubidium as Influenced by Soil Moisture Tension, *Can. J. Soil Sci.*, **43**: 44–51 (1963).

Pauw, F. van der. Critical Remarks Concerning the Validity of the Mitscherlich Effect Law, *Plant Soil*, **4**: 97–106 (1952).

Pauw, F. van der. Notes: Meaning of the Great German Soil Fertility Survey—A Reply to Criticism, *Soil Sci.*, **80**: 253–254 (1955).

Pauw, F. van der. Calibration of Soil Test Methods for the Determination of Phosphate and Potash Status, *Plant Soil*, **8**: 105–125 (1956).

Pauw, F. van der. Periodic Fluctuations of Soil Fertility, Crop Yields and Responses to Fertilization Effected by Alternating Periods of Low or High Rainfall, *Plant Soil* **17**(2): 155–182 (1962).

Pauw, F. van der, and J. Ris. A New Potassium Index for Cultivated Sand and Reclaimed Peat Soils, *Landbouwvoorlichting*, **71**: 719–725 (1960).

Pauw, F. van der, and J. Ris. The Usefulness of K Number for Arable Highly Organic Soils, *Landbouwvoorlichting*, **19**: 265–268 (1962).

Paver, H. and C. E. Marshall. The Role of Aluminum in the Reactions of the Clays, *Chemistry and Industry 1934*, pp. 750–760.

Pearsall, W. H. The Investigation of Wet Soils and its Agricultural Implications, *Empire J. Exptl. Agr.*, **18**: 289–298.

Pearson, R. W. Potassium Supplying Power of Eight Alabama Soils, *Soil Sci.*, **74**: 301–309 (1952).

Pearson, R. W., H. V. Jordan, O. L. Bernet, et al. Residual Effects of Fall- and Spring-Applied Nitrogen Fertilizers on Crop Yields in the Southwestern U.S., *USDA Tech. Bull. 1254*, 1961.

Peech, M. Determination of Exchangeable Cations and Exchange Capacity of Soils. Rapid Micromethods Utilizing Centrifuge and Spectrophotometer, *Soil Sci.*, **59**: 25–38 (1945).

Pekáry, K. Experimental Results on the Effects of Nitrogen in the Autumn, *Agrokém. Talajtan*, **9**: 357–364 (1960).

Petersburghskii, A. V., and B. Debreczeni. Availability of Phosphates of Compound and Simple Fertilizers to Plants, *Izv. Timiryazev. Sel'skokhog. Akad.* **1**: 76–91 (1961).

Petersburghskii, A. V., and B. Debreczeni. Availability to Oats of the Phosphates of Compound and Simple Fertilizers on Acid and Limed Soil, *Izv. Timiryazev. Sel'skokhoz. Akad.*, **1**(5): 112–120 (1961a).

Peterburgskii, A. V., and G. I. Nelubova. Anion Exchange in Roots, *J. Sci. Food Agr.*, **14**(3): 186–187 (1963).

Peterburghskii, A. V., and G. I. Nelubova. Procedure for Determining the Total Anion-Exchange Absorptive Power of Roots by the Adsorption Method, *Soviet Plant Physiol.*, **9**(5): 520–521 (1963a).

Petterson, S. Ion Absorption in Young Sunflower Plants. II. The Sulphate Uptake in the Apparent Free Space, *Physiol. Plantarum,* **14**: 124–132 (1961).

Pfeiffer, Th., E. Blanck, and K. Friske. Der Einfluss verschiedener Vegetationsfaktoren, namentlich des Wassers, auf die Erzielung von Maximalerträgen in Vegetationsgefässen, *Landwirtsch. Versuchsstat.*, **82**: 237 (1913).

Phillips, R. E. Soil Compaction and Corn Growth, *Dissertation Abstr.*, **20**: 20 (1959).

Pierre, W. H. Hydrogen Ion Concentration, Aluminum Concentration in the Soil Solution and Percentage Base Saturation as Factors Affecting Plant Growth on Acid Soils, *Soil Sci.*, **31**: 183–207 (1931).

Pierre, W. H., and F. W. Parker. The Use of Collodion Sacks in Obtaining Clear Soil Extracts for Determination of the Water Soluble Constituents, *Soil Sci.*, **23**: 13–32 (1927).

Pinkas, L., and Smith, L. H. Physiological Basis of Differential Strontium Accumulation in Two Barley Genotypes, *Plant Physiol.*, **41**: 1471–1475 (1966).

Place, G. A., and S. Barber. The Effect of Soil Moisture and Rubidium Concentration on Diffusion and Uptake of Rubidium 86, *Soil Sci. Soc. Am. Proc.*, **28**: 239–243 (1964).

Plant, W., J. O. Jones, and D. J. D. Nicholas. The Diagnosis of Mineral Deficiencies in Crops by Means of Chemical Tissue Tests, Ann. Rept. Long. Ashton, 1944.

Plesset, M. S., F. Hefferich, and J. N. Franklin. Ion Exchange Kinetics. A Non-Linear Diffusion Problem. II. Particle Diffusion Controlled Exchange of Univalent and Bivalent Ions, *J. Chem. Phy.*, **29**: 1064–1069 (1958).

Ponnamperuma, F. N. The Chemistry of Submersed Soils in Relation to the Growth and Yield of Rice, Thesis, Cornell University, 1955.

Porter, L. K., W. D. Kemper, R. D. Jackson, and B. A. Stewart. Chloride Diffusion in Soils as Influenced by Moisture Content, *Soil Sci. Soc. Am. Proc.*, **24**(6): 460–463 (1960).

Possingham, J. V., and R. Brown. The Nuclear Incorporation of Iron and Its Significance in Growth, *J. Exptl. Botany,* **9**: 277–284 (1958).

Pratt, P. F., and G. Benois. Potassium Fixation in Soil of a Long-Term Fertility Trial with Citrus, *Soil Sci.*, **84**: 225–232 (1951).

Pratt, P. F., and B. L. Grover. Monovalent-Divalent Cation Exchange Equilibria in Soils in Relation to Organic Matter and Type of Clay, *Soil Sci. Soc. Am. Proc.*, **28**: 32–35 (1964).

Pratt, P. F., H. D. Chapman, and M. J. Garber. Gains and Losses of Nitrogen and Depth Distribution of Nitrogen and Organic Carbon in the Soil of a Lysimeter Investigation, *Soil Sci.*, **90**: 293–297 (1960).

Prevot, P., and P. M. Ollagnier. Application du diagnostic foliaire a l'Arachide, *Oléagineux,* No. 6, pp. 329–337 (F) 1951.

Prevot, P., and M. Ollagnier, Mineral Fertilizers and Tropical Oil Crops, *Oléagineux*, 8e Ann. No. 2: 843–851 (1953a).

Prevot, P., M. Ollagnier, and P. Giller. Blatt-Test bei Arachis und Gleichgewicht der Düngebestandteile im Senegal, *Z. Pflanzeneranehr. Düng. Bodenk.*, **63**: 176 (1953b).

Prigogine, I. *Thermodynamics of Irreversible Processes*, C. C. Thomas, Springfield, Illinois, p. 115, 1955.

Primrost, E., and G. Rittmeyer. The Effect of Split Application of Nitrogen at Increased Rates on the Grain Yield and Protein and Gluten Contents of Winter Wheat, *Bodenkultur*, **13**: 98–117 (1962).

Prince, A. B. Residual Effects of Superphosphate Application on Soil Phosphate Level and Growth of Crimson Clover as Measured by Yield and Phosphorus Uptake, *Soil Sci.*, **75**: 51–57 (1953).

Prince, A. L. Determination of Total Nitrogen Ammonia, Nitrates and Nitrites in Soils, *Soil Sci.*, **59**: 47–52 (1945).

Prummel, J. Placement of Fertilizers, *Trans. Intern. Comp. Soil Sci. Paris*, Vol. D: 167–171 (1956).

Prummel, J. Fertilizer Placement Experiments, *Plant Soil*, **8**: 231–253 (1957).

Prün, H. Nitrogen Fertilizing Assures Rape-Seed Production, *MID Dtnh. Lanw. Ger.*, **74**: 114–116 (1959).

Purdue University, Fertilizer Recommendation Pointers, *Extension Circ. 491*, 1961.

Rabinowitch, E. I. *Photosynthesis and Related Processes*, Interscience Publishers, New York, Vol. 1, 1945; Vol. 2, Pt. 1, 1951; Vol. 2, Pt. 2, 1956.

Rader, L. F., L. M. White, and C. W. Whittaker. The Salt Index—a Measure of the Effect of Fertilizer on the Concentration of the Soil Solution, *Soil Sci.*, **55**: 201–218 (1943).

Rains, D. W., W. E. Schmid, and E. Epstein, Absorption of Cations by Roots, *Plant Physiol.*, **39**: 274–278 (1964).

Rakhno, P. Kh., and O. O. Ryys. The Use of Azotobacter Preparations, *Microbiology* (USSR) (English Transl.), **32**: 474–476 (1963).

Ramamoorthy, B., S. V. Desai, and S. P. Ray Chandhuri. Potassium Fixation in Soils with Special Reference to India, *Indian J. Agr. Sci.*, **22**: 49–62 (1952).

Reichman, G. A., D. L. Grunes, C. W. Carlson, and J. Alessi. N and P Composition and Yield of Corn as Affected by Fertilization, *Agron. J.*, **51**: 575–578 (1959).

Reitemeier, R. F. Effect of Moisture Content on the Dissolved and Exchangeable Ions of Soils and Arid Regions, *Soil Sci.*, **61**(3): 195–214 (1946).

Reitemeier, R. F. Soil Potassium, *Advan. Agron.*, **3**: 113–162 (1951).

Remy, T. Fertilization and Its Relationship to the Course of Nutrient Absorption by Plants, *Soil Sci.*, **46**: 187–209 (1938).

Rennie, D. A., and R. B. McKercher. Adsorption of Phosphorus by Four Saskatchewan Soils, *Can. J. Soil Sci.*, **39**: 64–75 (1959).

Rennie, D. A., and R. J. Soper. The Effect of Nitrogen Additions on Fertilizer Phosphorus Availability II, *J. Soil Sci.*, **9**(1): 155–167 (1958).

Rennie, D. A., and E. D. Spratt. The Influence of Fertilizer Placement on "A" Values, *Proc. 7th Intern. Congr. Soil Sci.*, Madison, Wisc., 535 (1960).

Reyes, E. D., J. G. Davide, L. C. Orara, and R. A. Calirinan. Nitrogen, Phosphorus and Potassium Uptake by a Lowland Rice Variety at Different Stages of Growth, Univ. of Penna. Publ. A., *Centre Expt. Sta. Contrib. No. 2841*, Proj. 671, 1962 or 1963.

Rheinwald, M. *Praktische Düngerlehre für den landwritschaftlichen Betrieb,* Paul Barey, Berlin, 1948.

Rible, J. M., and L. E. Davis. Ion Exchange in Soil Columns, *Soil Sci.,* **79**(1): 41–47 (1955).

Rich, C. I., and S. S. Obershain. Chemical and Clay Mineral Properties of a Red Yellow Podzolic Soil Derived from Muscovite Schist, *Soil Sci. Soc. Am. Proc.,* **19**: 334–339 (1955).

Richards, L. A. A Pressure-Membrane Extraction Apparatus for Soil Solution, *Soil Sci.,* **51**: 377–386 (1941).

Richards, L. A. (Editor). *Diagnosis and Improvement of Saline and Alkali Soils,* U.S. Dept. Agr. Handbook NO. 60, 160 pp., 1954.

Richardson, H. L. The Nitrogen Cycle in Grassland Soils, With Special References to the Rothamsted Park Grass Experiment, *J. Agr. Sci.,* **28**: 73–121 (1938).

Riehm, H. Bestimmung der lactatlöslichen Phosphorsäure in Carbonathaltigen Böden, *Phosphorsäure,* F: 167 (1942).

Riehm, H. Neues über die Lactatmethode, *Z. Pflanzenernähr. Düng. Bodenk.,* **40**(85): 152–156 (1948).

Riehm, H. Ergebnisse der Bodenuntersuchungen im Bundesgebiet 1936–1953, *Phosphorsäure,* **13**: 329–331 (1953).

Riehm, H. Phosphatuntersuchungen der Böden des Bundesgebietes in den Jahren 1936–1955, *Phosphorsäure,* **16**: 160–166 (1956).

Riehm, H., and A. Hofmann. Investigations on Changes in the Nutrient Content of Soils During Several Years, *Landwirtsch. Forsch.,* **7**: 169–178 (1955).

Rippel, A. Das Ertragsgesetz, *Handb. Pflanzenernaehr. Düng.* (Honcamp), **1**: 602–622 (1931).

Robinson, R. A., and R. H. Stokes. *Electrolyte Solutions,* Butterworths, London, p. 559, 1959.

Roelofsen, P. A. Nitrogen in Rain Water, *Natuurw. Tijdschr., (Ghent),* **10**: 179–180 (1941).

Roemer, T., and F. Scheffer. Untersuchungen nach der Keimpflanzenmethode Neubauer unter Anwendung von Reis als Versuchsfrucht, *Ernähr. Pflanze.,* **252**: 532–535 (1929).

Rohde, G. Einfache Methode zur Bestimmung des Nährstoffzustandes der Böden auf Grund der Löslichkeit ihrer Nährstoffverbindungen in kochendem Wasser, *Z. Pflanzenernaehr. Düng. Bodenk.,* A **30**: 331–344 (1933).

Romkens, M. J. M., and Bruce, R. R. Nitrate Diffusivity in Relation to Moisture Content of Non-Adsorbing Porous Media, *Soil Sci.,* **98**: 332–337 (1964).

Roo, H. C. de. Root Growth in Connecticut Tobacco Soils, *Conn. Agr. Expt. Sta. Bull. 608,* 1957.

Roschach, H. Reproducibility of the Aspergillus Method in the Quantitative Determination of Zinc, *Landwirtsch. Forsch.,* **8**: 227–234 (1956).

Roschach, H. The Nitrogen Balance in Buckwheat, *Z. Pflanzenernaehr. Düng. Bodenk.,* **88**: 18–35 (1960).

Rosenberg, T. On Accumulation and Active Transport in Biological Systems. I. Thermodynamic Considerations, *Acta Chem. Scand.,* **2**: 14–33 (1948).

Rossiter, R. C. *Aust. J. Agr. Res.,* **3**: 7 (1952).

Rotini, O. T., and N. Guerrucci. The Action of Cyanamide Compounds on Plants, *Agrochimica* **6**: 92–100 (1961).

Rubins, E. J. Residual Phosphorus of Heavily Fertilized Acid Soils, *Soil Sci.,* **75**: 59–67 (1953).

Russell, E. J. *Soil Conditions and Plant Growth,* 8th ed. Longmans, Green & Co., London, pp. 606, 1950.

Russell, E. W. *Soil Conditions and Plant Growth.* Longmans, Green & Co., London, 1961.

Russell, G. C., and P. F. Low. Reaction of Phosphate with Kaolinite in Dilute Solutions *Soil Sci. Soc. Am. Proc.,* **18**: 22–25 (1954).

Russell, M. B. Soil Aeration and Plant Growth, Am. Soc. Agron., *Agronomy* (a series of monographs), **2**: 253–301 (1952).

Russell, M. B., and L. A. Richards. The Determination of Soil Moisture Energy Relations by Centrifugation, *Soil Sci. Soc. Amer. Proc.,* **3**: 65–69 (1938).

Russell, R. S. The Relationship Between Metabolism and the Accumulation of Ions by Plants, *Proc. Soc. Exptl. Biol.,* **8**: 343–366 (1954).

Russell, R. S., and M. J. Ayland. Exchange Reactions in the Entry of Cations into Plant Tissue, *Nature,* **175**: 204–205 (1955).

Russell, R. S., and D. A. Barber. The Relationship Between Salt Uptake and the Absorption of Water by Intact Plants, *Ann. Rev. Plant Physiol.,* **11**: 127–140 (1960).

Russell, R. S., and R. P. Martin. A Study of the Absorption and Utilization of Phosphate by Young Barley Plants. I. The Effect of External Concentration on the Distribution of Absorbed Phosphate Between Roots and Shoots, *J. Exptl. Botany,* **4**(10): 108–127 (1953).

Russell, R. S., and V. M. Shorrocks. The Relationship Between Transpiration and the Absorption of Inorganic Ions by Intact Plants, *J. Exptl. Botany,* **10**(29): 301–316 (1959).

Russell, R. S., J. B. Rickson, and S. N. Adams. Isotopic Equilibria Between Phosphates in Soil and Their Significance in the Assessment of Fertility by Tracer Methods, *J. Soil Sci.,* **5**(1): 85–105 (1954).

Russell, R. S., E. W. Russell, and P. G. Marais. Factors Affecting the Ability of Plants to Absorb Phosphate from Soils. I. The Relationship Between Labile Phosphate and Absorption, *J. Soil Sci.,* **8**: 248–267 (1957).

Russell, R. S., E. W. Russell, P. G. Marais, and W. N. M. Foster. Factors Affecting the Availability to Plants of Soil Phosphates, *1st UN Intern. Conf. Peaceful Uses Atomic Energy, Geneva,* P/460, **12**: 103–108 (1956).

Salomon, M., and J. B. Smith. Residual Soil Phosphorus from Various Fertilizer Phosphates Extracted by Different Solvents, *Soil Sci. Soc. Am. Proc.,* **20**: 33–36 (1956).

Samuels, G., and H. C. Dias. Influence of the Number of Fertilizer Applications for Pineapple Yields, *J. Agr. Univ. Puerto Rico,* **42**: 7–11 (1958).

Savioja, T., and J. K. Miettinen. Influence of Calcium Upon the Uptake of Orthophosphate by Pea Roots, *Suomen Kemistilehti,* B **33**: 78–80 (1960).

Sayre, C. B., and A. W. Clarke. Rates of Solution and Movement of Different Fertilizers in the Soil and the Effect of the Fertilizers on the Germination and Root Development, *N.Y. State (Geneva) Agr. Expt. Sta. Tech. Bull. 231,* 1935.

Sayre, J. D. Mineral Accumulation in Corn, *Plant Physiol.* **23**: 267–281 (1948).

Sayre, J. D., and V. H. Morris. Lithium Method of Measuring the Extent of Corn Root Systems, *Plant Physiol.,* **15**: 761–764 (1940).

Scarseth, G. D. The Mechanism of Phosphate Retention by Natural Alumino-Silicate Colloids, *J. Am. Soc. Agron.,* **27**: 596–617 (1935).

Scarseth, G. D. *et al., Purdue Univ. Agr. Expt. Sta. Bul. 482,* p. 40 (1944).

Schachtschabel, P. Aufnahme von nichtaustauschbarem Kali durch die Pflanzen, *Bodenk. Pflanzenernähr.,* **3**(48): 107–133 (1937).

Schachtschabel, P. Die Bestimmung des Kalibedarfes im Boden, *Bodenk. Pflanzenernähr.,* **24**(69): 371–384 (1941).

Schachtschabel, P. Das pflanzenverfügbare Magnesium des Bodens und seine Bestimmung, *Z. Pflanzenernähr. Düng. Bodenk.*, **67**(112): 9–23 (1954).

Schachtschabel, P. Magnesium in Soil and Plants, *Landwirtsch. Versuchs- Untersuchungswesen,* **2**: 507–523 (1956).

Schanderl, H. *Botanische Bakteriologie und Stickstoffhaushalt der Pflanzen auf neuerer Grundlage*, Stuttgart, p. 133, 1947.

Scharrer, K., and G. Lemme. Untersuchungen über die Brauchbarkeit der Blattanalyse von Lundegårdh zur Ermittlung des Düngerbedürfnisses der Böden *Z. Pflanzenernaehr. Düng. Bodenk.*, **60**(105): 125–148 (1952).

Scheffer, F., and B. Ulrich. Physical-Chemical Basis of Nutrient Dynamics in the Soil, *Acad. Rep. Populare Romine Probl. Act. Biol. Stiinte Agr.*, 663–670, 1960.

Scheffer, F., E. Welte, and C. Heinemann. Über die "Lactatlöslichkeit" von bodenbürtiger und Mineraldünger-Phosphorsäure in carbonatreichen Böden, *Z. Pflanzenernaehr. Ern. Düng. Bodenk.*, **58**(103): 40–59 (1952).

Scheffer, F., B. Ulrich, P. Benecke, and W. Sendler. The Kinetics of Isotopic Exchange of Phosphates. I. Methods of Investigation and Graphical and Numerical Interpretation. *Z. Pflanzenernaehr. Ern. Düng. Bodenk.*, **91**: 224–232 (1960).

Schell, W. R., and J. V. Jordan. Anion-Exchange Studies of Pure Clays, *Plant Soil,* **10**: 303–318 (1959).

Scheuring, D. G., and R. Overstreet. Sodium Uptake by Excised Barley Roots from Sodium Betonite Suspensions and from Their Equilibrium Filtrates, *Soil Sci.*, **92**(3): 166–171 (1961).

Schmid, K. Grundsätzliches zur Düngung im Qualitätstabakbau, *Der Deut. Tabakbauer,* **4**: 27–29 (1951).

Schmitz, G. W., and P. F. Pratt. Exchangeable and Non-Exchangeable Potassium as Indexes to Yield Increases and Potassium Absorption by Corn in the Greenhouse, *Soil Sci.*, **76**: 345–354 (1953).

Schofield, R. K. A Ratio Law Governing the Equilibrium of Cations in the Soil Solution, *Proc. 11th Intern. Congr. Pure and Appl. Chem. (London)*, **3**: 257–261 (1947).

Schofield, R. K. Can a Precise Meaning be Given to "Available" Soil Phosphorus?, *Soil Fertilizers,* **18**: 373–375 (1955).

Scholl, W., G. W. Schmidt, and H. P. Toland. Consumption of Commercial Fertilizers and Primary Plant Nutrients in the United States, Year Ended June 30, 1963. *Commercial Fertilizers,* **108**(5): 23–27 (1964a).

Scholl, W., G. W. Schmidt, and H. P. Toland. Consumption of Commercial Fertilizers and Primary Plant Nutrients in the U.S., Year Ended June 30, 1963, U.S. Fertilizer Laboratory USDA, Beltsville, Md., F-55, Preliminary Report, 8 pp., 1964b.

Schollenberger, C. J., and R. M. Simon. Determination of Exchange Capacity and Exchangeable Bases in Soil-Ammonium Acetate Method, *Soil Sci.*, **59**: 13–24 (1945).

Schwertman, U. and M. L. Jackson. Hydrogen-Aluminum Clays: A Third Buffer Range Appearing in Potentiometric Titration, *Science,* **139**: 1052–1053, (1963).

Sears, P. D., and L. T. Evans. Pasture Growth and Soil Fertility. III. The Influence of Red and White Clovers, Superphosphate Lime and Dung and Urine on Soil Composition and on Earthgrown and Grass-Grub Population, *N.Z. J. Sci. Tech.*, **35**(A), (Suppl. 1): 42 (1953).

Seatz, L. F. Phosphate Activity Measurements in Soils, *Soil Sci.*, **77**: 43–52 (1954).

Seatz, L. F., and C. O. Stanberry. Advances in Phosphate Fertilization, *Fert. Tech. Usage Symp.* Soil Sci. Soc. Am., Madison, Wisc., 1963.

Selke, W. Supplemental Late Application of Nitrogen as a Means of Further Increasing the Yield and Quality of Cereals, *Dent. Handw. W*, pp. 191–197 (1959).

Semb, G., and G. Uhlen. A Comparison of Different Analytical Methods for the Determination of Potassium and Phosphorus Soil Based on Field Experiments, *Acta Agr. Scandinavia*, **5**(1): 44–68 (1955).

Semma, V. G. The Effectiveness of Different Dates and Methods of Mechanized Application of Fertilizers to Potatoes, *Udobr. Urozhai*, **3**: 33–36 (1959).

Sen, A. N. Nitrogen Economy of Soil Under Rahar (Cajanus cajan.), *J. Indian Soc. Soil Sci.*, **6**: 171 (1958).

Sen Gupta, M. B., and A. H. Cornfield. Phosphorus in Calcareous Soils. III. "Available" Phosphate in Calcareous Soils as Measured by Five Chemical Methods and Phosphate Uptake by Ryegrass in a Pot Test, *J. Sci. Food Agr.*, **14**: 563–567 (1963).

Shalhevet, J., and P. J. Zwerman. Nitrogen Response of Corn under Variable Conditions of Drainage—A Lysimeter Study, *Soil Sci.*, **93**: 172–182 (1962).

Shapiro, R. E. Effect of Organic Matter and Flooding on Availability of Soil and Synthetic Phosphates, *Soil Sci.*, **85**: 267–272 (1958).

Shapiro, R. E., and M. Fried. Relative Release and Retentiveness of Soil Phosphates, *Soil Sci. Soc. Am. Proc.*, **23**(3): 195–198 (1959).

Shapiro, R. E., W. H. Armiger, and M. Fried. The Effect of Soil Water Movement vs. Phosphate Diffusion on Growth and Phosphorus Content of Corn and Soy Beans, *Soil Sci. Soc. Am. Proc.*, **24**: 161–164 (1960).

Sharp, J. C., and R. C. Powell. *Advances in Nitrogen Fixation Fertilizer Technology and Usage, Soil Sci.* Soc. Am., Madison, Wisc., 1963.

Shaw, E., and L. A. Dean. Use of Dithizone as an Extractant to Estimate the Zinc Nutrient Status of Soil, *Soil Sci.*, **73**: 341–347 (1952).

Shaw, E., R. G. Menzel, and L. A. Dean. Plant Uptake of Zinc-65 from Soils and Fertilizers in the Greenhouse, *Soil Sci.*, **77**: 205–214 (1954).

Shaw, K. Loss of Mineral Nitrogen from Soil, *J. Agric. Sci.*, **58**: 145–151 (1962).

Shaw, W. M., and B. Robinson. Reaction Efficiencies of Liming Materials as Indicated by Leachate Composition, *Plant Physiol.*, **89**: 209–218 (1960).

Shealtiel, M. Specific Exchange of Rubidium Ion in Barley Roots, *Compt. Rend.*, **247**: 1647–1649 (1958).

Shen, R. C. Fertilizer Contaminants. Rate of Biuret Formation from Urea, *J. Agr. Food Chem.*, **7**: 762–763 (1959).

Shilova, Y. I., and L. V. Korovkina. Seasonal Dynamics of the Chemical Composition of Lysimeter Water in Podzolic Fine Clay Loam Soils, *Pochvovedenie* (English Translation), **3**: 263–272 (1961).

Shilova, Y. I., and L. V. Korovkina. Comparative Study of the Composition of Soil Solutions and Lysimetric Water from Sod-podzolic Soils, *Soviet Soil Sci.*, (English translation), **8**: 796–805 (1962).

Shiori, M., and T. Tanaka. *The Chemistry of Paddy Soils in Japan*, Ministry of Agr. Japan & Tokyo, 1959.

Sigmond, A. A. J. v. Die Bestimmung der in Salzsäure löslichen Mineral—und Nährstoffe des Bodens und die Bewertung der Befunde des Salzsäurenauszuges, *Handb. Bodenlehre* (Blanck), **18**: 148–174 (1931).

Simpson, J. R. Mineral Nitrogen Fluctuations in Soils Under Improved Pasture in Southern New South Wales, *Australian J. Agr. Res.*, **13**: 1059–1072 (1962).

Singh, J. N., and N. Murayama. Analytical Studies on the Productive Efficiency of Nitrogen in Rice, *Soil Sci. Plant Nutr.* (*Tokyo*), **9**: 25–35 (1963).

Skinner, S. O. Fertilizer Burn, *Cane Gr. Quart. Bull. 25*, p. 87, 1962.

Skogley, E. D., and C. B. McCants. Ammonium Influences on Rubidium Absorption and Distribution by Tobacco Seedlings, *Soil Sci. Soc. Am. Proc.*, **27**: 549–552 (1963).

Skyring, G. W., and B. J. Callow. The Physiology and Biochemistry of Nitrification and Denitrification, *Bull. 46,* Commonwealth Bur. Pasture and Field Crops, Hurley, Berkshire, England, 1962.

Smith, C. M., and J. T. Pesek. Comparing Measurements of the Effect of Residual Fertilizer Phosphorus in Some Iowa Soils, *Soil Sci. Soc. Am. Proc.*, **26**(6): 563–566 (1962).

Smith, D. H., J. M. Blume, and C. W. Whittaker. Radio-chemical Measurement of Reaction Rates, *Agr. Food Chem.*, **1**: 67 (1963).

Smith, G. E. Soil Fertility and Corn Production, *Missouri Agr. Expt. Sta. Bull. 583,* 1952.

Smith, H. V. A Lysimeter Study of the Nitrogen Balance in Irrigated Soils, *Ariz. Agr. Expt. Sta. Tech. Bull. 102, p.* 257, 1944.

Smith, R. L., and A. Wallace. Cation-Exchange Capacity of Roots and Its Relation to Calcium and Potassium Content of Plants, *Soil Sci.*, **81**(2): 97–109 (1956).

Smith, R. L., and A. Wallace. Influence of Nitrogen Fertilization, Cation Concentration and Root Cation-Exchange Capacity on Calcium and Potassium Uptake by Plants, *Soil Sci.*, **82**(2): 165–172 (1965a).

Smith-Brun, T. Kobberets Binding i Humus, *Bergens Museums Åzbok,* **6**: 1–21 (1945).

Sokolov, A. V. Determination of the Availability of Soil Phosphates and Fertilizers with the Aid of Radioactive Isotopes of Phosphorus, *Proc. 1st UN Intern. Conf. Peaceful Uses Atomic Energy,* Geneva, P/695, **12**: 118–122 (1956).

Soubiès, L., R. Gadet, and P. Maury. Migration Hivernale de l'Azote Nitrique dans un Sol Limoneaux de la Région Toulousaine, *Ann. Agron.*, **3**: 365–383 (1952).

Spicher, G. The Question of the Reliability of Micro-biological Methods for Determining Trace Elements and Magnesium in Soil, *Zbl. Bakter.*, II, **108**: 259–267 (1954).

Spiegler, K. S., and C. D. Coryell. Electromigration in Cation-Exchange Resins III, *J. Phys. Chem.*, **57**: 687–690 (1953).

Spillman, W. J., and E. Lang. *The Law of Diminishing Returns,* Yonkers-on-Hudson, 178 pp., 1924.

Sprengel, C. *Die Lehre vom Dünger* 2. Auflage, Leipzig, 1845.

Spurway, C. H. A Test for Water-Soluble Phosphorus; Studies on Water-Soluble Phosphorus in Field Soils, *Mich. State Tech. Bull. 101,* (25), 1929.

Stahlberg, S. Soil and Fertility, Abstract No. 1249 (1955), *Voxtnärrings-Nytt,* **10**: 81–85 (1954).

Stahlberg, S. Studies on the Release of Bases from Minerals and Soils. V. The Uptake of Potassium by Clover from Different Mineral-Peat and Soil-Peat Mixtures, *Acta Agr. Scand.*, **13**(4): 392–403 (1963).

Stanberry, C. O., C. D. Converse, H. R. Haise, and O. J. Kelley. Effect of Moisture and Phosphate Variables on Alfalfa Hay Production on the Yuma Mesa, *Soil Sci. Soc. Am. Proc.*, **19**(3): 303–310 (1955).

Stanberry, C. O., W. H. Fuller, and N. R. Crawford. Comparisons of Phosphate Sources for Alfalfa on a Calcareous Soil, *Soil Sci. Soc. Am. Proc.*, **27**: 364–366 (1960).

Stanford, G., and D. R. Bouldin. Biological and Chemical Availability of Phosphate-

Soil Reaction Products, *Proc. 7th Intern. Soil Sci. Congr.*, Madison, Wisc., 1960, **2**:388–396 (1960).

Stanford, G, and J. D. DeMent. A Method for Measuring Short-Term Nutrient Absorption by Plants. I. Phosphorus, *Soil Sci. Soc. Am. Proc.*, **21**: 612–617 (1957).

Stanford, G., and J. D. DeMent. A Method for Measuring Short-Term Nutrient Absorption by Plants. II. Potassium, *Soil Sci. Soc. Am. Proc.*, **23**: 47–50 (1959).

Stanford, G., and J. D. DeMent. A Method for Measuring Short-Term Nutrient Absorption by Plants. III. Nitrogen, *Soil Sci. Soc. Am. Proc.*, **23**: 371–374 (1959a).

Stanford, G., and W. H. Pierre. The Relation of Potassium Fixation to Ammonium Fixation, *Soil Sci. Soc. Am. Proc.*, **11**: 155–160 (1946).

Stanford, G., J. Hanway, and H. R. Meldrum. Effectiveness and Recovery of Phosphorus and Potassium Fertilizers Topdressed on Meadows, *Soil Sci. Soc. Am. Proc.*, **17**: 378–382 (1953).

Stanford, G., J. Hanway, and H. R. Meldrum. Effectiveness and Recovery of Initial and Subsequent Fertilizer Applications on Oats and the Succeeding Meadows, *Agron. J.*, **47**: 25–31 (1955).

Stapp, C., and C. Wetter. Contribution to the Quantitative Estimation of Magnesium, Zinc, Iron, Molybdenum and Copper in Soil, *Landwirtsch. Forsch.*, **5**: 167–180 (1953).

Starkey, R. L. Some Influences of the Development of Higher Plants Upon the Microorganisms in the Soil. V. Effects of Plants Upon Distribution of Nitrates, *Soil Sci.*, **32**: 395–404 (1931).

Starostka, R. W., and W. L. Hill. Influence of Soluble Salts on the Solubility of and Plant Responses to Dicalcium Phosphate, *Soil Sci. Soc. Am. Proc.*, **19**: 193–198 (1955).

Steenbjerg, F. Undersgelser over Manganindholdet i Dansk Jord. I. Det Ombyttelige Mangan, *Tidsskr. Planteavl*, **39**: 401–436 (1933).

Steenbjerg, F. Yield Curves and Chemical Plant Analyses, *Plant Soil*, **3**: 97–109 (1951).

Steenbjerg, F. Verification of Chemical Soil and Plant Analyses: General Considerations, *Trans. Intern. Soc. Soil Sci.*, **1**: 309–322 (Dublin) (1952).

Steenbjerg, F., and S. T. Jakobsen. Some Approaches to Experimental Investigations into the Correlation Between the Slope and the Sigmoidal Shape of Yield Curves, *Plant Soil*, **10**: 284–295 (1959).

Steenbjerg, F., and S. T. Jacobsen. Plant Nutrition and Yield Curves, *Soil Sci.*, **95**: 69–88 (1963).

Stephens, R. C., and J. S. Ward. Pot Experiments on Urea as Fertilizer. III., *Plant Soil*, **19**(2): 184–192 (1963).

Stevenson, G. Use of N-15 in the Study of Fixation of Atmospheric Nitrogen by Non-nodulated Seed Plants, *Proc. UN Intern. Conf. Peaceful Uses of Atomic Energy*, pp. 51–57, 1958.

Steward, F. C. (Editor). *Plant Physiology—A Treatise*. Academic Press, Inc., New York and London, pp. 758, 1959.

Stewart, B. A., D. D. Johnson, and L. K. Porter. The Availability of Fertilizer Nitrogen Immobilized During Decomposition of Straw, *Soil Sci. Soc. Am. Proc.*, **27**: 656–659 (1963).

Stewart, B. A., L. K. Porter, and D. D. Johnson. Immobilization and Mineralization of Nitrogen in Several Organic Fractions of the Soil, *Soil Sci. Soc. Am. Proc.,* **27**: 302–304 (1963a).

Stewart, D. P., and G. Bond. The Effect of Ammonium Nitrogen on Fixation of Elemental N in Alnus and Myrica, *Plant Soil,* **14**(4): 347–359 (1961).

Stewart, I., and C. D. Leonard. Effect of Various Salts on the Availability of Zn and Mn to Citrus, *Soil Sci.,* **95**: 149–154 (1962).

Stiles, W. *Trace Elements in Plants,* 3rd ed., University Press, Cambridge, 249 pp., 1961.

Stock, J. Kulturversuche mit Aspergillus Niger als Indikator für die Düngerbedüftigkeit, *Boston. Archiv.,* **35**: 1 (1933).

Stornier, R. R. The Availability of Mineral Nitrogen in Wheat Soil from Southern New South Wales, *Austrialian J. Exptl. Agr. Animal Husbandry,* **2**: 185–192 (1962).

Stout, P. R., and D. R. Hoagland. Upward and Lateral Movement of Salt in Certain Plants as Indicated by Radioactive Isotopes of K, Na, and P Absorbed by Roots, *Am. J. Botany,* **26**: 320–324 (1939).

Strang, J., and J. Weir. The Effect of Biuret on Pasture, *Agr. Gaz. N. S. Wales,* **72**: 424–425 (1961).

Street, N. The Surface Conductance of Kaolinite, *Australian J. Chem.,* **9**: 333–346 (1956).

Street, N. Surface Conductance of Suspended Particles, *J. Phys. Chem.,* **64**: 173–174 (1960).

Street, N. On the Isoconductivity of Clays, *Soil Sci.,* **95**: 367 (1963).

Struchtemeyer, R. A., C. E. Cunningham, and P. N. Carpenter. Utilization of Residual Fertility by Potatoes, *Soil Sci. Soc. Am. Proc.,* **19**: 212–218 (1955).

Sumner, M. E., and G. H. Bolt. Isotopic Exchange of Potassium in an Illite under Equilibrium Conditions, *Soil Sci. Soc. Am. Proc.,* **26**(6): 541 (1962).

Sutton, C. D., and S. Larsen. Pyrophosphate as a Source of Phosphorus for Plants, *Soil Sci.,* **97**: 196–201 (1964).

Sutton, P., and W. A. Seay. Relationship Between the Potassium Removed by Millet and Red Clover and the Potassium Extracted by Four Chemical Methods from Six Kentucky Soils, *Soil Sci. Soc. Am. Proc.,* **22**: 110–115 (1958).

Suzuki, A., K. Lawton, and E. C. Doll. Phosphorus Uptake and Soil Tests as Related to Forms of Phosphorus in Some Michigan Soils, *Soil Sci. Soc. Am. Proc.,* **27**(4): 401–403 (1963).

Swaby, R. J., and B. I. Passey. Availability of Trace Elements from Rocks and Minerals, *Australian J. Agr. Res.,* **4**: 292–304 (1953).

Swanback, T. R. Granite Stone Meal as a Source of Potash for Tobacco, *Comm. Agr. Exptl. Sta. Bull. 536,* p. 17, 1950.

Swarajosunghkam, S., L. T. Alexander, J. G. Cady, and M. G. Cline. Laterite, *Advan. Agron.,* **14**: 1–60 (1962).

Swenson, R. M., C. V. Cole, and D. H. Sieling. Fixation of Phosphate by Iron and Aluminum and Replacement by Organic and Inorganic Ions, *Soil Science,* **67**: 3–23 (1949).

Swerzey, A. W., and G. O. Turner. Crop Experiments on the Effect of 2-Chloro-6-(Trichloromethyl) Pyridine for the Control of Nitrification of Ammonium and Urea Fertilizer, *Agron, J.,* **54**: 532–535 (1962).

Tabikh, A. A., I. Barshad, and R. Overstreet. Cation-Exchange Hysteresis in Clay Minerals, *Soil Sci.,* **90**(4): 41–53 (1960).

Talibudeen, O. The Determination of Isotopically Exchangeable Phosphorus in

Some Rothamsted Soils, *Proc. 2nd Conf. Radioisotopes,* Oxford, pp. 405–411, 1954.

Talibudeen, O. Isotopically Exchangeable Phosphorus in Soils. II. Factors Influencing the Estimation of Labile Phosphorus, *J. Soil Sci.,* **8**: 86–96 (1957).

Tanada, T. Effects of Ultraviolet Radiation and Calcium and Their Interaction of Salt Absorption by Excised Mung Bean Roots, *Plant Physiol.,* **30**(3): 221–225 (1955).

Tanada, T. Effect of Ribonulease on Salt Absorption by Excised Mung Bean Roots, *Plant Physiol.,* **31**(3): 251–253 (1956).

Tanada, T. Localization and Mechanism of Calcium Stimulation of Rubidium Absorption in the Mung Bean Root, *Am. J. Botany,* **49**(10): 1068–1072 (1962).

Tanada, T. Localization and Mechanism of Calcium Stimulation of Rubidium Absorption in the Mung Beet Root, *Am. J. Botany,* **49**: 1068–1072 (1962).

Tanada, T. Kinetics of Rb Absorption by Excised Barley Roots Under Changing Rb Concentrations. I. Effects of Other Cations on Rb Uptake, *Plant Physiol.,* **38**(4): 422–425 (1963).

Tanaka, A., S. Patnaik, and C. T. Abichandani. Studies on the Nutrition of Rice Plant. III Partial Efficiency of Nitrogen Absorbed by Rice Plant at Different Stages of Growth in Relation to Yield of Rice, *Proc. Indian Acad. Sci.,* **49**: 207–226 (1959a).

Tanaka, A., S. Patnaik, and C. T. Abichandani. Studies on the Nutrition of Rice Plant. VI Utilization of Ammonium and Nitrate Nitrogen by Rice Plant under Water Logged Soil State, *Proc. Indian Acad. Sci.,* **508**: 61–74 (1959b).

Tarabrin, C. A. Absorption Capacity of Plant Roots and Utilization by Them of Ions Absorbed by the Soil, *Dokl. Akad. Nank Timiryazeva,* **70**: 111–115 (1961).

Tchan, Y. T. Study of Soil Algae. III. Bioassay of Soil Fertility by Algae, *Plant Soil,* **10**(3): 220–232 (1959).

Tchan, Y. T., L. N. Balaam, R. Hawkes, and F. Draetta. Study of Soil Algae. IV. Estimation of the Nutrient Status of Soil Using an Algal Growth Method with Special Reference to Nitrogen and Phosphorus, *Plant Soil,* **14**(2): 147–158 (1961).

Tchan, Y. T., L. N. Balaam, and F. Draetta. Study of Soil Algae. V. Estimation of the Nutrient Status of S in Soil Using an Algal Growth Method, *Plant Soil,* **19**(2): 233–240 (1963).

Teakle, L. J. H. Phosphate in the Soil Solution as Affected by Reaction and Cation Concentration, *Soil Sci.,* **25**: 143 (1928).

Templeman, W. G. Urea as a Fertilizer, *J. Agr. Sci.,* **57**: 237–239 (1961).

Tepe, W., and E. Leidenfrost. Ein Vergleich zwischen pflanzenphysiologischen, kinetishen und statischen Bodenuntersuchungswerten. I. Die Kinetik der Bodenionen gemessen mit Ionenaustauschern, *Landwirtsch. Forsch.,* **11**: 217–230 (1958).

Terman, G. L. Yield Response in Experiments with Phosphorous Fertilizer in Relation to: I. Meaningful Differences Among Sources in Acid Soils of the Southeastern States, *Soil Sci. Soc. Am. Proc.,* **24**: 356–360 (1960).

Terman, G. L. Yield Response in Experiments with Phosphorus Fertilizer in Relation to: II. Variability and Differences Among Sources in Soils of Northern and Western States, *Soil Sci. Soc. Am. Proc.,* **25**: 49–52 (1961a).

Terman, G. L. Crop Response to Granular or Finely Divided Fertilizer, *Agr. Chem.,* Feb. 1961b.

Terman, G. L., J. D. De Ment, L. B. Clements, et al. Crop Response to Ammoniated Superphosphates and Dicalcium Phosphate as Affected by Granule Size Water Solubility and Time of Reaction with Soil, *J. Agr. Food Chem.*, **8**: 13–18 (1960).

Terman, G. L., D. R. Bouldin, and J. R. Webb. Evaluation of Fertilizers by Biological Methods (USA), *Advan. Agron.*, **14**: 265–319 (1962).

Terman, G. L., A. Hawkins, C. E. Cunningham, and P. N. Carpenter. Rate Placement and Source of Phosphorus Fertilizers for Potatoes in Maine, *Maine Agr. Expt. Sta. Bull. 506*, p. 24, 1952.

Terrien, J. *Light, Vegetation, and Chlorophyll*, Philosophical Library, New York, p. 228, 1957.

Thomas, H. C., and G. L. Gaines. Clays and Clay Minerals, *2nd Conf. Natl. Acad. Sci., Publ. 327*, p. 398, (1954).

Thomas, W. Foliar Diagnosis. Principle and Practice, *Plant Physiol.*, **12**: 571 (1937).

Thomas, W. Present Status of Diagnosis of Mineral Requirements of Plants by Means of Leaf Analysis, *Soil Sci.*, **59**: 353–374 (1945).

Thompson, E. J., A. L. F. Oliveira, U. S. Moser, and C. A. Black. Evaluation of Laboratory Indexes of Absorption of Soil Phosphorus by Plants II, *Plant Soil*, **13**(1): 28–38 (1960).

Thompson, L. M., and C. A. Black. The Mineralization of Organic Phosphorus, Nitrogen and Carbon in Clarion and Webster Soils, *Soil Sci. Soc. Am. Proc.*, **14**: 147–151 (1950).

Thornton, S. F. Soil and Fertilizer Studies by Means of the Neubauer Method, *Indian Agr. Expt. Sta. Bull. 399*, p. 38, 1935.

Thornton, S. F., S. D. Conner, and R. R. Fraser. The Use of Rapid Chemical Tests on Soils and Plants as Aids in Determining Fertilizer Needs, *Circ. Purdue Agr. Expt. Sta. Bull. 204*, 1934.

Thun, R. Untersuchungen über die Neubauer-Methode sowie bezüglich des leicht aufnehmbaren Stickstoffs, *Z. Pflanzenernaehr. Düng. Bodenk. A*, **16**: 257–283 (1930).

Tisdale, S. L., and W. L. Nelson. *Soil Fertility and Fertilizers*, The Macmillan Co., New York, 430 pp., 1956.

Torii, K., and Laties, G. G. Mechanisms of Ion Uptake in Relation to Vacuolation of Corn Roots, *Plant Physiol.*, **41**: 863–870 (1966).

Tromp, J. Interactions in the Absorption of Ammonium, Potassium, and Sodium Ions by Wheat Roots, *Acta Botan. Neerland.*, **11**: 147–192 (1962).

Troughton, A. The Underground Organs of Herbage Grasses, *Comm. Agr. Bur. Bull. 44*, 1957.

Truog, E. The Utilzation of Phosphate by Agricultural Crops, Including a New Theory Regarding the Feeding Power of Plants, *Wisconsin Univ. Agr. Expt. Sta. Res. Bull. 41*, 1916.

Truog, E. The Determination of the Readily Available Phosphorus of Soils, *J. Am. Soc. Agr.*, **22**: 879 (1930).

Truog, E. Availability of Essential Soil Elements—A Relative Matter, *Ann. Arbor.*, **1**: 135 (1937).

Truog, E. Present Status and Future Soil Testing and Soil Analysis, *Ann. Arbor.*, **2**: 191 (1938).

Truog, E. Determination of Total and Available Boron in Soils, *Soil Sci.*, **59**: 85–90 (1945).

Tshapek, M. The Ion Sieving Phenomenon in Soils, *Z. Pflanzenernaehr. Düng. Bodenk.*, **102**: 193–203 (1963).

Tucker, T. C., and L. T. Kirk. A Comparison of Several Chemical Methods with the Bio-Assay Procedure for Extracting Zinc from Soils, *Soil Sci. Soc. Am. Proc.*, **19**: 477–481 (1955).

Tucker, T. C., L. T. Kurtz, and D. L. Lunch. Zinc Status of Some Illinois Soils as Estimated by an Aspergillus niger Method, *Soil Sci. Soc. Am. Proc.*, **17**: 111–114, (1953).

Tyner, E. H. The Relation of Corn Yields to Leaf Nitrogen, Phosphorus and Potassium Content, *Soil Sci. Soc. Am. Proc.*, **11**: 317–343 (1946).

Uexküll, H. V. Fertilizer Use, Nutrition and Manuring of Tropical Crops, *Verlagsges. Ackerbau*, Hanover, Germany, 1963.

Ulrich, A. Plant Analysis as a Diagnostic Procedure, *Soil Sci.*, **55**: 101–112 (1943).

Ulrich, B. *Soil and Plant: Their Interrelationships—Physical-Chemical View*, Ferd. Enke Verlag, Stuttgart, pp. 1–114, 1961.

U.S. Salinity Laboratory Staff. *Diagnosis and Improvement of Saline and Alkali Soils, USDA Agricultural Handbook No. 60*, 156 pp., 1954.

Vageler, von P. *An Introduction to Tropical Soils*, The Macmillan Co., New York, 1933.

Vageler, von P., and F. Alten. Böden des Nil und Gash II, IV, *Z. Pflanzenernaehr. Düng. Bodenk. A*, **22**: 21, 191 (1931).

Vageleri, von P., and J. Woltersdorf. Beiträge zur Frage des Basenaustausches und der Aziditäten. *Z. Pflanzenernähr. Düng. Bodenk.*, **15**:329–342 (1930).

Vaidyanathan, L. V., and O. Talibudeen. Alteration in the Surface Properties of Soils by Ion Exchange Resins, *Nature*, **194**: 897 (1962).

Van Bladel, R., and H. Laudelout. Thermodynamic and Thermochemical Study of Ammonium-Calcium Exchange in Clays (in preparation).

Vandecaveye, S. C. *Biological Methods of Determining Nutrients in Soil Diagnostic Techniques for Soils and Crops*. The American Potash Institute, Washington, D.C., 1948.

van den Honert, T. H. *The Phosphate Absorption by Sugar Cane*. Verslag 13e Bijeenkomst van de Vereniging van Proefstations-Personeel, pp. 7–20, 1933.

van den Honert, T. H. *Limiting Factors in Phosphate Absorption*, Verslag van de Vergadering van de Vereniging van Proefstation-Personeel, Djember, October: 85–93 (1936).

van den Honert, T. H. Over eigenschappen van plantenwortels welke een rol spelen bij de opname van voedingszouten, *Natuurw. Tijdschr. Ned. Indie*, **97**: 150–162 (1937).

van den Honert, T. H., and J. J. M. Hooymans. On the Absorption of Nitrate by Maize in Water Culture, *Acta Botan. Neerl.*, **4**: 376–384 (1955).

van den Honert, T. H., and J. J. Hooymans. Diffusion and Absorption of Ions in Plant Tissue, I. Observations on the Absorption of Ammonium by Cut Discs of Potato Tuber as Compared to Maize Roots, *Acta Botan. Neerl.*, **10**: 261–273 (1961).

van den Honert, T. H., J. J. M. Hooymans, and W. S. Volkers. Experiments on the Relation Between Water Absorption and Mineral Uptake by Plant Roots, *Acta Botan. Neerl.*, **4**(1): 139–155 (1955).

van der Pauw, F. Periodic Fluctuations of Soil Fertility, Crop Yield and Responses to Fertilization Effected by Alternating Periods of Low or High Rainfall, *Plant Soil*, **17**: 155–182 (1962).

Van Olphen H. Surface Conductance of Various Ion Forms of Bentonite in Water and the Electrical Double Layer, *J. Phys. Chem.*, **61**: 1276 (1957).

Van Olphen H., and M. H. Waksman. Surface Conductance of Sodium Bentonite, *Proc. Fifth Natl. Conf. Clays and Clay Minerals Publ. 566,* p. 61–80 (1958).

Vanselow, A. P. Equilibria of the Base-Exchange Reactions of Bentonites, Permutites, Soil Colloids and Zeolites, *Soil Sci.*, **33**: 95–113 (1932).

Vanstallen, R. The Variability of Soluble Nitrogen in Loam Soil, *Agricultura Louvain,* **7**: 45–60 (1959).

Vasey, E. H., and S. A. Barber. Effect of Placement on the Absorption of Rb^{86} and P^{32} from Soil by Corn Roots, *Soil Sci. Soc. Am. Proc.*, **27**(2): 193–197 (1963).

Vialard-Goudou, A., and C. Richard. Physico-Chemical and Economic Studies of Rainfall in Saigon (1950–1954), *Agron. Trop. Paris,* **11**: 74 (1956).

Vincente-Chandler, J., R. Caro-Costas, and J. Figarella. The Effects of Two Heights of Cutting and Three Fertility Levels on the Yield, Protein Content, and Species Composition of a Tropical Kudzu and Molasses-Grasses Pasture, *Agron. J.*, **20**: 45–397 (1953).

Vickar, M. H., G. L. Bridges, and L. B. Nelson. Fertilizer Technology and Usage, *Soil Sci. Soc. Am. Proc.*, Madison, Wisc., 1963.

Viets, F. G., C. E. Nelson, and C. L. Crawford. The Relationship Among Corn, Yield Leaf Composition and Fertilizers Applied, *Soil Sci. Soc. Am. Proc.*, **18**: 297–301 (1954).

Viets, F. G. Calcium and Other Polyvalent Cations as Accelerators of Ion Accumulation by Excised Barley Roots, *Plant Physiol.*, **19**(3): 466–480 (1944).

Vilenskii, D. G. *Soil Science* (Pochvovedenie), 3rd enlarged ed., Moscow, 1957. English translation published for the National Science Foundation, Wash., D.C. and the Department of Agriculture, USA, by the Israel Program for Scientific Translations, 477 pp. Printed in Jerusalem by S. Monson, PST Cat. No. 188, 1960.

Viro, P. J. Use of Ethylenediaminetraacetic Acid in Soil Analysis. I. Experimental II. Determination of Soil Fertility, *Soil Sci.*, **79**: 459–465 and **80**: 69–74 (1955).

Vlamis, J. Acid Soil Infertility as Related to Soil-Solution and Solid-Phase Effects, *Soil Sci.*, **75**: 383–394 (1953).

Voigtländer, G. Experiments on Different Distributions of Nitrogen on Pastures of the Swabian Alb, *Z. Acker. Pflanzenbau,* **113**: 263–279 (1961).

Volk, G. M. Volatile Loss of Ammonia Following Surface Application of Urea to Turf or Bare Soils, *Agron. J.*, **51**: 746–749 (1959).

Volk, G. M. Gaseous Loss of Ammonia From Surface Applied Nitrogeneous Fertilizer, *J. Agr. Food Chem.*, **9**: 280–283 (1961).

Vose, P. B. The Cation Content of Perennial Ryegrass *Lolium Perenne L.* in Relation to Intraspecific Variability and Nitrogen/Potassium Interaction, *Plant Soil,* **19**(1): 49–64 (1963).

Wagner, R. E. Über die Bestimmung des aufnehmbaren Stickstoffs im Boden durch Gräserkeimpflanzen, *Wiss. Arch. Landwirtsch. Abt. A,* **5**: 166–200 (1930).

Wagner, R. E. Legume Nitrogen Versus Fertilizer Nitrogen in Protein Production of Forage, *Agron. J.*, **46**: 233 (1954).

Waisel, Y. The Absorption of Li and Ca by Barley Roots, *Acta Botan. Neerl.*, **11**: 56–68 (1962).

Waisel, Y. The Effect of Ca on the Uptake of Monovalent Ions by Excised Barley Roots, *Physiol. Plantarum,* **15**: 709–724 (1962a).

Wakefield, R. C., D. A. Schallock, M. Solomon, and C. B. Olney. Yield and Chemical Composition of Ladino Clover as Affected by Fertilizer Treatments, *Agron. J.*, **49**: 374–377 (1957).

Waksman, S. A. *Soil Microbiology*. John Wiley & Sons, Inc., pp. 1–345, 1952.

Walker, J. M., and S. A. Barber. Ion Uptake by Living Plant Roots, *Science*, **133**: 881–882 (1961).

Walker, J. M., and S. A. Barber. Absorption of Potassium and Rubidium from the Soil by Corn Roots, *Plant Soil*, **17**: 243–259 (1962).

Walker, T. W., A. F. R. Adams, and H. D. Orchiston. Fate of Labelled Nitrate and Ammonium Nitrogen When Applied to Grass and Clover Grown Separately and Together, *Soil Sci.*, **81**: 339–351 (1956).

Walker, T. W., B. K. Thapa, and A. F. R. Adams. Studies on Soil Organic Matter. 3. Accumulation of Carbon Nitrogen, Sulfur, Organic and Total Phosphorus in Improved Grassland Soils, *Soil Sci.*, **87**: 135 (1959).

Wallace, A. Solute Uptake by Intact Plants, Los Angeles, Calif., 1963.

Wallace, A., R. T. Mueller, O. R. Lunt, R. T. Ashcroft, and L. M. Shamon. Comparisons of Fine Chelating Agents in Soils, Nutrient Solutions, and in Plant Responses, *Soil Sci.*, **80**: 101–108 (1955).

Walsh, T., and S. J. Culliman. The Effect of Wetting and Drying on Potash-Fixation in Soils, *Empire J. Exptl. Agr.*, **13**: 203–212 (1945).

Walsh, T., and E. Quinn. Use of Potash Rich Soils as a Source of Potash for Crops, *J. Dept. Agr. Dublin*, **53**: 40–43 (1957).

Ward, R. C., E. J. Langin, and R. A. Olson. Factors Responsible for Poor Response of Corn and Grain Sorghum to Phosphorus Fertilization III, *Soil Sci. Soc. Am. Proc.*, **27**: 326–330 (1963).

Waring, S. A., and J. M. Bremner. Ammonium Production in Soil under Waterlogged Conditions, as an Index of Nitrogen Availability, *Nature*, **201**(4922): 951–952 (1964).

Washburn, E. W. (Editor). *International Critical Tables of Numerical Data, Physics, Chemistry, and Technology*, McGraw-Hill Book Co., Inc., New York, pp. 63–69, 1929.

Washko, J. B., and R. P. Pennington. Forage and Protein Production of Nitrogen-Fertilized Grasses Compared with Grass-Legume Associations, *Pa. Agr. Expt. Sta. Bull. 611*, p. 22, 1956.

Watanabe, A., S. Nishigaki, and C. Konishi. Effect of Nitrogen Fixing Blue-Green Algae on the Growth of Rice Plants, *Nature* (London), **168**: 748–749 (1951).

Wear, J. I., J. E. Steckel, M. Fried, and J. L. White. Clay Mineral Models: Construction and Implications, *Soil Sci.*, **66**: 111–117 (1948).

Weaver, J. E. *Root Development of Field Crops*, McGraw-Hill Book Co., New York, 1926.

Weaver, J. E. Summary and Interpretation of Underground Development in Natural Grassland Communities, *Ecol. Monographs*, **28**: 55–78 (1958).

Webb, J. R., Eik, K. and J. T. Pesch. An Evaluation of Phosphorus Fertilizers Applied Broadcast on Calcareous Soils for Corn, *Soil Sci. Soc. Am. Proc.*, **25**: 232–236 (1961).

Webster, G. R., and M. E. Harward. Hydrogen and Calcium Ion Ratios in Dilute Equilibrium Solutions as Related to Cation Saturation, *Soil Sci. Soc. Am. Proc.*, **23**(6): 446–451 (1959).

Webster, R., and J. K. R. Gasser. Soil Nitrogen V—Leaching of Nitrate from Soils in Laboratory Experiments, *J. Sci. Food Agric.*, **10**: 584–585 (1957).

Weigl, J. Die Bedeutung der energiereichen Phosphate bei der Ionenaufnahme durch Wurzeln, *Planta,* **60**: 307–321 (1963).

Weigl, J., and U. Luettge. Microradiography of the Uptake of S^{35} Sulfate by Roots of *Zea Mays.* The Function of the Primary Epidermis, *Planta,* **59**: 15–28 (1962).

Weigl, J., and Ziegler, H. Die Räumliche Verteilung von ^{35}S und die Art der Markierten Verbindungen in Spinatblättern nach Begasung mit $^{35}SO_2$, *Planta,* **58**: 435–447 (1962).

Welte, E., and U. Marckwordt. Die Wechselbeziehungen einiger Kationen bei der Anreicherung durch Gerstenkeimpflanzen, *Agrochimica,* **7**(2): 161–172 (1963).

Wetter, C. A Procedure for Quantitative Determination of Soil Manganese by Means of Aspergillus niger, *Landwirtsch. Forsch.,* **6**: 114–119 (1954).

White, J. W., F. J. Holben, and A. C. Richer. Correlation of Microbiological and Chemical Soil Data with Crop Yields of the Jordan Soil Fertility Plots, *Soil Sci.,* **67**: 279–285 (1949).

White, L. M., and W. H. Ross. Influence of Fertilizers on the Concentration of the Soil Solution, *Soil Sci. Soc. Am. Proc.,* **1**: 181–186 (1936).

White, R. E., and P. H. T. Beckett. Studies on the Phosphate Potentials of Soils. Part I. The Measurement of Phosphate Potential, *Plant Soil,* **20**: 1–6 (1964).

White, R. F., O. Kempthorne, C. A. Black, and J. R. Webb. Fertilizer Evaluation. II. Estimation of Availability Coefficients, *Soil Sci. Soc. Am. Proc.,* **20**: 179–186 (1956).

White, W. C., and J. Pesek. Nature of Residual Nitrogen in Iowa Soils, *Soil Sci. Soc. Am. Proc.,* **23**: 39–42 (1959).

Whitehead, D. C. Soil and Plant Nutrition—Aspects of the Sulfur Cycle, *Soils Fertilizers,* **37**: 1–8 (1964).

Whitney, R. S., and R. Gardner. Notes on Estimating Available Phosphorus by Extracting Soils with a Potassium Carbonate Solution, *Soil Sci.,* **41**: 33 (1936).

Widdowson, F. V., A. Penny, and R. J. B. Williams. Autumn Nitrogen for Winter Wheat, *J. Agr. Sci.,* **57**: 329–337 (1961).

Wiegand, C. L., and E. R. Lemon. A Field Study of Some Plant-Soil Relations in Aeration, *Soil Sci. Soc. Am. Proc.,* **22**: 216–221 (1958).

Wiersum, L. K. Density of Root Branching as Affected by Substrate and Separate Ions, *Acta Botan. Neerl.* **7**: 174–190 (1958).

Wiersum, L. K. Bodemstructuur, Beworteling en Planten Voeding, *Landbouwk. Tijdschr.,* **74**(22): 961–972 (1962).

Wiersum, L. K., and K. Bakema. Competitive Adaptation of the Cation Exchange Capacity of Roots, *Plant Soil,* **11**(3): 287–292 (1959).

Wijler, J., and C. C. Delwiche. Investigations on the Denitrifying Process in Soil, *Plant Soil,* **5**: 155 (1954).

Wiklander, L. Fixation of Potassium by Clays Saturation with Different Cations, *Soil Sci.,* **69**: 261–268 (1950a).

Wiklander, L. Kinetics of Phosphate Exchange in Soils, *Ann. Royal Agr. Coll. Sweden,* **17**: 407–424 (1950b).

Wiklander, L. Forms of Potassium in the Soil, *Potassium Symp. Intern. Potash Inst.,* Bern, pp. 109–121, 1954.

Wiklander, L., and G. Hallgreng. Studies on Gyttja Soils, *Kungl. Lantsburkhögsk. Ann.,* **16**:811–827 (1949).

Wild, A. The Retention of Phosphate by Soil, *J. Soil Sci.,* **1**: 221–238 (1950).

Wild, A. The Determination of Calcium Chloride-Soluble Phosphate in Soils, *Z. Pflanzenernähr. Düng. Bodenk.*, **84** (129):220–224 (1959).

Wild, A. Soluble Phosphate in Soil and Uptake by Plants, *Nature,* **203**(4942): 326–327 (1964).

Wilde, S., and R. O. Rosendahl. Value of Potassium Feldspar as a Fertilizer in Forest Nurseries, *J. Forestry,* **43**: 466–367 (1945).

Wilkinson, S. R., and A. J. Ohlrogge. Fertilizer Nutrient Uptake as Related to Root Development in the Fertilizer Band. Influence of Nitrogen and Phosphorus Fertilizer on Endogenous Auxin Content of Soybean Roots, *Proc. 7th Intern. Congr. Soil Sci.,* **4**:234–242 (1960).

Wilkinson, S. R., and A. J. Ohlrogge. Influence of Biuret and Urea Fertilizers Containing Biuret on Corn Plant Growth and Development, *Agron. J.,* **52**: 560–562 (1960a).

Wilkinson, S. R., and A. J. Ohlrogge. Principles of Nutrient Uptake from Fertilizer Bands. I. Mechanism Responsible for Intensive Root Development in Fertilized Zones, *Agron. J.,* **54**: 288–291 (1962).

Willcox, O. W. Evaluation of Fertilizer Tests. I. The Scientific Basis. II. Evaluation of a Multiple Test. III. The Strip Test, *Sugar,* **49**(2): 46–50; (3), 51–56; (4), 40–43 (1954).

Willcox, O. W. Quantitative Agrobiology. I. The Inverse Yield Nitrogen Law. II. The Nitrogen Constant. III. The Mitscherlich Equation and its Constants. IV. Apparent Exceptions to the Mitscherlich Law, *Agron. J.,* **46**: 315–328 (1954a).

Williams, C. H., and A. Steinbergs. The Evaluation of Plant-Available Sulfur in Soils. I. The Chemical Nature of Sulfate in Some Australian Soils, *Plant Soil,* **17**: 279–294 (1962).

Williams, D. E. The Absorption of Potassium Influenced by Its Concentration in the Nutrient Medium, *Plant Soil,* **15**: 387–399 (1961).

Williams, D. E., and N. T. Coleman. Cation-Exchange Properties of Plant Root Surfaces, *Plant Soil,* **2**(2): 243–256 (1950).

Williams, E. G. Chemical Soil Tests as an Aid to Increased Productivity, *Trans. Intern. Soil Conf.,* New Zealand, pp. 820–834, 1962.

Williams, E. G., and W. M. H. Saunders. Significance of Particle-Size Fractions in Readily Soluble Phosphorus Extractions by the Acetic, Truog, and Lactate Methods, *J. Soil Sci.,* **7**(2): 189–202 (1956).

Williamson, R. E. The Effect of Root Aeration on Plant Growth, *Soil Sci. Soc. Am. Proc.,* **28**: 86–90 (1964).

Willis, A. J., and E. W. Yemm. Respiration of Barley Plants. VIII. Nitrogen Assimilation and the Respiration of the Root System, *New Phytol.,* **54**: 163–181 (1955).

Wilson, P. W., and R. H. Burris. Biological Nitrogen Fixation—A Reappraisal, *Bactiol. Rev.,* **11**: 41–73 (1947).

Winogradsky, S. Etudes sur la microbiologie du Sol: II. Sur les microbes fixateurs d'arote, *Ann. Inst. Pasteur,* **40**: 455–520 (1925).

Wit, C. T. de. A Physical Theory on Placement of Fertilizer, Thesis, Wageningen Agr. Univ., 1953a.

Wit, C. T. de. A Physical Theory on Placement of Fertilizers, *Versl. Landbouwk. Onderzoek.,* **59**(4): 71(E) (1953b).

Wit, C. T. de, W. Dijkshoorn, and J. C. Noggle. Ionic Balance and Growth of Plants, *Verslag. Landbouwk. Onderzock.,* **69**: 15 (1963).

Woodruff, C. M. Ionic Equilibria Between Clay and Dilute Salt Solutions, *Soil Sci. Soc. Am. Proc.,* **19**: 36–40 (1955).

Yemm, E. W., and A. J. Willis. The Respiration of Barley Plants. IX. The Metabolism of Roots During the Assimilation of Nitrogen, *New Phytol.,* **55**: 229–252 (1956).

Zink, F. W. Rate of Growth and Nutrient Absorption of Late Garlic, *Proc. Am. Soc. Hort. Sci.,* **83**: 579–584 (1963).

Zuev, I. A., and P. F. Golubeva. Comparative Action of Nitrogen Phosphorus and Potassium Deficiency on the Absorption and Metabolism of Phosphorus by Winter Wheat in Light and Darkness, *Soviet Plant Physiol.,* 9(1): 29–33 (1962).

Author Index

Numbers in italics indicate the pages on which the complete references are listed.

A

B

E

F

G

L

M

N

O

P

Q

R

S

T

U

V

Y

Z

Subject Index